NOVEL PRODUCTION METHODS FOR ETHYLENE, LIGHT HYDROCARBONS, AND AROMATICS

CHEMICAL INDUSTRIES

A Series of Reference Books and Textbooks

Consulting Editor
HEINZ HEINEMANN
Heinz Heinemann, Inc.,
Berkeley, California

Volume 1: Fluid Catalytic Cracking with Zeolite Catalysts, *Paul B. Venuto and E. Thomas Habib, Jr.*

Volume 2: Ethylene: Keystone to the Petrochemical Industry, *Ludwig Kniel, Olaf Winter, and Karl Stork*

Volume 3: The Chemistry and Technology of Petroleum, *James G. Speight*

Volume 4: The Desulfurization of Heavy Oils and Residua, *James G. Speight*

Volume 5: Catalysis of Organic Reactions, *edited by William R. Moser*

Volume 6: Acetylene-Based Chemicals from Coal and Other Natural Resources, *Robert J. Tedeschi*

Volume 7: Chemically Resistant Masonry, *Walter Lee Sheppard, Jr.*

Volume 8: Compressors and Expanders: Selection and Application for the Process Industry, *Heinz P. Bloch, Joseph A. Cameron, Frank M. Danowski, Jr., Ralph James, Jr., Judson S. Swearingen, and Marilyn E. Weightman*

Volume 9: Metering Pumps: Selection and Application, *James P. Poynton*

Volume 10: Hydrocarbons from Methanol,
Clarence D. Chang

Volume 11: Foam Flotation: Theory and Applications,
Ann N. Clarke and David J. Wilson

Volume 12: The Chemistry and Technology of Coal,
James G. Speight

Volume 13: Pneumatic and Hydraulic Conveying of Solids,
O. A. Williams

Volume 14: Catalyst Manufacture: Laboratory and Commercial Preparations, *Alvin B. Stiles*

Volume 15: Characterization of Heterogeneous Catalysts,
edited by Francis Delannay

Volume 16: BASIC Programs for Chemical Engineering Design,
James H. Weber

Volume 17: Catalyst Poisoning,
L. Louis Hegedus and Robert W. McCabe

Volume 18: Catalysis of Organic Reactions,
edited by John R. Kosak

Volume 19: Adsorption Technology: A Step-by-Step Approach to Process Evaluation and Application, *edited by Frank L. Slejko*

Volume 20: Deactivation and Poisoning of Catalysts,
edited by Jacques Oudar and Henry Wise

Volume 21: Catalysis and Surface Science: Developments in Chemicals from Methanol, Hydrotreating of Hydrocarbons, Catalyst Preparation, Monomers and Polymers, Photocatalysis and Photovoltaics, *edited by Heinz Heinemann and Gabor A. Somorjai*

Volume 22: Catalysis of Organic Reactions,
edited by Robert L. Augustine

Volume 23: Modern Control Techniques for the Processing Industries, *T. H. Tsai, J. W. Lane, and C. S. Lin*

Volume 24: Temperature-Programmed Reduction for Solid Materials Characterization, *Alan Jones and Brian McNicol*

Volume 25: Catalytic Cracking: Catalysts, Chemistry, and Kinetics, *Bohdan W. Wojciechowski and Avelino Corma*

Volume 26: Chemical Reaction and Reactor Engineering, *edited by J. J. Carberry and A. Varma*

Volume 27: Filtration: Principles and Practices, second edition, *edited by Michael J. Matteson and Clyde Orr*

Volume 28: Corrosion Mechanisms, *edited by Florian Mansfeld*

Volume 29: Catalysis and Surface Properties of Liquid Metals and Alloys, *Yoshisada Ogino*

Volume 30: Catalyst Deactivation, *edited by Eugene E. Petersen and Alexis T. Bell*

Volume 31: Hydrogen Effects in Catalysis: Fundamentals and Practical Applications, *edited by Zoltán Paál and P. G. Menon*

Volume 32: Flow Management for Engineers and Scientists, *Nicholas P. Cheremisinoff and Paul N. Cheremisinoff*

Volume 33: Catalysis of Organic Reactions, *edited by Paul N. Rylander, Harold Greenfield, and Robert L. Augustine*

Volume 34: Powder and Bulk Solids Handling Processes: Instrumentation and Control, *Koichi Iinoya, Hiroaki Masuda, and Kinnosuke Watanabe*

Volume 35: Reverse Osmosis Technology: Applications for High-Purity-Water Production, *edited by Bipin S. Parekh*

Volume 36: Shape Selective Catalysis in Industrial Applications, *N. Y. Chen, William E. Garwood, and Frank G. Dwyer*

Volume 37: Alpha Olefins Applications Handbook, *edited by George R. Lappin and Joseph L. Sauer*

Volume 38: Process Modeling and Control in Chemical Industries, *edited by Kaddour Najim*

Volume 39: Clathrate Hydrates of Natural Gases, *E. Dendy Sloan, Jr.*

Volume 40: Catalysis of Organic Reactions, *edited by Dale W. Blackburn*

Volume 41: Fuel Science and Technology Handbook, *edited by James G. Speight*

Volume 42: Octane-Enhancing Zeolitic FCC Catalysts *Julius Scherzer*

Volume 43: Oxygen in Catalysis, *Adam Bielański and Jerzy Haber*

Volume 44: The Chemistry and Technology of Petroleum, Second Edition, Revised and Expanded, *James G. Speight*

Volume 45: Industrial Drying Equipment: Selection and Application *C. M. van 't Land*

Volume 46: Novel Production Methods for Ethylene, Light Hydrocarbons, and Aromatics, *edited by Lyle F. Albright, Billy L. Crynes, and Siegfried Nowak*

Additional Volumes in Preparation

Catalysis of Organic Reactions, *edited by William E. Pascoe*

NOVEL PRODUCTION METHODS FOR ETHYLENE, LIGHT HYDROCARBONS, AND AROMATICS

edited by

Lyle F. Albright

School of Chemical Engineering
Purdue University
West Lafayette, Indiana

Billy L. Crynes

School of Chemical Engineering
The University of Oklahoma
Norman, Oklahoma

Siegfried Nowak

Institut für Chemische Technologie
Berlin, Germany

Marcel Dekker, Inc. New York • Basel • Hong Kong

ISBN 0-8247-8588-6

This book is printed on acid-free paper.

MARCEL DEKKER, INC.
270 Madison Avenue, New York, New York 10016

Current printing (last digit):
10 9 8 7 6 5 4 3 2 1

PRINTED IN THE UNITED STATES OF AMERICA

Preface

The objective of this book is to present, in an organized format, current results from 29 international studies on novel production methods of ethylene, light hydrocarbons, and aromatics. The book materialized as a result of the remarkably positive response to our call for papers for a symposium for the 199th American Chemical Society meeting, Boston, Massachusetts, April 22-27, 1990.

This book is for the most part based on the proceedings of the papers presented at the Boston symposium, which was sponsored by the Division of Industrial and Engineering Chemistry of the American Chemical Society. Experimental and other results obtained since the symposium have, however, been incorporated in several papers. Five papers are included that were not a part of the official program at Boston. Three of these papers were, however, presented in an impromptu or unofficial manner there. Two additional papers were submitted and accepted by us subsequent to the meeting; each provides important contributions to the subject material. Each chapter has been reviewed and edited as needed.

This book is considered complementary to three other books prepared by either one or two of the current editors. In the previous books, conventional pyrolysis processes producing olefins and aromatics and/or coking problems associated with conventional (pyrolysis) processes were discussed. In the previous books and also in the current book, the intent is to present important information for the scientific, engineering, and business communities.

A unique feature of the book is its international flavor. Chapters were prepared in 11 different countries located on 4 continents. Eleven chapters were prepared in the United States and 18 in other countries. It is rare to find such an international and comprehensive collection of papers reflecting different cultures and perspectives on novel methods of producing light hydrocarbons. With such a relatively broad subject and material from a range of international interests, considerable effort has been given to organizing the material for use by a wide range of readers. The challenge was to meld the chapters discussing mechanistic and fundamental studies, design and practical investigations, and studies addressing cost and business elements into a cohesive text. A wide variety of processes and feedstocks is discussed.

In most of the initial 13 chapters, use of methane as a feedstock to produce ethylene and other light hydrocarbons is emphasized. In the two exceptions, ethane and butane were used as feedstocks, but the technology involved was highly related to technology using methane. In these chapters, the following types of technology are employed: catalytic processes using oxygen, processes using chlorinated methane, and thermal pyrolysis of methane. Over the years, methane has held the attention of numerous institutes and businesses. Its high chemical stability and yet great availability attract those wishing to obtain mechanistic, technical, and economic information on its upgrade. New technology proposed suggest economic processes may soon be developed. Methods that minimize energy requirements are presented.

Subjects covered in Chapters 14, 15, and 16 include catalytic hydrogenation of carbon monoxide (or use of synthesis gas) to produce light olefins, catalytic dehydrogenation of propane, and catalytic pyrolysis. A dry pyrolysis process is discussed in Chapter 17. In Chapter 18, a novel method of reducing or even eliminating coke formation in an ethylene furnace is reported. The proposed technique has already caught the attention of more than one industrial company.

The use of oxygenated hydrocarbons or compounds as feedstocks is addressed in Chapters 19-27. Feedstocks considered include methanol, ethanol, "bioacids," "bioacetone," waste materials, and wood. The promise of plentiful feedstocks employing wastes and biomaterials is confounded by difficulties in the separation and the recovery of valuable products. Chapters 28 and 29 report on novel methods to produce aromatic hydrocarbons.

We are indebted to the authors in meeting deadlines and considering revisions in the interest of a more cohesive text. Our appreciation can never be fully expressed to all of our colleagues, who over the years provided valuable advice, assistance, and constructive debate as well as encouragement. We dare not attempt a listing of these many friends and colleagues here. Without the help of the Petroleum Research Fund through a generous travel assistance grant for some of the speakers at the Boston ACS symposium, we could not have achieved the broad international scope of this book. Finally, our thanks to Mrs. Phyllis Beck at Purdue University for outstanding secretarial assistance.

Lyle F. Albright
Billy L. Crynes
Siegfried Nowak

Contents

1

CATALYTIC PRODUCTION OF ETHYLENE AT PHILLIPS. I. OVERVIEW AND METHANE COUPLING

J.H. Kolts, J.B. Kimble and R.A. Porter

Research and Development
Phillips Petroleum Company
Bartlesville, Oklahoma 74004

Scientists at Phillips Petroleum Company have for many years investigated potential catalytic routes to produce ethylene that may have economic advantages over the currently used thermal cracking processes. For example, with ethane an oxidative dehydrogenation process would remove the equilibrium restrictions that hinders current thermal dehydrogenation methods. In the case of propane and butane, a viable catalytic cracking process that substantially improves selectivity to ethylene or even allows the choice of ethylene or propylene as the preferred product could have substantial economic advantage, if perfected. In the case of methane, whose only large scale uses are as a combustion fuel and feed for synthesis gas production, the development of a catalytic route to convert methane to higher molecular weight hydrocarbons in high yield has almost unlimited potential.

The catalytic conversion of methane (1-5 and references therein), ethane (6-11) and to a lesser extent propane and butane (12) has received a considerable amount of attention in the open literature. Indeed the publications of Keller and Bhasin (1), and Hinsen and Baerns (2) have led to a virtual flood of research on the direct conversion of methane to higher hydrocarbons. Although the order of presentation for this series of papers begins with methane and ends with butane, the actual research began with the oxidative conversion of ethane to ethylene, and that work was followed by the selective catalytic conversion of methane, propane and butane. These later successes had as their foundation discoveries made during the initial ethane research. Valuable basic knowledge was acquired, for example, during the early Phillips work using alkali metal/manganese (10,13) and alkali metal/cobalt (9) catalysts for the cyclic oxidative dehydrogenation of ethane and during the work using lithium supported on magnesium oxide (14) for the continuous oxidative dehydrogenation of ethane.

Mechanistically, all of the catalysts, to be discussed here and in companion papers (15,16), appear to operate by first activating the hydrocarbon, then releasing a free-radical to the gas phase where it sequentially reacts to form the final products. The initial formation of free-radicals and their subsequent release to the gas-phase appears to be independent of the hydrocarbon used or whether the process is oxidative or non-oxidative in nature.

In remaining portions of this paper methane coupling results in three areas will be presented: 1) a summary of methane coupling data for several different catalysts, 2) a summary of results using thermal gravimetric analysis (TGA) and X-ray diffraction (XRD) on the Li/MgO and Li/La_2O_3 catalysts, and 3) kinetic modeling results for the continuous oxidative coupling of methane.

EXPERIMENTAL

Catalysts used in this study were prepared by first milling the solid support material to a fine powder. To this was added an appropriate amount of alkali metal carbonate and just enough water to form a thick paste. In some preparations, the source of the alkali promoter was the nitrate salt; in this case a solution of this salt was used as the slurry water. The resulting paste was dried at 150°C for at least 2 hours in a forced draft oven. The resulting solid mass was then calcined in an open porcelain dish by raising the temperature slowly to 300°C, holding at this temperature for 4 hours. The temperature was then raised to 850°C over 4 hours. The material was held at the high temperature for 4 hours, then allowed to cool to room temperature. This solid material was ground and sieved to the desired mesh (usually -20/+40 for testing). Catalyst testing was carried out in quartz reactors of 25 cc volume. Unless otherwise noted all experiments used 20 cc of catalyst having an approximate bed depth of 4.5 inches. Temperatures were monitored using a standard Chromel-Alumel thermocouple in a quartz thermowell which extended axially through the bed. The composition of the feed was controlled by independently controlling the flow of pure gases and was electronically monitored throughout the experiments. All gases used were either Phillips or Matheson CP grade and were used without further purification.

Effluent from the reactor was analyzed for C1 through C2 hydrocarbons plus CO, CO_2, N_2, and O_2 on an HP5880 chromatograph using 18 ft of 27% Bis 2-EEA, 6 ft of 80% Porapak N + 20% Porapak Q and 10 ft of 13X molecular sieve. The chromatograph was calibrated using a certified standard blend from Air Products. Absolute material balances were better than 96%. Reported selectivities are based upon the moles of feed carbon converted to a particular product.

Thermogravimetric measurements were made on a Cahn/Ventron Model R-100 microbalance equipped with a gas flow system. Data from this instrument were collected and analyzed on a DEC MINC-11 computer. X-ray powder diffraction (XRD) patterns were taken using a Philips diffractometer using Cu-Ka radiation.

METHANE COUPLING RESULTS

Shown in Table 1 are conversion and selectivities for several catalysts that have been tested. As the table shows, all of the materials tested had some activity for methane conversion. Data taken over quartz chips had no measurable activity at the conditions shown, however, this result is very dependent upon the prior history of the quartz. In the authors experience, any reactor which had been previously used with any other catalyst or had been contaminated with any metal or halogen may have substantial methane conversion activity. None of the selectivity trends shown in Table 1 are surprising when compared to the extensive literature available for these catalysts. In general all of the supports tested had moderate activity for methane coupling. For the MgO, CaO, La_2O_3 and ZnO catalysts the effects of Li, Na, and K as promoters decreased from Li to K.

All of the alkali metal containing catalysts tested

TABLE 1. *Methane coupling catalyst summary*

				Selectivity				
Catalyst	CH_4\\O_2*	Temp	Conv	C_2	C_2=	C_3's	CO	CO_2
Quartz	4.4/1	740	nil					
MgO	4.4/1	705	16	9	7		30	55
3%Li/MgO	4.4/1	700	23	24	40	5		31
3%Na/MgO	4.4/1	710	21	23	31	4	3	40
CaO	4.4/1	714	9	36	21		4	38
3%Li/CaO	4.4/1	711	20	27	39	5	3	25
3%Na/CaO	4.4/1	707	15	37	36	6	2	18
3%K/CaO	5.0/1	702	10	39	19	3	3	36
La_2O_3	4.4/1	710	16	28	23	3	10	35
3%Li/La_2O_3	4.4/1	710	21	22	39	5	4	29
3%Na/La_2O_3	5 0/1	700	16	30	32	4	2	31
3%K/La_2O_3	5.0/1	701	16	38	26	5	1	29
3%Li/ZnO	5.0/1	720	13	42	32	3		23
3%Na/ZnO	5.0/1	701	11	37	16	2		45
3%K/ZnO	5.0/1	702	13	20	6			74
3%Li/CeO_2	5.0/1	711	15	29	16			55
3%Li/SnO_2	5.0/1	703	13	10	22	1	3	65
3%Li/SiO_2	5.0/1	721	3		11		62	27
Mg_2TiO_4	5.0/1	708	10				33	67
3%Li/Mg_2TiO_4	5.0/1	708	3	56	14		15	16
Ca_2TiO_4	5.0/1	706	12	4			27	69
3%Li/Ca_2TiO_4	5.0/1	707	17	31	40	5	2	22
Sr_2TiO_4	5.0/1	709	13	11	4		25	60
3%Li/Sr_2TiO_4	5.0/1	710	15	34	32	4	2	28

* Total gas flow was 150 cc/min, oxygen was added as dry air. GHSV has been kept low to minimize hot spot formation. Reactor temperatures across the bed were within plus or minus 10C.

TABLE 2. *Change in Li concentration and surface area before and after use on CaO support.**

Li Conc Start	Li Conc After	Nitrogen BeET SA
0.0		10.3
0.50	0.20	1.4
3.00	0.93	1.3
8.00	1.01	1.3
15.00	0.98	1.3

* CaO starting material was 110 m^2/gram before preparation and between 50 and 60 after preparation but before testing. Li added as the nitrate. Test period was approximately 48 hours.

in quartz reactors tended to lose some of the initially present alkali promoter during the methane coupling reaction. The rate of alkali metal loss being highest for lithium and least for potassium. As shown in Table 2, catalysts prepared between 3.0 and 15 wt % Li on CaO had actual Li concentrations of approximately 1% after 48 hours of testing. In addition, the catalyst surface areas before and after testing had been substantially reduced. As shown in Fig. 1, a lithium on lanthanum oxide catalyst showed initially high selectivities to C2+ hydrocarbons which dropped off under our test conditions after about 140 minutes of continuous testing. This initial high selectivity is apparently due to absorption of carbon dioxide product rather than an actual selectivity increase and will be discussed in more detail in the next section.

Several experiments have been conducted to maximize catalyst performance. Some of the important variables

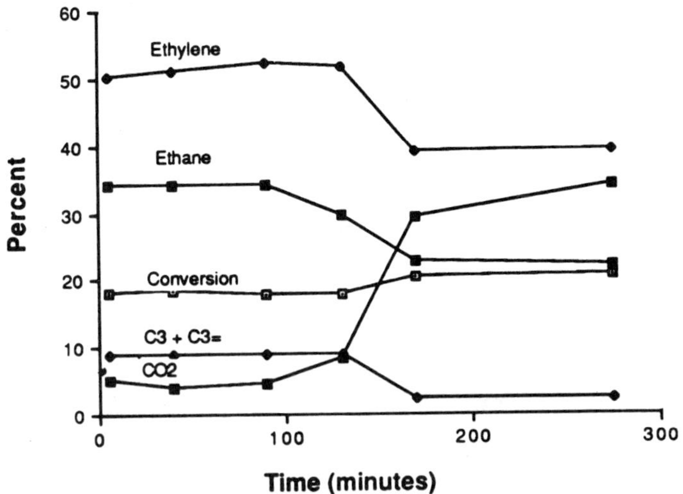

FIG. 1 *Conversion and selectivity changes as a function of reaction time for 3%Li on* La_2O_3.

which have been shown to improve performance are summarized below:

1) Substitution of water for nitrogen diluent has been shown to increase ethylene and ethane selectivity, and increase methane conversion for MgO based catalysts (4). This result appears to be in conflict with the findings of Korf et al. (17).

2) Carbon oxides selectivity actually decrease with increased temperature (see Fig. 2). The optimum temperature for any given catalyst being dependent upon space velocity, reactor geometry, and methane to oxygen ratio.

3) Selectivity to ethane and ethylene, in general, increases as the mole percent of oxygen in the feed decreases.

4) Reducing the residence time of the hydrocarbon products in the hot zone of the reactor improves product yields.

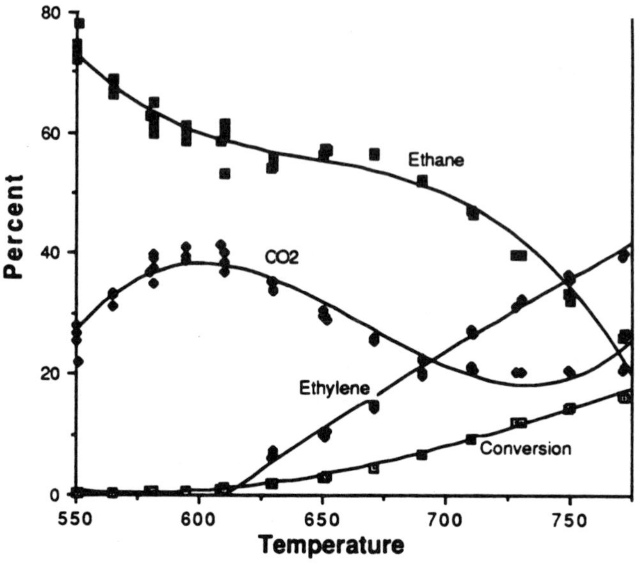

FIG. 2 *Conversion and selectivity over 3%Li/MgO at 8 to 1 methane to oxygen as a function of reactor temperature.*

TGA AND XRD RESULTS

Figure 3 compares the selectivity to C_2's and CO_2 for catalysts prepared as Li_2CO_3 on MgO to that for Li_2CO_3 on La_2O_3. Initially, Li on La_2O_3 shows lower conversion and much higher selectivity to C_2's and almost no CO_2. However, after 2.0 hours, both the conversion and selectivity to CO_2 increase and the C_2's selectivity decreases to be more in line with that seen with Li on MgO. Catalysts prepared from $LiNO_3$, with no prior exposure to CO_2 before testing, behave very similarly to that observed for those prepared as Li_2CO_3 on La_2O_3. In an effort to better understand what species are being formed and what the differences are between the MgO and La_2O_3 catalysts, a TGA and XRD study of the catalysts,

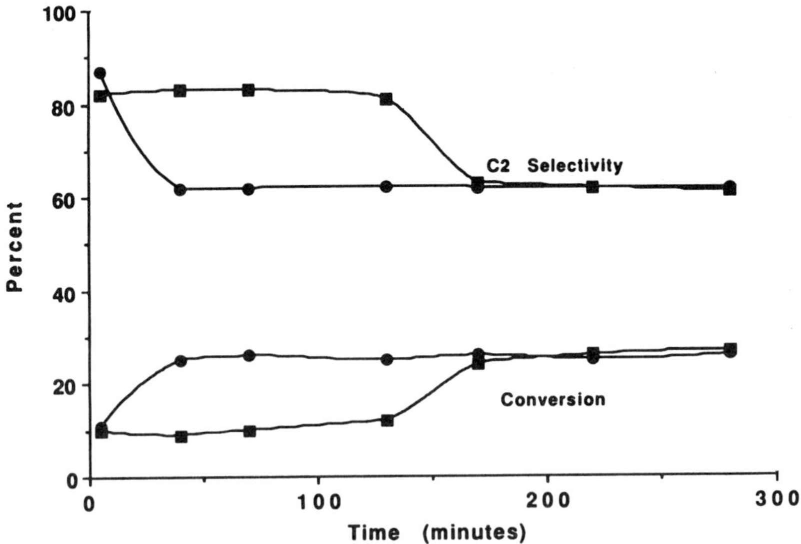

FIG. 3 *Comparison of conversion and selectivity for Li/MgO (circles) and Li/La_2O_3.*

in the presence of CH_4/air, CO_2/N_2 and CO/N_2, was conducted.

Methane/air reaction The reaction of a mixture of methane and air with a catalyst prepared from $LiNO_3$ and supported on MgO results in the quantitative formation of Li_2CO_3 as evidenced by the weight gain shown in Fig. 4. When this compound was heated to 900°C in flowing nitrogen, it began to decompose at approximately 710-720°C, see Fig. 4, the final form being Li_2O. This is in good agreement with tabulated literature values (18). Calculations based on the weight gain and loss data imply that the Li was initially present as LiOH. The Li_2O species formed after heating to 900°C was rapidly rehydrated to LiOH when exposed to water vapor at 675°C.

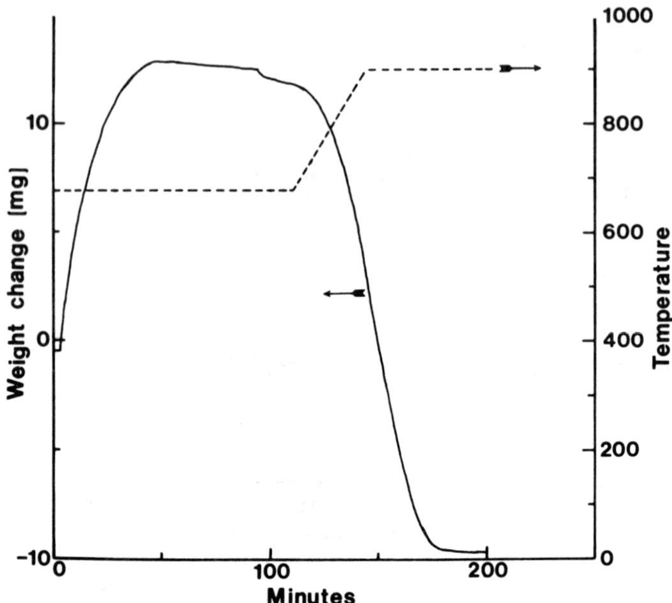

FIG. 4 *3.9 wt% Li/MgO reacted with 1/1 methane to air diluted to 10% in nitrogen at 675°C. At 90 minutes pure nitrogen was flowed over the sample and ramped to 900°C at 10 deg/min.*

Li supported on La_2O_3 behaved similarly to Li on MgO when exposed to the methane/air mixture. Lithium was initially in the form of LiOH and $LiLaO_2$, which formed the carbonate under methane/air, then Li_2O after heating in nitrogen. For both Li supported on MgO and La_2O_3 the presence of LiOH, Li_2CO_3 and Li_2O at each step was confirmed using X-ray powder patterns. $LiLaO_2$ was also observed on the Li on La_2O_3 catalyst.

Reaction with carbon dioxide The reaction of Li/MgO with 10% CO_2 in nitrogen results in much more rapid carbonate formation compared to the CH_4/air mixtures. This may be due to the increased partial pressure of CO_2 in this experiment. The weight increase is also somewhat greater than would be expected for the stoichiometric formation of Li_2CO_3. Unpromoted MgO alone shows no formation of $MgCO_3$ when exposed to CO_2 at these temperatures. One possible explanation could be the formation of a carbonate phase that involves both Li and Mg that is more stable than $MgCO_3$.

At this higher CO_2 partial pressure the Li on La_2O_3 behaves differently than Li on MgO in that CO_2 begins reacting with the La_2O_3 support, see Fig. 5. The rate of reaction with La_2O_3 appears to be slower than with LiOH, $LiLaO_2$ or Li_2O supported on La_2O_3. Pure La_2O_3 run under the same conditions also formed the carbonate, but at a slower rate when compared to Li promoted La_2O_3. As soon as the flow of CO_2 was stopped, at 675°C, the

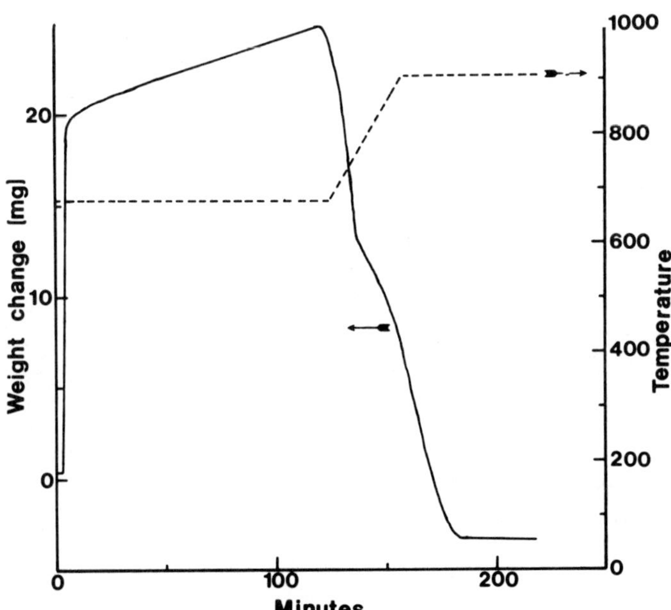

FIG. 5 *3 wt% Li/La_2O_3 reacted with 10% CO_2 in N_2 at 675°C then switched to pure N_2 at 120 minutes and ramped to 900°C at 10 deg/min.*

sample began to lose weight and returned to pure La_2O_3 starting material.

Reaction with CO Li on MgO was found to react reversibly with CO at 675°C. However, in this case, we do not believe the product to be Li_2CO_3, but rather a C_6O_6 oligomer adsorbed on the catalyst. The decomposition of the CO species showed two distinct decomposition rates which we ascribe to the C_6O_6 oligomer and LiOH decomposition. Evidence to support this conclusion is shown by the fact that only one decomposition rate was observed if the catalyst was pretreated to decompose the LiOH prior to contact with CO. No stoichiometric compound formation was achieved in this reaction, even after 13 hours of exposure to CO. This sample was also examined by Scanning Electron Microscopy and found to be coke-like. It appears that the C_6O_6 can form reversibly and is stable at 675°C for short periods of time but that prolonged heating converts this material to coke.

KINETIC MODELING RESULTS

In previous work from this laboratory, (4) the oxidative coupling of methane was kinetically modeled using the method of Bulrisch and Stoer (19). This first attempt at modeling the methane coupling reaction sequence only considered hydrocarbon species and totally neglected the carbon oxide forming reactions as well as carbon-oxygen intermediate species. Even though the reaction sequence was limited, the modeled hydrocarbon product distributions were surprisingly close to those observed experimentally.

In the present study the set of gas phase chemical reactions used was substantially expanded to include the formation of hydrocarbons up to propane and propylene,

TABLE 3. *Gas-phase methane coupling model reactions.*

Reactions of hydrocarbon species.					
$2CH_3$	→	C_2H_6	$H+C_3H_8$	→	$H_2+1-C_3H_7$
$CH_3+C_2H_6$	→	$CH_4+C_2H_5$	$H+C_3H_8$	→	$H_2+2-C_3H_7$
C_2H_5	→	$H+C_2H_4$	$1-C_3H_7$	→	$CH_3+C_2H_4$
$H+C_2H_6$	→	$H_2+C_2H_5$	$2-C_3H_7$	→	$H+C_3H_6$
$CH_3+C_2H_5$	→	C_3H_8	$C_2H_5+C_3H_8$	→	$C_2H_6+1-C_3H_7$
$CH_3+C_3H_8$	→	$CH_4+1-C_3H_7$	$C_2H_5+C_3H_8$	→	$C_2H_6+2-C_3H_7$
$CH_3+C_3H_8$	→	$CH_4+2-C_3H_7$	$H+CH_4$	→	H_2+CH_3
Reactions of hydrocarbons with oxygen species.					
CH_4+O2	→	CH_3+HO_2	CH_3+HO_2	→	CH_4+O_2
CH_4+HO	→	CH_3+H_2O	CH_3+CH_2O	→	CH_4+HCO
CH_4+O	→	CH_3+HO	C_2H_6+HO	→	C_2H_5+H2O
CH_4+HO2	→	$CH_3+H_2O_2$	C_2H_6+O	→	C_2H_6+HO
CH_3+O2	→	CH_3O+O	$C_2H_5+O_2$	→	$C_2H_4+HO_2$
CH_3+HO	→	H_2+CH_2O	C_2H_4+HO	→	CH_3+CH_2O
CH_3+O	→	$H+CH_2O$	C_2H_4+O	→	CH_3+HCO
CH_3+HO2	→	CH_3O+HO			
Reactions of carbon/oxygen and non-carbon species.					
CH_3O	→	$H+CH_2O$	$HCO+O_2$	→	$CO+HO_2$
CH_3O+O_2	→	CH_2O+HO_2	HCO	→	$H+CO$
CH_2O+HO	→	$HCO+H_2O$	$2HCO$	→	$CO+CH_2O$
CH_2O+H	→	$HCO+H_2$	$HCO+HO_2$	→	CH_2O+O_2
CH_2O+O	→	$HCO+HO$	$CO+HO$	→	CO_2+H
CH_2O+HO_2	→	$HCO+H_2O_2$	$CO+HO_2$	→	CO_2+HO
$HCO+HO$	→	$CO+H_2O$	$CO+O$	→	CO_2
$HCO+CH_3$	→	$CO+CH_4$			
Reactions of non-carbon species.					
H_2+O_2	→	$2HO$	$H+O_2$	→	$HO+O$
H_2+O	→	$H+HO$	$H+HO_2$	→	$2HO$
H_2+HO	→	$H+H_2O$	H_2O_2	→	$2HO$

and to include the formation of oxygen intermediates as well as CO and CO_2. The complete set of gas phase reactions are summarized in Table 3. The literature rate constants used (20,21) were those that had been measured at temperatures near the actual methane coupling temperature. In cases where several rate constants had been measured or calculated at the desired temperature an average value was used in the modeling program. All of the gas phase rate constants were used as given and were not altered to provide a better fit to the data. The reaction sequence used to form methyl radicals is shown in Table 4. The particular reactions shown follow the mechanism proposed by Lunsford and co-workers (22,23). Rate constants used for the methyl radical formation steps were empirically derived by adjusting them until the correct level of methane conversion was achieved. The methane activation step at the surface was also forced to be the slow step in the sequence, following the recent isotopic work of Cant et al. (24). Although the particular surface initiation reactions chosen follow a specific mechanism, a model of this type cannot confirm that they have any real significance in the actual reaction sequence.

Results of the model using only the reactions shown in Tables 3 and 4 are summarized in Table 5. As can be seen in the table, there are some serious inconsistencies between calculated and observed product distributions. Specifically, selectivities to ethane

TABLE 4. *Methane activation reactions used to supply methyl radicals into the gas-phase reaction sequence.*

O2+Catalyst	→ 2Cat-O	Oxygen activation
CH4+Cat-O	→ CH3+Cat-OH	Methyl radical formation
2Cat-OH	→ H2O+Cat+Cat-O	Water desorption

18. Mellor, J. W., *A Comprehensive Treatise on Inorganic and Theoretical Chemistry*, "Alkali Metals Part 1", p252, Vol II, Supp II; John Wiley and Sons, New York,1961.

19. Bulrisch, R., and Stoer, J., *Numerisch Mathimatik*, 1966, *8*, 1.

20. Allara, D. L., and Shaw, R., *J. Phys. Chem. Ref. Data*, 1980, *9*, 523.

21. Tsang, W., and Hampson, R. F., *J. Phys. Chem. Ref. Data*, 1986, *15*, 1087, and references therein.

22. Driscoll, D. J., Martir, W. M., Wang, J., and Lunsford, J. H., *J. Am. Chem. Soc.*, 1985, *107*, 58.

23. Ito, T., and Lunsford, J. H., *Nature*, 1985, *314*, 721.

24. Cant, N. W., Lukey, C. A., Nelson, P. F., and Tyler, R. J., *J. Chem. Soc., Chem. Commun.*, 1988, *12*, 766.

2

CATALYTIC PRODUCTION OF ETHYLENE AT PHILLIPS. II. ETHANE OXIDATIVE DEHYDROGENATION

A. D Eastman, J. H. Kolts and J. B. Kimble

Research and Development
Phillips Petroleum Company
Bartlesville, Oklahoma 74004

Ethylene's place as a key building block in the chemical industry has been unchallenged. However, fluctuations in the economics and supply of ethylene precursors have encouraged research on novel ways to produce ethylene. Phillips successful commercialization of the Butene OXD (oxidative dehydrogenation) process to produce 1,3-butadiene from mixed butenes and subsequent extension of the chemistry to n-butane suggested that perhaps ethylene might be synthesized by oxidative dehydrogenation of ethane.

Simplistically, oxidative dehydrogenation can be considered the combination of a simple dehydrogenation coupled with combustion of the produced hydrogen:

$$C_nH_{2n+2} \rightarrow C_nH_{2n} + H_2 \qquad \text{Endothermic}$$

$$H_2 + 1/2\ O_2 \rightarrow H_2O \qquad \text{Highly exothermic}$$

$$C_nH_{2n+2} + 1/2\ O_2 \rightarrow C_nH_{2n} + H_2O \qquad \text{Exothermic}$$

The problem is to keep the hydrocarbon from being completely oxidized. The difficulty of this problem is illustrated by a schematic energy profile of possible reactions of ethane with oxygen, shown as Figure 1.

Activation of the alkane, region 1 in the figure, usually involves breaking a C-H bond. Bond cleavage becomes more difficult (and the activation energy correspondingly greater) as the carbon number decreases. The small valley in the energy plot (at 2) corresponds to olefin formation; olefins are activated much more easily than paraffins. Finally, combustion to CO_x and H_2O is energetically quite favorable. The key problem of oxidative dehydrogenation is to furnish enough energy for activation (region 1) without passing region 2.

Oxidative dehydrogenation possesses some advantages relative to conventional dehydrogenation. Conventional dehydrogenation is equilibrium limited. For example, equilibrium conversion of ethane to ethylene is only 50%

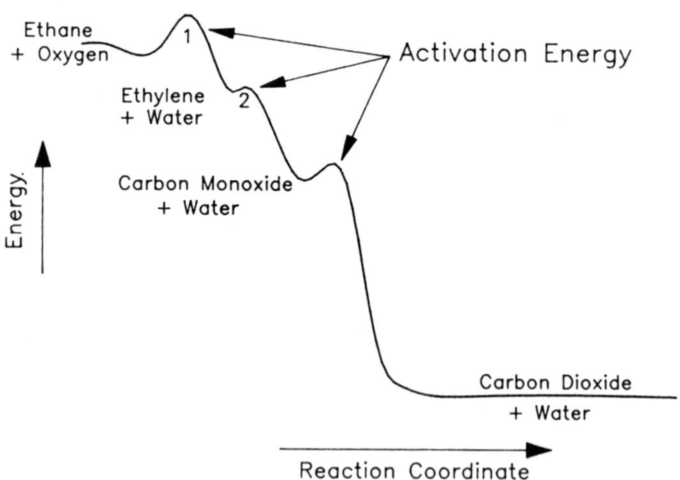

FIG. 1 *Schematic Energy Diagram of Ethane/Oxygen Reaction*

at 1341°F and one atmosphere pressure. Because hydrogen is removed in the OXD reaction, equilibrium conversion is near 100% for OXD reactions at the same conditions. Second, conventional dehydrogenation is endothermic at most temperatures. OXD reactions would be net energy contributors to a commercial installation, though at the cost of hydrogen loss.

The literature describes several processes for producing ethylene via oxidative dehydrogenation of ethane; none of them has been commercially successful. Imperial Oil of Canada, for example, developed a process using a cadmium/alumina catalyst and hydrogen sulfide.(1) They believe the active sulfur from reaction between H_2S and O_2 in the presence of the catalyst promotes the OXD reaction.

Ethyl was granted a series of patents (2-6) on halogenated catalysts for ethane OXD. Ethyl's process uses catalytic amounts of HCl in combination with copper or iron chloride plus rare earth and alkali metal or alkaline earth chlorides supported on a fluidizable carrier. In one version (5), the ethylene product is passed over a supported copper halide catalyst in the presence of excess HCl to give vinyl chloride and dichloroethane. There is a similar Russian patent. (7)

Allied Chemical holds patents (8-9) on a process in which oxygen and chlorine are introduced into a stream of flowing ethane via an axial jet. They claim 88.8% conversion of ethane and 78.8% yield of ethylene at 960°C in the absence of any catalyst.

Many papers (10-19) have dealt with the use of N_2O as a probe to investigate the mechanism of the OXD reaction. No commercial process has emerged from those investigations.

The closest approaches to commercial operation of ethane OXD stem from work at Union Carbide (20-28), using a Mo/V/Nb/Sb catalyst and Phillips Petroleum (29-36), using promoted Co or Mn/Li/Ti oxide catalysts. The Phillips research is the basis of this paper.

In addition to the work cited, several other groups have investigated ethane OXD. Shchukin and his group at the Tolyatti Polytechnic Institute in the U.S.S.R. have looked at the kinetics and mechanism of partial oxidation of ethane (37), and have developed a calcium chloride-based OXD process (38). Another Russian group (39) and an Italian group (40) have examined ethane OXD over Cd-doped zeolite A. Exxon researchers looked at ethane OXD over supported silver catalysts (41), while Sohio has patented both fixed-bed V-P oxide catalysts (42) and an electrocatalytic reaction in a fuel cell (43).

EXPERIMENTAL

CATALYST PREPARATION

Unsupported (bulk) catalysts such as the Co/Sn series were prepared by mixing metal carbonates, nitrates, or chlorides with sodium hydroxide or phosphate, and aqueous ammonia to form slurries. The slurries were dried in air at 125°C, then calcined in air for 3-5 hours at 1500°F (815°C).

Supports for the supported catalysts were prepared by slurrying together in minimum distilled water the appropriate amounts of metal carbonates or oxides. The slurries were dried in air at 125°C, then calcined 3 hours at 1100°F (593°C). Promoters were added by the method of incipient wetness, followed by drying and calcination at 815°C for 3 hours.

REACTION AND ANALYSIS

Prepared catalysts were ground and screened to a convenient size, usually -20/+40 mesh. Four cubic

TABLE 1 *Details of cyclic operation*

Segment	Minutes	Reagent	GHSV
Process	3	Ethane	500
		Nitrogen	500
Purge	3	Nitrogen	500
Regeneration	6	Air	1200
Purge	3	Nitrogen	500

centimeters of catalyst was placed in a quartz reactor tube, which was placed into a tube furnace and heated to 1300°F (704°C).

In continuous flow (co-feed) mode, ethane, air, and diluent nitrogen were passed simultaneously over the catalyst bed. Samples were taken at regular intervals for GC analysis.

Cyclic operation involved a process segment, a nitrogen purge, regeneration in air, and another nitrogen purge; the entire cycle lasted 15 minutes as shown in Table 1

The automated catalyst screening unit employed for this research has been previously described. (44) Furnace temperature was usually 1300°F.

The effluent from 12 consecutive cycles was collected in a large bomb; a portion of that sample was introduced into a dual-column gas chromatograph for computer-controlled analysis.

RESULTS AND DISCUSSION

CYCLIC PROCESS AND CATALYSTS

Promoted cobalt oxide and manganese oxide on lithium/titanium spinel are active and selective catalysts for the oxidative dehydrogenation of ethane in cyclic reaction. Both cobalt and manganese have stable and isostructural oxides of the forms MO and M_3O_4. In this reaction mode, the catalysts were observed to cycle between the two oxides.

Promoted Cobalt Oxide Cobalt oxide gave fairly good yield of ethylene in cyclic operation, but only if promoted with such oxidation suppressors as phosphorus and alkali metals. Because empirical data indicated that the best catalysts contained non-reducible oxides such as that of tin, a balanced incomplete block experiment was conducted to determine the importance of each promoter and to give some indication whether the effect of the promoters was cumulative or if there was interaction among the promoters.

Results of the study are shown in Fig. 2 and 3. Figure 2 shows that lithium, sodium and potassium gave

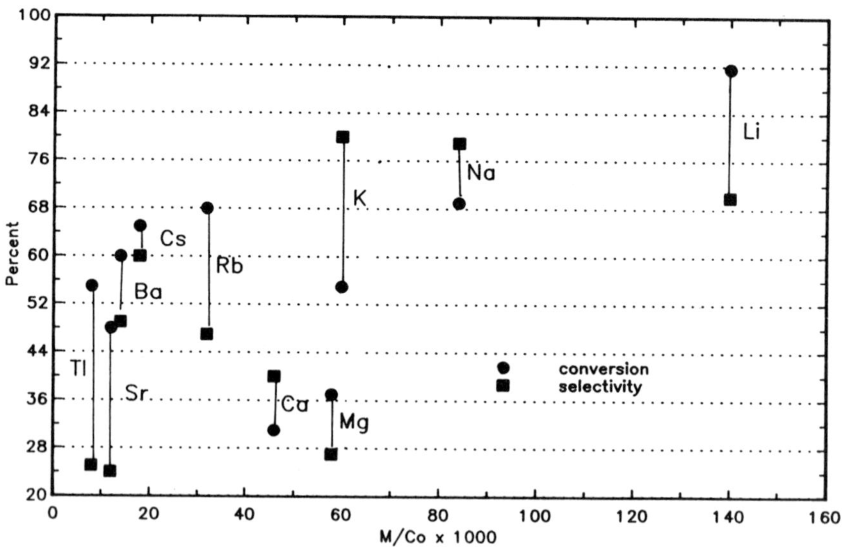

FIG. 2 *Effect of Group I and Group II Metals on Ethane OXD Activity*

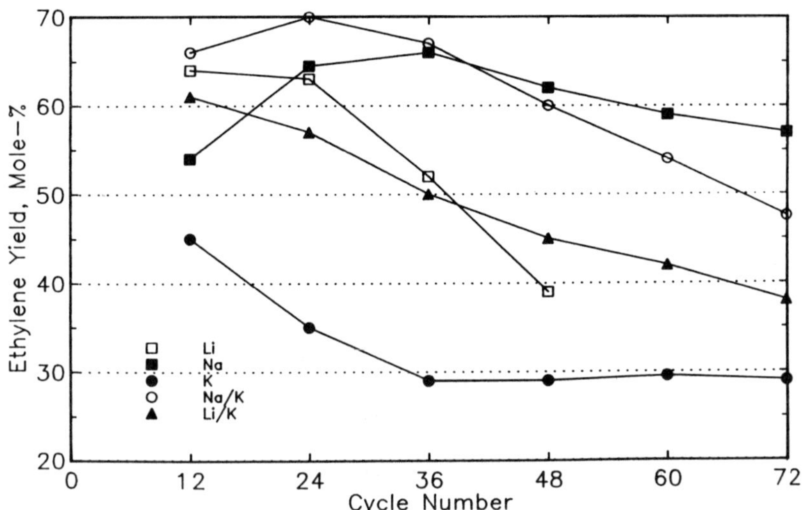

FIG. 3 *Effect of Li, Na, K on Ethane OXD Activity*

the best results. Sodium gives higher conversion but slightly lower selectivity than potassium. Lithium gave very high conversion, but selectivity to ethylene was not as high. Ions larger or smaller than K and Na gave either low conversion (like Ca) or low selectivity (like Sr). Further testing of the alkali metals, shown in Fig. 3, indicated that catalyst behavior varied widely as a function of time on stream for the various metals. Sodium gave the best ethylene yield overall and the most stable performance.

Phosphorus or similar promoters and tin or zirconium were shown to be essential to best catalyst performance. Tin and zirconium are believed to function as moderators or inhibitors of reduction of the cobalt oxide, as shown in Fig. 4. That figure shows the differential thermal analysis traces of cobalt oxide spinel (Co_3O_4) with various promoters in an ethane atmosphere. Reduction to Co metal is rapid and complete

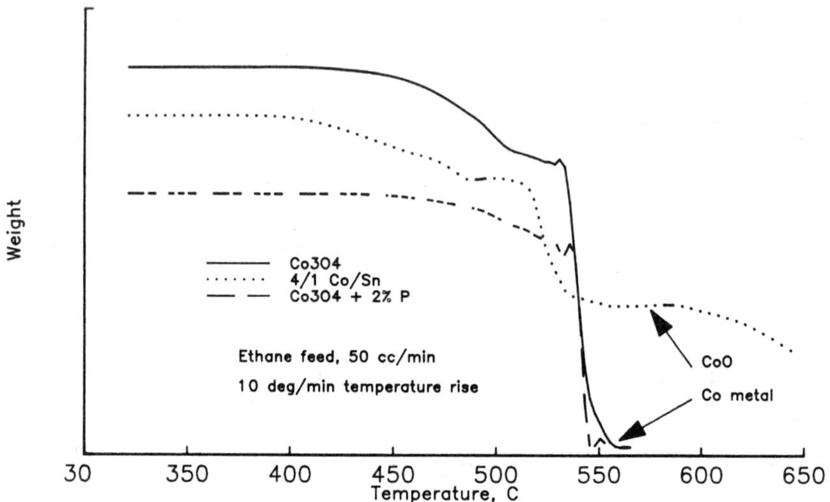

FIG. 4 *Reduction of Cobalt Oxide by Ethane - TGA Evidence*

for both unpromoted Co_3O_4 and Co_3O_4 promoted with phosphorus. However, Co_3O_4 promoted with 25% tin reduces first to CoO, then to cobalt metal, but at a higher temperature than the other samples.

The role of phosphorus in the catalyst seems to be related to ease of re-oxidation of the reduced cobalt oxide. Catalysts promoted with phosphorus had longer lifetimes under cyclic operation than unpromoted materials.

Fresh catalyst showed only Co_3O_4 to x-ray diffraction; as the OXD cycle progressed, increasing amounts of CoO were observed until no more Co_3O_4 was present. Continued contact with hydrocarbon produced cobalt metal. Successful regeneration of the catalyst to Co_3O_4 depended on avoiding reduction to the metal. The crystal structures of CoO and Co_3O_4 are related. Both are based on a cubic-close-packed oxygen lattice; the differences lie in the location of the cations in the lattice. The unit cells for the two oxides are close in size (8.084Å for Co_3O_4 and 8.520Å for the same number of oxygens--twice the unit cell--for CoO), so cycling between the two involves a change in lattice size of roughly 5%.

Efforts to limit reduction of the catalyst did not achieve lifetimes greater than a few hundred cycles with the Co/Sn/P/Na catalyst. Reduction to cobalt metal and catalyst powdering were observed. Shorter cycle time may prevent reduction to the metal, but powdering is probably a consequence of the constant expansion and contraction of the oxide lattice during the redox cycle, and may not be avoidable.

Addition of a small amount of halogen increased both catalyst lifetime and selectivity. The Co/Sn/P/Na catalyst, with the addition of 0.5% chlorine, was able to produce ethane conversions near 60% with selectivity to ethylene greater than 90%, as shown in Fig. 5.

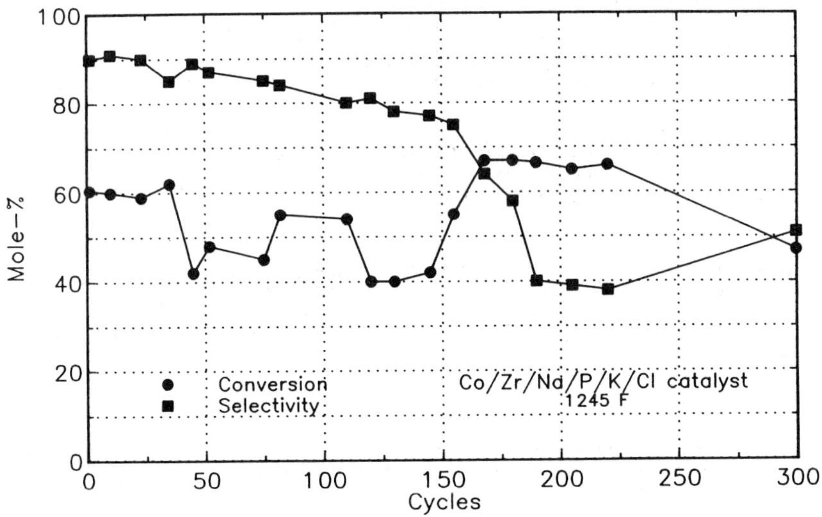

FIG. 5 *Lifetime of Promoted Cobalt Ethane OXD Catalyst*

Promoted Manganese Oxide Manganese oxide supported on silica, alumina, magnesium oxide, or zinc aluminate spinel and promoted with sodium and phosphorus is an active and selective ethane OXD catalyst in cyclic operation. Unsupported catalysts were also active and selective, but their activity declined rapidly with the number of cycles. The supported catalysts maintained their initial activity for more than 2000 cycles, as shown in Fig. 6. Analogous to the cobalt system, manganese oxide alternates in cyclic operation between Mn_3O_4 and MnO. The catalyst as prepared is Mn_2O_3, but the first cycle reduces it to MnO; regeneration produces Mn_3O_4 and regular cycling commences.

In contrast to the cobalt system, tin and zirconium had little effect on manganese oxide. This is probably because, unlike cobalt, manganese can be reduced to the metal only with difficulty; tin and zirconium apparently block reduction of cobalt to the metal.

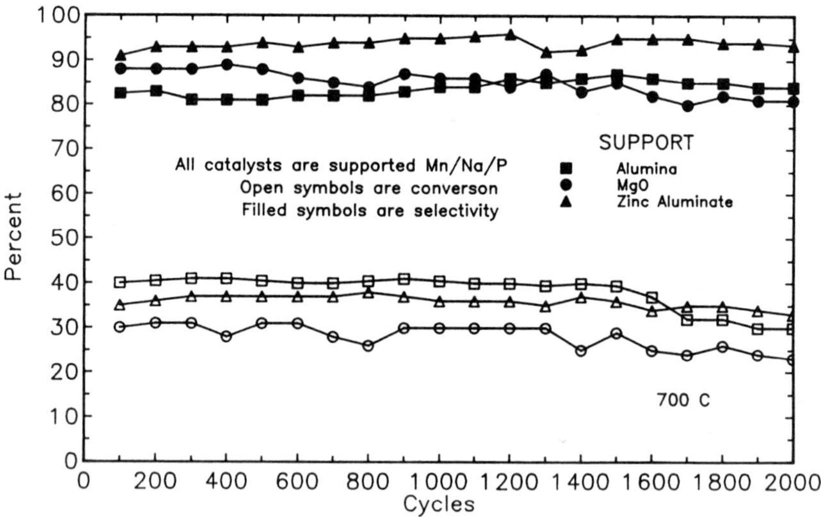

FIG. 6 *Activity of Supported Catalysts for Ethane OXD*

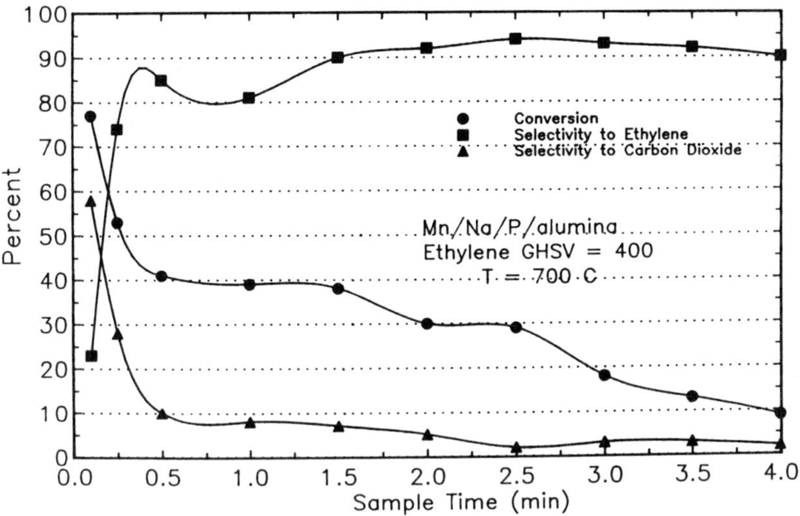

FIG. 7 *Ethane OXD Activity vs Sample Time*

The manganese and cobalt systems behave differently during reaction cycles, particularly with respect to selectivity to ethylene as a function of time. With cobalt, CO_2 formation is relatively independent of time during the cycle. Manganese, however, gives high initial CO_2 production which decreases with catalyst reduction, as shown in Fig. 7

Promoted manganese on a lithium-titanium spinel ($Li_{4/3}Ti_{5/3}O_4$, LTS) or a mixed oxide of stoichiometry $Li_2Mg_2TiO_5$ (LMT) is an active and selective catalyst for OXD of ethane in cyclic operation. Conversions of around 25% at selectivity to ethylene above 90% can be achieved for extremely long periods of time without noticeable catalyst degradation. In contrast to the other catalysts tested, both hydrogen and carbon oxides can be detected in the reaction effluents, suggesting that both oxidative and non-oxidative processes are operative.

CONTINUOUS PROCESSES AND CATALYSTS

Supported Manganese Catalysts Catalyst lifetime and physical stability were improved by passing oxygen and the hydrocarbon concurrently rather than alternately over the catalyst. Supported manganese catalysts give good conversion of ethane and selectivity to ethylene, as shown in Table 2.

Addition of manganese increases selectivity to ethylene, compared to the supports alone. However, the only significant product other than ethylene is CO_2.

Li/MgO Catalysts Mixed lithium/magnesium oxides showed surprising activity for ethane OXD, as demonstrated in Table 3 (45).

Li/TiO_2 Catalysts Mixed lithium/titanium oxides with Li/Ti ratios greater than two show ethane OXD activity. Ethane conversions on the order of 25% with selectivities to ethylene in the 90-95% range are attainable. Catalyst activity is stable with time on stream. Activity, however, is adversely effected by the presence of CO_2 in the feedstock.

TABLE 2 *Manganese-based Co-feed ethane OXD*

Catalyst	Hours	Conversion	Selectivity
MgO	2.7	48.7%	24.1%
MgO	3.2	40.8	55.7
$Li_2Mg_2TiO_5$ (LMT)	1.6	57.7	28.5
2.5% Mn on LMT	92.7	56.0	64.4
2.5% Mn/MgO	4.7	50.1	61.9
Quartz chips	0.01	12.8	41.9

All runs at 700°C, with GHSV (Ethane) = 600
GHSV (Air) = 1800

TABLE 3 *Lithium/Magnesium oxide catalysts*

Catalyst	Hours	Conversion	Selectivity
MgO	1	30.6	55.9
MgO	11	25.7	54.1
0.5% Li/MgO	1	55.6	92.1
0.5% Li/MgO	16	32.5	92.6

The catalytic activity of the mixed Li/Ti oxide cannot be ascribed to any particular compound, as mixtures of such materials as $Li_4Ti_5O_{12}$, $Li_{0.8}Ti_{2.2}O_4$, $LiTi_2O_4$, and $LiTiO_2$ are seen in X-ray diffraction patterns from the active catalysts. All the oxides seen, however, are based on the spinel structure, and so may share a common surface configuration.

In addition to the activity shown by the mixed oxides, metals supported on the Li/Ti mixed oxides are active and selective ethane OXD catalysts. Figure 8 shows results from a number of catalysts made by

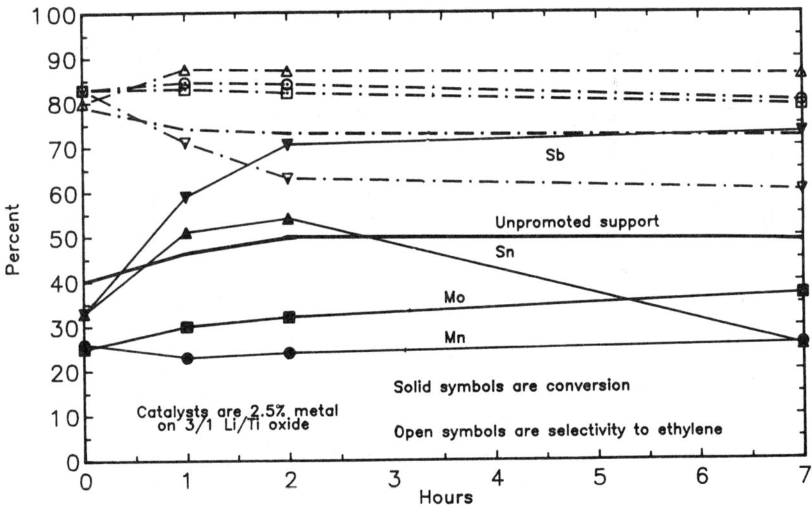

FIG. 8 *Ethane/Air over Metals on 3/1 Li/Ti Oxide*

impregnating a 3/1 Li/Ti mixed oxide with 2.5% metal. Antimony gave the highest conversion, but with selectivity to ethylene of only about 60%. Tin had good initial activity, but declined after a few hours on stream. Molybdenum and manganese both showed 25-30% conversion, and selectivities above 80%; both had activity which was stable with time on stream.

Promoted Co/Zr Catalysts Promoted unsupported cobalt/zirconium catalysts similar to those used as cyclic catalysts are active and selective for co-feed ethane OXD. Conversion of ethane decreases from 95% early in the reaction sequence to 40% after 24 hours on stream. Selectivity to ethylene changes from 85% initially to 95% during the same period.

Catalytic activity of aged catalyst can be restored by brief exposure to a chlorine-containing gas such as methyl chloride. In addition, catalyst performance can be improved by following chlorine exposure with a short

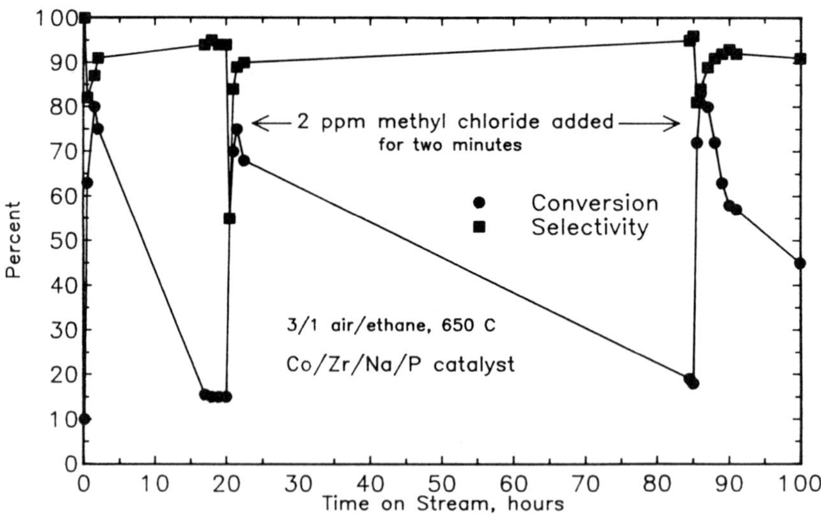

FIG. 9 *Effect of Chlorine on Ethane OXD Activity*

(<1 hour) period of exposure to ethane in the absence of air. Addition of ppm quantities of chlorine to the reaction feed gas raises conversion and retards (but does not completely prevent) deactivation, as shown in Fig. 9.

MECHANISMS

Ethane OXD was at first thought to proceed via a relatively simple Mars-Van Krevelen mechanism in which oxygen for the oxidation reactions is furnished from the transition metal oxide lattice. Pulse reactor studies performed at Phillips on a Co/Sn oxide catalyst suggested the following reaction mechanism:

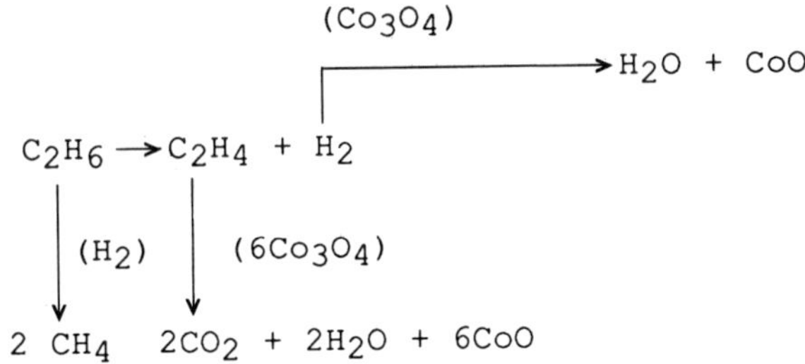

The first step in the original mechanism is the reversible catalytic dehydrogenation of ethane. Ethylene and hydrogen then compete for lattice oxygen. Since ethane does not, in this scheme, react directly with lattice oxygen, the highest selectivities should be obtained at the lowest conversion levels and selectivity should improve with lattice oxygen removal. Both effects were seen in the pulse experiments. In addition, ethane is relatively insensitive to the presence of gas-phase oxygen. Experiments using pulses containing as much as 20% oxygen still gave good selectivity (>80%) to ethylene.

However, experimental anomalies suggest that one or

more other mechanisms may also be operative. For example, the activity of lithium/magnesium or lithium/titanium oxides does not fit the Mars-Van Krevelen scheme, since no easily reducible oxide is present in either case.

The activity of non-reducible oxides suggests a free-radical reaction mechanism similar to that of methane oxidative coupling, which has been the object of much work at Phillips and elsewhere. The initial reaction step, whatever the catalyst, appears to be activation of the catalyst by formation of a catalyst-oxygen complex, with oxygen possibly present as O^-. Ethane reacts with this species to form ethyl radicals, which then desorb and react further in the gas phase, as in the analogous methane reaction.

For non-reducible metals, oxygen is supplied from the gas phase. With reducible metals, however, oxygen is also provided by the catalyst lattice. It is likely that alkali or alkaline-earth metals provide a locus for the activated oxygen, as reducible metals without those promoters give only carbon oxides in the presence of ethane. Thus, reducible metals may function as an oxygen pump, absorbing oxygen from the gas phase and channeling it into active surface sites.

In the methane coupling reaction, deep oxidation to carbon oxides is believed to occur mostly on the catalyst surface, though there is some gas-phase reaction as well. The same pattern also appears to apply to ethane OXD.

Morales and Lunsford (46) recently studied ethane OXD over the Li/TiO_2 mixed oxide catalyst, concluding that the primary mechanism involved is generation of ethyl radicals, followed by desorption into the gas phase.

The effect of chlorine on ethane OXD is not

completely clear. However, chlorine appears to increase ethane conversion as an apparent source of free radicals, increasing the rate of ethane radical formation. In addition, chlorine may aid in regeneration of the reduced metal oxides in cyclic systems by facilitating metal atom transfer through formation of volatile metal chlorides.

CONCLUSIONS

Oxidative dehydrogenation of ethane has been shown to occur over a number of different catalysts in both cyclic (redox) and co-feed reaction modes. The most likely reaction model is a free-radical mechanism similar to that proposed for the methane coupling reaction. For reducible metal oxides in cyclic operation, some lattice oxygen participates in the reaction. Despite the promise of escaping equilibrium restrictions on ethane conversion, ethane OXD is not commercially viable.

ACKNOWLEDGMENTS

We are grateful to Sue Gaughan, Larry Potts, Helen Vanderveen, and Vernon Lakey, who did much of the lab work on ethane OXD.

REFERENCES

1. Gaspar, N.J.; Pasternak, I.S.; Can. J. Chem. Eng., 49(2), 248-251, 1971.
2. Beard, William Q., Jr., U.S. Patent 3 644 560, 1972, to Ethyl Corp.; Chem. Abstr. 76(17):99066e.
3. Beard, William Q., Jr., U.S. Patent 3 644 561, 1972, to Ethyl Corp.; Chem. Abstr. 76(17):99067f.
4. Beard, William Q., Jr., U.S. Patent 3 658 933, 1972, to Ethyl Corp.; Chem. Abstr. 77(6):35210k.

5. Beard, William Q., Jr., U.S. Patent 3 658 934, 1972, to Ethyl Corp.; Chem. Abstr. 77(6):35211m.

6. Beard, William Q., Jr., U.S. Patent 3 769 362, 1973, to Ethyl Corp.; Chem. Abstr. 80(1):3054h.

7. Luk'yanova, T.E.; Shchukin, V.P.; Obrubov, V.A.; Avergukh, A. Ya.; U.S.S.R. SU 667 229, 1979; Chem. Abstr. 93(17):167288g.

8. Kurtz, Bruce E.; Ger. Offen. DE 2 645 766, 1977, to Allied Chemical Corp.; Chem. Abstr. 87(2):6637f.

9. Kurtz, Bruce E.; Smalley, Edmund W.; Ger. Offen. DE 2 654 424, 1977, to Allied Chemical Corp.; Chem. Abstr. 87(6):40056a.

10. Ward, M.B.; Lin, M.J.; Lunsford, J.H.,; J. Catal., 50(2), 306-318, 1977.

11. Lunsford, Jack H.; Ward, Mark B.;Yang, Tsong-Jen; Chem. Uses Molybdenum, Proc. Int.Conf., 3rd, 166-169. Ed. Barry, H.F.; Mitchell, P.C.H., Climax Molybdenum Co., Ann Arbor, MI, 1979.

12. Yang, Tsong-Jen; Lunsford, Jack H.; J. Catal., 63(2), 505-509, 1980.

13. Iwamoto, Masakazu; Taga, Toshiyuki; Kagawa, Shuichi; Chem. Lett. (9), 1469-1472, 1982.

14. Aika, Kenichi; Tajima, Masaki; Onishi, Takaharu; Chem. Lett. (11), 1783-1786, 1983.

15. Mendelkovici, Leonardo; Lunsford, Jack H.; J. Catal., 94(1), 37-50, 1985.

16. Aika, K.; Tajima, M.; Isobe, M.; Onishi,T.;Int. Congr. Catal., [Proc.], 8th, 1984, Vol. 3, III 335-III 346, Verlag Chemie; Weinheim, FRG, 1985.

17. Iwamatsu, Eiji; Aika, Kenichi; Onishi, Takaharu; Bull. Chem. Soc., Jpn., 59(6), 1665-1669, 1986.

18. Aika, Kenichi; Isobe, Makoto; Kido, Kazuhiro, Moriyama, Takeshi; Onishi, Takaharu; J. Chem. Soc., Faraday Trans. 1 82(10), 3139-3148, 1987.

19. Vereshchagin, S.N.; Baikalova, L.I.; Anshits, A.G.; Izv. Akad. Nauk SSSR, Ser. Khim. (8), 1718-1722, 1988; Chem. Abstr. 110(16):135673m.

20. Young, Frank Glynn; Thorsteinson, Erling Magnus; Ger. Offen. DE 2 644 107. 1977.

21. Thorsteinson, E.M.; Wilson, T.P.; Young, F.G.; Kasai, P.H.; J. Catal., 52(1), 116-132, 1978.

22. U.S. Patent 4 250 346, 1981, to Union Carbide Corp; Chem. Abstr. 94(21):174288f.

23. McCain, James H.; U.S. Patent 4 524 236A, 1985, to Union Carbide Corp.; Chem. Abstr. 104(2):6298r.

24. McCain, James Herndon; Eur. Pat. Appl. EP 166 438A2, 1986, to Union Carbide Corp.; Chem. Abstr. 104(21):185967c.

25. Brockwell, Jonathan L.; Kendall, John E.; Arabian J. Sci. Eng. 10(4), 351-360, 1985.

26. Manyik, Robert M.; McCain, James H.; U.S. Patent 4 596 787A, 1986, to Union Carbide Corp.; Chem. Abstr. 105(12):98107z.

27. Manyik, Robert Michael; Brockwell, Jonathan Lester; Kendall, John Edward; Eur. Pat. Appl. EP 261 264 A1, 1988; Chem. Abstr. 109(2):7116v.

28. Japan. Kokai Tokkyo Koho JP 63/96140 A2, 1988; Chem. Abstr. 109(22):191042s.

29. Marsheck. Robert Milton, UK Pat. Appl. GB 2 050 188, 1981; to Phillips Petroleum Co.; Chem Abs. 95(5):42321x.

30. Eastman, Alan D.; Kolts, John H.; U.S.Patent 4 310 717, 1982; to Phillips Petroleum Co.; Chem. Abstr. 96(21):180742g.

31. Eastman, Alan D.; U.S.Patent 4 368 346, 1983; to Phillips Petroleum Co.; Chem. Abstr. 98(16):126786c.

32. Eastman, Alan D.; Kimble, James B.; U.S.Patent 4 450 313, 1984; to Phillips Petroleum Co.; Chem. Abstr. 101(10):74759e.

33. Kimble, James B; U.S.Patent 4 476 344, 1984; to Phillips Petroleum Co.; Chem. Abstr. 102(4):25199j.

34. Eastman, Alan D.; Guillory, Jack P.; Cook, Charles F.; Kimble, James B.; U.S. Patent 4 497 971, 1985; to Phillips Petroleum Co.; Chem. Abstr. 102(16):133973r.

35. Kolts, John Henry; Guillory, Jack Paul; Eur. Pat. Appl. EP 205 765, 1986; to Phillips Petroleum Co.; Chem. Abstr. 106(16):120370z.

36. Japan Kokai Tokkyo Koho JP 61/282322 A2, 1986; to Phillips Petroleum Co.; Chem. Abstr. 107(10):79935t.

37. Averbukh, A. Ya.; Shchukin, B.P.; Evpova, T.I., Rutkovskii, V.I.; Kinet. Mater.-Vses. Konf. Kinet. Katal. Reakts., 2nd, Vol. I, 70-73. Akad. Nauk SSSR, Sib. Otd., Inst. Katal.: Novosibirsk, USSR, 1975; Chem. Abstr. 86(3):16048h.

38. Luk'yanova, T.E.; Shchukin, V.P.; Obrubov, V.A.; Averbukh, A. Ya.; U.S.S.R. SU 667 229, 1979; Cham. Abstr. 91(15):123383p.

39. Gaziev, R.G.; Berman, A.D.; Yanovskii, M.I.; Elinek, A.V.; Krylov, O.V.; Tkachenko, O.P.; Shpiro, E.S.; Kinet. Katal., 29(2), 242-430, 1988; Chem. Abstr. 109(21):189645x.

40. Ottana, Rosaria; Parmaliana, Adolfo; Zipelli, Cesare; Giordana, Nicola; Ann. Chim. (Rome), 73(7-8), 403-410, 1983.

41. Wachs, Isreal E.; Keleman, Simor R.; Prepr. Div. Pet. Chem., Am. Chem. Soc., 25(2), 332-333, 1980.

42. Blum, Patricia, R.; Milberger, Ernest c.; U.S.Patent 4 410 752, 1983; Chem. Abstr. 100(1):5826s.

43. Mazanec, Terry J.; Cable, Thomas L.;UK Pat. Appl. GB 2 203 446 A1, 1988; Chem. Abstr. 110(16):135899k.

44. Hogan, R.J.; Doane, E.P.; Preprints, Div. Petr. Chem, ACS, 16(4), 035, 1960.

45. Kolts, John H.; Guillory, Jack P.; U.S.Patent 4 672 145, 1987.

46. Morales, Edrick; Lunsford, Jack H.; J. Catal. 118, 255-265, 1989.

3

CATALYTIC PRODUCTION OF ETHYLENE AT PHILLIPS. III. ETHYLENE-FROM-BUTANE CRACKING CATALYSTS

G. A. Delzer and J. H. Kolts

Research and Development
Phillips Petroleum Company
Bartlesville, Oklahoma 74004

Our investigation of the catalytic cracking reaction of butane to make ethylene started with the observation that n-butane cracks preferentially to ethylene rather than to propylene over an Mn/MgO catalyst. The initial work with Mn/MgO, and subsequently with Fe/MgO and Mn/La_2O_3, showed that ethane is also a major product, thus demonstrating the uniqueness of this series of cracking catalysts.

Figure 1 illustrates what is observed when n-butane is cracked over an Mn/MgO or an Fe/MgO catalyst, as compared to thermal pyrolysis over quartz chips. It can be seen that conversion is enhanced by 100%. Ethylene and ethane are also doubled. Propylene selectivity is down by 50%. The observed selectivity is not what would be expected for a free radical pyrolysis reaction nor is it what would be seen with traditional acid cracking

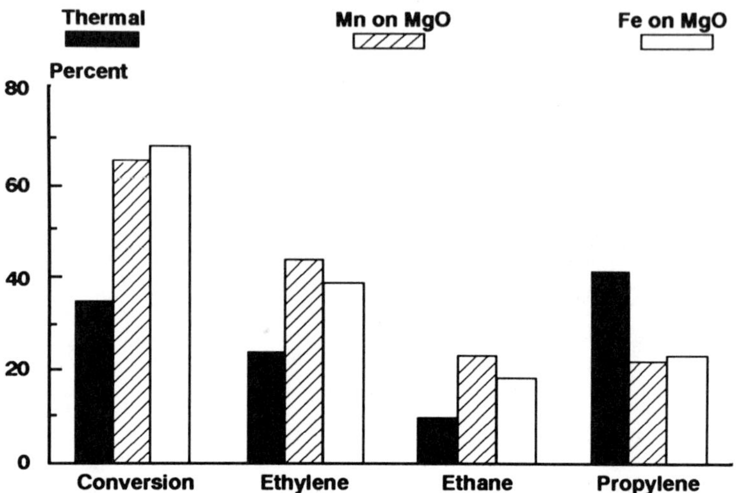

FIG. 1 *Comparison of thermal and catalytic cracking of n-butane. The temperature is 725 C.*

catalysts which tend to go through a carbenium ion mechanism (1). Free radical cracking and acid cracking would also be expected to give increased selectivities to propylene and methane.

Several techniques have been used to characterize the butane cracking catalysts to help understand their unique behavior. Reactor studies with active and related, but inactive, catalysts were conducted. In addition, physical measurements such as thermal gravimetric analysis, x-ray diffraction, electron spin resonance, and x-ray photoelectron spectroscopy were used. These were applied to both the oxidized and the reduced forms of the catalyst since it is likely that the catalyst is in a reduced form under our reaction conditions. The principal findings confirm that the ethylene-from-butane cracking catalysts are a unique series of materials.

EXPERIMENTAL

Catalyst Testing The catalysts were prepared either from the oxide or from the hydroxide as the base. Manganese was added as either the acetate salt or by means of a 50% manganese nitrate solution. Iron was added as the nitrate salt. In one preparation method, incipient wetness was used to add the desired level of manganese or iron (expressed as weight percent of the metal relative to the oxide form of the support). Either water or methanol was used as the solvent in the impregnation. In another method, the hydroxide support was slurried with a small amount of water, and the nitrate salt was dissolved in water and added to the slurried support in a blender. This mixture was blended for five minutes, the sides scraped down, and blended for five more minutes. Drying was done at 125C for five hours. The solid mass was broken up, heated slowly in a forced draft oven to 350-400C, and held for three hours. A further calcination was then performed by heating slowly to 750-800C and holding for 4-5 hours. The resulting solid was ground and sieved, usually to -20/+40 mesh for the catalytic studies. All samples were analyzed for iron or manganese content, as appropriate. No differences in catalytic activity were observed because of the particular method of preparation.

The testing reactor was constructed of 15mm ID quartz tubing with a thermocouple well centered axially along the catalyst bed. During testing the catalyst was supported by plugs of quartz wool. Feed was preheated by passing it over 5 cc of -16/+25 mesh quartz chips placed directly upstream of the catalyst bed. Hydrocarbon, air, hydrogen, and nitrogen were individually metered using mass flow controllers and were measured using calibrated electronic flow sensors. Steam was added by saturating flow to the reactor with water heated to the temperature required for the desired partial pressure. In some experiments, liquid water was pumped at a suitable rate into the reactor with an ISCO Model 314 metering pump. Hydrocarbons used were Phillips pure or research grade; all other gases were Linde CP grade and were used without further purification.

Gaseous products from the reactor were analyzed with a Hewlett-Packard 5880 gas chromatograph using a stacked column with 5 ft. of 2.4% Carbowax 1540 on Porasil C (80/100) and 18 ft. of 27% bis-2-EEA, 6.5% DC-200 (500 cSt), and 3.5% Carbowax 1540 on Chromosorb P AW (60/80), and two additional columns with 6 ft. of Porapak Q (80/100) and 10 ft. of 13X molecular sieve (42/60) at 50C. Hydrogen analysis was done on a Carle BGC using nitrogen carrier gas. The chromatographs were calibrated using a certified standard blend from Air Products. Reported selectivities are based on moles of feed converted to a particular product. Coke was not factored into the selectivity calculations.

Unless otherwise noted, catalyst treatment for reaction testing was carried out using the following series of steps:

1) heat to reaction temperature in flowing nitrogen for 60 minutes;
2) regenerate in air for 20 minutes;
3) purge with nitrogen for 5 minutes;
4) reduce with hydrogen for 10 minutes; and
5) flow hydrocarbon for the desired time interval.

All new catalysts were cycled through steps 2-5 a minimum of five times before analytical data were taken. In this case the feed time was 30 minutes. When steam was to be used during an experiment, it was added to all flows. The flow rates of air, nitrogen, and hydrogen were set to be the same as that for the hydrocarbon. Sample times were based upon the total minutes of feed to which the catalyst had been exposed since the last air regeneration.

X-Ray Photoelectron Spectroscopy Oxidized or reduced catalyst samples were prepared in a microreactor, quenched, and sealed under nitrogen. Samples were then loaded onto the XPS sample holder in a glove box and transported to the spectrometer in an air-tight carrier. XPS measurements were performed using a Physical Electronics Model 548 electron spectrometer with AlKα radiation (1486.6 eV). The spectrometer was interfaced to a Hewlett-Packard 21MX computer for data acquisition and manipulation, including curve resolving. The instrument was operated at about 2×10^{-9} torr, with the samples being introduced into the ultrahigh vacuum from a prechamber evacuated to about 1×10^{-6} torr. Binding energies were referenced to carbon 1s at 284.6 eV. Referencing to gold gave inconsistent results. The relative surface coverage was determined by using both Scofield's cross sections (2) and the method of calculation described by Carter et al. (3).

Matrix Isolation of Radicals The method of Lunsford (4) was used for matrix isolation electron spin resonance experiments to measure radicals formed by passing butane over a reduced catalyst. These were done using a Varian Model V4500 ESR spectrometer. Figure 2 indicates the arrangement of the reactor, matrix isolation, and gas analyzer systems. The reactor for the matrix experiments was constructed of fused quartz (2.5 cm ID, 35.8 cm length) and had a thermocouple well centered along its axis which allowed measurement of the temperature

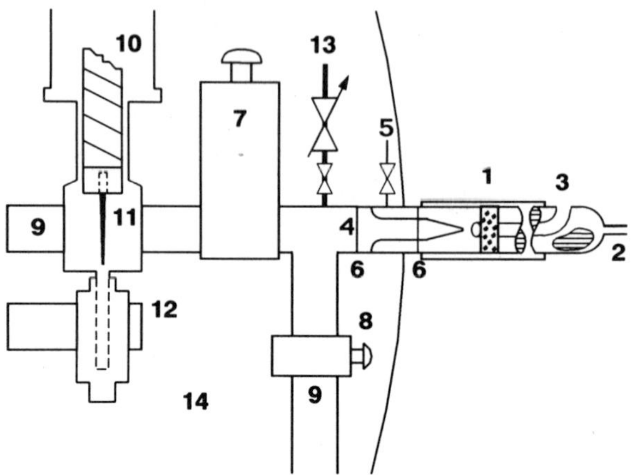

FIG. 2 *Schematic for the matrix isolation and gas analyzer systems coupled to the catalytic reactor: (1) catalyst, (2) gas inlet, (3) thermocouple well, (4) pressure probe, (5) metal valve, (6) "O" ring joints, (7) gate valve, (8) butterfly valve, (9) to vacuum pumps, (10) vacuum shroud, (11) sapphire rod, (12) microwave cavity, (13) quadrupole mass spectrometer inlet, and (14) ESR magnet.*

along the entire reactor length. The reactor was resistively heated over a 23.5 cm length and contained a 9.0 cm reaction zone and a 14.5 cm preheater zone. A temperature profile of the reaction zone showed a maximum variation of $\pm$2C. A gas leak was placed between the exit of the reactor and the collection system, creating a pressure gradient. Pressure in the reaction zone was approximately 0.7 torr; in the collection region, the pressure was typically 2×10^{-5} torr. The reactions were generally run at 600C.

To conduct a matrix isolation experiment, a regenerated catalyst was heated under vacuum at 600C. After 15-20 minutes, evacuation was discontinued, and gas flows were started. The catalyst was reduced for several minutes in hydrocarbon flow before radical collection was begun. Radicals were collected and analyzed in a solid argon matrix on a sapphire rod cooled to about 13K. The collection period for most samples was 25 minutes. Analysis of the radicals was accomplished by lowering the sapphire rod, under vacuum at 13K, into a TE102 microwave cavity and recording the spectrum. Spin concentrations were compared on a relative basis and were not calibrated. This was suitable as long as comparison was made only for spins that had the same spectral shape.

Other Analytical Tests Thermal gravimetric studies (TGA) were done on a DuPont 1090 Thermal Analyzer. Catalysts were sequentially reduced with hydrogen and oxidized in air. The powder diffraction (XRD) measurements were made on a Philips instrument using both oxidized and reduced samples of several catalysts. Additional electron spin resonance (ESR) measurements on several catalyst samples were done with a Varian E-4 spectrometer.

RESULTS

Table 1 shows the results obtained from reactor tests with a number of catalyst systems. Each material was run at a temperature to give 50% conversion of the *n*-butane to products, excluding coke. It can be seen that selectivity to C_2 products is especially good for manganese supported on magnesium oxide and on lanthanum oxide. Iron supported on magnesium oxide is also an effective catalyst. Intermediate selectivities result from manganese or iron on calcium oxide and from iron on lanthanum oxide. Manganese on silica exhibits little advantage over pyrolysis on quartz chips.

The effects of manganese concentration (up to 1%)

Table 1. *Activity Data for Cracking of n-Butane over Various Catalysts*

Catalyst	Temp C	Conv %	Selectivity, %			$\frac{C2= + C2}{C3}$
			C2=	C3=	C2	
Quartz chips	720	50	30	39	7	0.9
4%Mn/MgO	659	50	35	27	19	2.0
5%Mn/CaO	718	50	32	32	11	1.3
3%Ca/5%Mn/MgO	652	50	36	25	24	2.5
1%Al/3%Ca/4%Mn/MgO	675	48	40	19	28	3.6
3%Ca/3.5%Mn/SiO_2	702	50	32	39	7	1.0
5%Fe/MgO	675	50	31	26	19	1.9
5%Fe/CaO	705	50	30	34	10	1.2
4%Mn/La_2O_3	660	50	27	24	21	2.0
5%Fe/La_2O_3	672	50	18	21	14	1.5

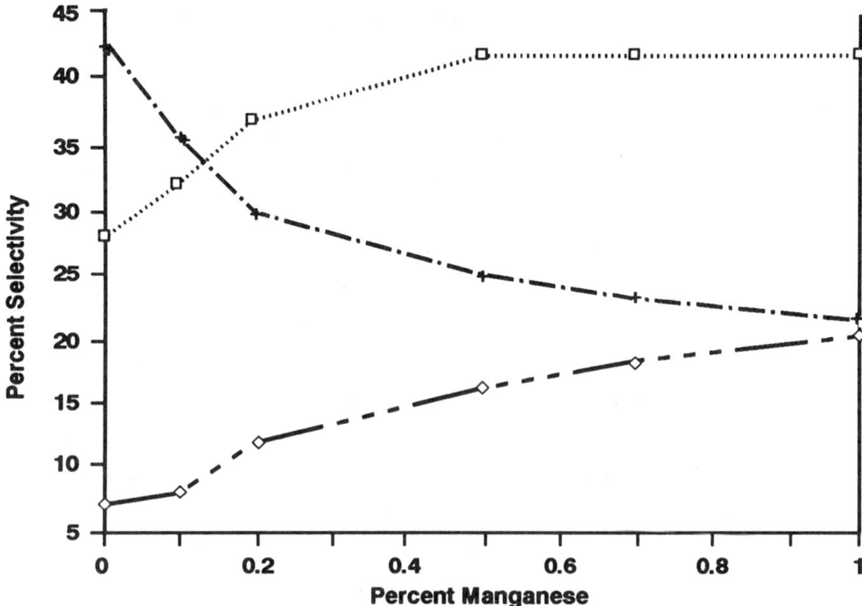

FIG. 3 *Variation of selectivity for ethylene (□), ethane (♦), and methane (+) in the cracking of n-butane over Mn/MgO at increasing manganese concentration.*

on *n*-butane cracking are shown in Fig. 3. Small amounts of manganese rapidly increase the selectivity to ethylene, which reaches a maximum at 0.5% Mn. Ethane selectivity does not increase as rapidly; its maximum is reached at 1% Mn. The decline in propylene and methane follows inversely the changes in both ethylene and ethane selectivity. For manganese concentrations greater than 1% (up to 17%), the observed selectivities do not change within experimental error. Conversion also increases from 24% to over 50% within this same range in manganese concentration. As before, however, most of the increase occurs below 1% Mn.

Several other features of the catalytic cracking of butane over these catalysts are the following:

a) Hydrogen reduction of the Mn/MgO catalyst is not needed. The feed hydrocarbon adequately reduces the manganese.

b) Substantial improvement in the lifetime of the catalyst can be achieved by operating at a 2:1 steam to hydrocarbon ratio.

c) The addition of aluminum as a promoter increases the ethane selectivity.

d) Ethylene is not hydrogenated over the catalyst.

e) Whereas the presence of calcium improves the lifetime of the manganese catalyst when steam is co-fed with hydrocarbon, its addition to an Fe/MgO catalyst impairs conversion and selectivity.

f) Compared to pyrolysis, this type of catalytic cracking offers an advantage in that the C_5+ fraction is reduced. Separation costs in purifying the products, therefore, would be lower.

g) The reaction is first order in *n*-butane, and the apparent activation energy is 53 kcal/mole for 4%Mn-3%Ca/MgO. The corresponding energies for reaction over quartz and magnesium oxide are 47 and 59 kcal/mol, respectively, which may be within experimental error of the value for the catalyst.

Table 2 shows results from TGA studies where the catalysts were cycled between reduction and air oxidation. The weight loss for the Mn/MgO catalyst indicates that there was almost one oxygen atom change per manganese atom. Thus, it is likely that the oxidized form of manganese on magnesium oxide is in the 4+ oxidation state and that the reduced form is in the 2+ oxidation state. Similar results were found for Mn/La_2O_3. Mn/CaO also shows nearly the same change in oxidation state, but for reasons we don't understand, it is not catalytically active. The implication that manganese is stabilized in the higher oxidation state

Table 2. *Thermogravimetric analysis (Reduction-Oxidation) of selected manganese and iron catalysts.*

Catalyst	O/M* Reduction Cycle First	Second	Fourth
Manganese oxide	0.33	0.33	0.33
4%Mn/SiO_2	0.43	0.34	0.37
4%Mn/CaO	1.07	1.03	1.08
8%Mn/CaO	0.96	0.85	0.90
3%Mn/La_2O_3	1.17	0.96	0.88
4%Fe/MgO	0.94	0.52	0.44
4%Mn/Al_2O_3	0.28	0.23	0.22
6%Mn/MgO	1.04	0.86	0.86
4.5%Mn-3%Ca/MgO	1.04	1.03	0.95

*O/M is defined as the change in O atoms per metal atom when going from the oxidized to the reduced state. Calculated O/M values are shown below:

Oxidized	Reduced	O/M
MnO_2	MnO	1.00
MnO_2	Mn_3O_4	0.67
MnO_2	Mn_2O_3	0.50
Mn_2O_3	MnO	0.50
Mn_3O_4	MnO	0.33
Mn_2O_3	Mn_3O_4	0.17

has been previously noted in the papers of Cordischi (5) and Derouane (6). The values of O/M in Table 2 for most other materials that do not show the same catalytic properties is much less, and so it is likely that manganese exists on them as more nearly a bulk phase material.

XRD patterns from oxidized and reduced samples of the several catalysts showed that all supports, except for silica and alumina, had high crystallinity. Manganese components for the oxidized catalyst were easily found on the magnesium oxide, but less easily on lanthanum oxide. Oxidized Mn/MgO gave evidence of compound formation with the support, as indicated by the presence of Mg_6MnO_8. MnO was found on the reduced catalyst. This result supports the TGA finding. The MnO was, however, only detected at Mn levels greater than 10%; below this level no reflections for Mn-containing species were detected on reduced catalysts.

X-ray photoelectron spectroscopy (XPS) data on oxidized and reduced samples are given in Table 3. This technique also indicated Mn(IV) and Mn(II), respectively, as the predominant manganese oxidation states. In addition, XPS measurements on catalytically active Mn/MgO indicated that most of the Mn was uniformly distributed throughout the MgO lattice with only slight enrichment at the catalyst surface. Similar analysis of Mn/SiO_2 and Mn/Al_2O_3 showed the oxidized form to be Mn_3O_4 and the reduced form to be MnO, a difference corresponding to much less than one oxygen per manganese. As the samples were reduced, the relative surface coverage of manganese tended to increase. (There was one exception--manganese on calcium oxide showed lower surface coverage on the reduced catalyst than it did on the oxidized catalyst.) The magnitudes of the relative surface coverages also went along with expectations from changes in the surface areas (7). For example, when supported on silica, the manganese catalyst had the largest BET surface area. Accordingly, one would expect a more dilute coverage of the surface by manganese and, indeed, Mn/SiO_2 had the lowest relative surface coverage. The XPS data for iron catalysts are harder to interpret because, as long as iron metal is not formed, one would not expect to see significant changes in binding energy with oxidation state. The uniformly high values measured confirm this. It was shown that reduction with steam/H_2 did not reduce the surface iron as greatly as hydrogen alone did.

Table 3. *X-Ray Photoelectron Spectroscopy of Oxidized and Reduced Manganese and Iron Catalysts.**

Catalyst	Binding Energy, eV Oxidized		Binding Energy, eV Reduced	Δ, eV	Relative Surface Coverage Oxidized	Relative Surface Coverage Reduced
4%Mn/SiO_2	641.66		641.94	(0.28)	0.6	0.9
4%Mn/CaO	641.77		640.76	1.01	4.3	3.4
5%Mn/La_2O_3	641.69		641.01	0.68	11.8	11.6
5%Mn/Al_2O_3	641.95		642.01	(0.06)	2.6	2.7
6%Mn/MgO	642.66		641.14	1.52	4.1	4.7
4%Fe/MgO	711.33	H_2:	710.73	0.60-	---	---
		H_2/H_2O:	710.97	0.36	5.1	5.5

*tabulated values of binding energies ($2p_{3/2}$)

MnO_2	641.8 eV
Mn_2O_3	641.4
Mn_3O_4	641.0
MnO	640.2
Fe_3O_4	711.2
Fe_2O_3	711.2
FeO	710.3
Fe	706.8

In ESR measurements on oxidized and reduced samples of manganese catalysts supported on magnesium oxide, calcium oxide, and silica, only manganese(II) was detected, regardless of the sample treatment. On MgO, two forms of Mn(II) were detected, as shown in Fig. 4. The sharp six-line structure corresponds to a hyperfine splitting of Mn(II) in a highly ordered cubic environment. The broad underlying structure, shown as a dashed line in Fig. 4, corresponds to Mn(II) in an unsymmetrical environment, presumably MnO clusters. The decrease in the intensity of the hyperfine lines from left to right results from Mn(II) ions being in octahedral sites that are distorted due to their nearness to the particle surface (5,8). Comparison of

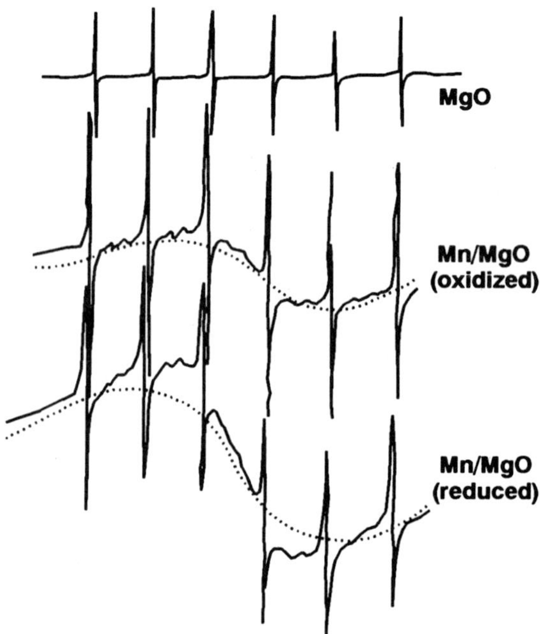

FIG. 4 *ESR spectra of impurity Mn in MgO and in Mn/MgO catalyst that has been oxidized or reduced.*

the line height ratios between the oxidized and reduced catalysts indicated that the manganese seen by ESR did not change its position relative to the surface when oxidized or reduced. ESR for manganese on silica showed only the broad structure when reduced and no signal when oxidized.

Manganese(IV) was not detected in any of the prepared samples. This follows reported literature results (6,9,10) in which Mn(IV) was only detected at very low concentrations or when a charge compensating ion was present. Mn(III), because of its high-spin state, has unpaired electrons, but, as indicated by Derouane (6), there have been no reports in the literature of an ESR spectrum being observed for this species.

In matrix isolation measurements, radicals produced over MgO, La_2O_3, Mn/MgO, and Mn/La_2O_3 were frozen out of the gas phase into a 13K argon matrix. Figure 5 shows the ESR spectra obtained after passing butane over Mn/La_2O_3 and, for reference purposes, that for the methyl radicals obtained by passing methane over MgO. Over either active catalyst, a mixed methyl-ethyl radical spectrum was obtained. Deconvolution of the combined signal indicated that 30-40% of the trapped radicals were ethyl. When *n*-butane was passed over MgO or La_2O_3, only 1-5% of the collected radicals could be attributed to ethyl radicals. In all matrix isolation experiments, only active catalysts for the production of ethylene and ethane showed enhanced ethyl radical concentrations in the gas phase.

Thus, the following are the principal findings that have been found for the active catalysts:

(a) manganese forms a mixed metal oxide which effectively disperses the manganese in the support;

(b) manganese can have an oxidation state which ranges

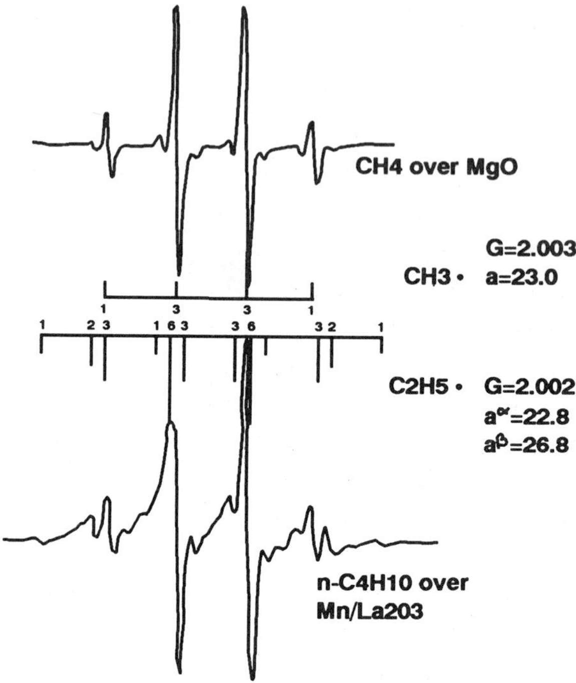

FIG. 5 *ESR spectra of radicals frozen out of the gas phase into a 13K argon matrix. Center labels correspond to theoretical line positions and relative intensities for methyl and ethyl radicals.*

from 4+ to 2+ between the oxidized and the reduced states (3+ and 2+, respectively, are the more typical oxidation states for manganese);

(c) manganese is, at least partially, in isolated clusters with a uniform octahedral symmetry; this manganese shows no tendency to migrate toward the surface under oxidizing or reducing conditions; and

(d) manganese generates ethyl radicals from *n*-butane far in excess of what would be expected over MgO or La_2O_3 alone.

Based on what we know of the catalytic and physical

properties of the catalyst system, it is likely that free radicals are generated by the catalyst surface. To give the product slate that is observed, a primary radical probably is involved.

MECHANISM OF REACTION

Speculation on the reaction sequence controlling cracking over the supported manganese and iron catalysts has produced the mechanism shown in Fig. 6. Although we do not have specific evidence, we propose that the active catalytic site may be isolated manganese or iron atoms stabilized by the support. A hydrocarbon molecule interacts with the surface forming a stabilized carbanion. (Garrone and Stone (11) have shown that pure MgO can form carbanions at the surface.) Electron

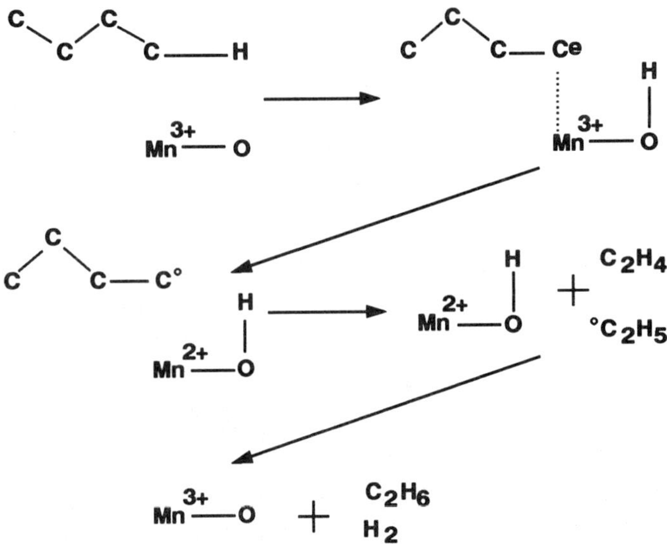

FIG. 6 *Proposed model for the catalytic cracking of n-butane over Mn/MgO.*

transfer then occurs with subsequent release of a free radical into the gas phase. The active catalytic site is regenerated via hydrogen atom recombination at the surface or by abstraction of the surface bound hydrogen by gas phase radicals.

Several aspects of the proposed mechanism are consistent with the experimental data. First, the initial formation of carbanions at the surface explains the observed products. The formation of carbanions directs the activation of hydrocarbon to the primary position as a result of inherent carbanion stabilities (primary>secondary>tertiary). Steric hindrance may also help direct activation to the primary position. Electron transfer from the butyl anion and subsequent desorption of the hydrocarbon radicals results in the directed formation of primary butyl radicals. Gas phase decomposition of these via accepted carbon-carbon bond scission pathways then produces ethylene and ethyl radicals, which can abstract hydrogen from the catalyst surface to form ethane.

The fact that butyl radicals were not trapped in the matrix isolation experiments is consistent with their extremely rapid unimolecular decomposition rate. They had already decomposed to ethyl radicals by the time they were trapped. Furthermore, it appears that most of the ethane is formed from reaction of ethyl radicals with hydrogen atoms on the catalyst surface rather than by gas phase hydrogen atom abstraction or catalytic hydrogenation of ethylene. If most of the ethane was formed by gas phase abstraction processes, the ethane/ethylene ratio would be near that observed for thermal conversion at comparable reaction conditions. Experiments in which ethylene and hydrogen were passed over the Mn/MgO catalyst showed it had virtually no hydrogenation activity.

CONCLUSIONS

It has been shown that manganese or iron on an appropriate support will give anomalously high selectivities to ethylene and ethane. The active manganese catalysts are unique in that the manganese is 4+ in the oxidized state. The hydrocarbon appears to react via primary carbanion formation at the surface and subsequent decomposition to the observed products.

ACKNOWLEDGMENTS

We are grateful to Larry Potts for the XPS work and to Helen Vanderveen for the catalytic runs and the XRD results.

REFERENCES

1. Haag, W. O. , and Dessau, R., M., in *Proceedings of the 8th International Congress on Catalysis*, G. Ertl, Ed. (Verlag Chemie, Weinheim, Federal Republic of Germany), 1984, *2*, 305.

2. Scofield, J. H., *J. Electron Spectrosc. Relat. Phenom.*, 1976, *8*, 129.

3. Carter, W. J., Schweitzer, G. K., and Carlson, T. A., *J. Electron Spectrosc. Relat. Phenom.*, 1974, *5*,827.

4. Lunsford, J. H., Martir, W., and Driscoll, D. J., *Prepr., Div. Pet. Chem., Am.Chem. Soc.*, 1984, *29*, 920;

   Driscoll, D. J., Wilson, M., Wang, J. X., and Lunsford, J. H., *J. Am. Chem. Soc.*, 1985, *107*, 58.

5. Cordischi, D., Nelson, R. L., and Tench, A. J., *Trans. Faraday Soc.*, 1969, *65*, 2740.

6. Derouane, E. G., and Indovina, V., *Bull. Soc. Chim. Belg.*, 1973, *82*, 645.

7. Kerkhof, F. P. J. M., and Moulijn, J. A., *J. Phys. Chem.*, 1979, *83*, 1612.

8. Derouane, E. G., Thelen, J., and Indovina, V., *Bull. Soc. Chim. Belg.*, 1973, *82*, 657.

9. Henderson, B., and Hall, T. P. P., *Proc. Phys. Soc.*, 1967, *90*, 511.

10. Henderson, B., and Wertz., J. E., *Adv. in Phys.*, 1968, *17*, 749.

11. Garrone, E., and Stone, F. S., in *Proceedings of the 8th International Congress on Catalysis*, G. Ertl, Ed. (Verlag Chemie, Weinheim, Federal Republic of Germany), 1984, *3*, 441.

4

TRANSIENT KINETIC ANALYSIS OF THE OXIDATIVE COUPLING OF METHANE OVER Sm_2O_3

KEVIN P. PEIL[1], GEORGE MARCELIN[2], JAMES G. GOODWIN, JR.[2,4], and ALAIN KIENNEMANN[3]

[1]The Dow Chemical Company, Midland, MI 48674

[2]Chemical and Petroleum Engineering Department
University of Pittsburgh, Pittsburgh, PA 15261

[3]Laboratoire de Chimie Organique Appliquée
Université de Strasbourg, Strasbourg, France

[4]Author to whom all correspondence should be addressed.

ABSTRACT

This paper details an investigation of the oxidative coupling of methane over cubic and monoclinic Sm_2O_3 using steady-state isotopic transient kinetic analysis (SSITKA). Oxygen exchange experiments in the absence of methane resulted in a quantification of the lattice oxygen diffusivity and total oxygen uptake. Drastic differences were seen in the lattice oxygen mobility for cubic and monoclinic Sm_2O_3. Using isotopic switches of oxygen and methane under steady-state reaction, the active intermediates along the carbon and oxygen reaction pathways were quantified. Lattice oxygen is suggested to play a significant role in the oxidation process under steady-state reaction. Exchange of the oxygen of the product gases occurred very readily. Quantification of carbon intermediates during steady-state reaction showed 57% of the cubic Sm_2O_3 surface active at any given time while only 23% of the monoclinic Sm_2O_3 surface was active.

INTRODUCTION

Considerable progress in recent years has been made in the search for active and selective catalysts for the oxidative coupling of methane. One of the most active and well studied catalysts has been Sm_2O_3.(1-3) However, only recently has detailed information on the surface of a working Sm_2O_3 catalyst and the overall carbon reaction pathway been presented.(4) The overall oxygen reaction network and role of the lattice oxygen in the selective and nonselective oxidation of methane is unknown.

The feasibility of using isotope switching experiments to study methane oxidation has recently been shown.(4-9) The unique feature of using isotopes is the ability to follow aspects of the reaction in such a way as to obtain detailed mechanistic information without perturbing the steady-state equilibrium.(10)

This paper presents results which delineate for the first time the nature of the surface reaction steps on cubic and monoclinic Sm_2O_3 and quantifies the working surface under steady-state reaction conditions.

EXPERIMENTAL

Steady-State Isotopic Transient Kinetic Analysis (SSITKA)

For a single, first order elementary reaction step with intermediates present at a coverage of θ_i and exhibiting a lifetime of τ_i on the surface, the reaction rate (TOF) can be expressed as:

$$TOF = \frac{1}{\tau_i} * \theta_i = k_i * \theta_i \qquad (1)$$

where k_i (sec^{-1}) is the intrinsic site turnover frequency (TOF_i).(11-14)

The evaluation and quantification of such reaction parameters are best conducted at steady-state working conditions.(10,15) The evaluation of such parameters can best be accomplished by means of steady-state isotopic transient kinetic analysis (SSITKA). This technique entails abrupt switches in the isotopic composition of one of the reactants accompanied by the continuous monitoring (e.g., by mass spectrometry) of the relaxation and evolution of labeled reactants and products (see Figure 1).

An example of the response of a single unidirectional reaction step to a step change in the isotopic concentration of one of the reactants is shown in Figure 2. The isotope switch is made at time $t=0$ and decays as:(16)

$$\frac{dN_A}{dt} = -\frac{N_A}{\tau} = -kN_A \qquad (2)$$

where N_A is the amount of isotopically labeled surface intermediates which form the labeled product. The decay in N_A is observed as a decay in the rate (R_A). If the assumption is made that there is no isotope effect, τ is independent of N_A and equation (2) can be integrated to yield:

$$R_A = Re^{-t/\tau} \qquad (3)$$

Since $R = R_A + R_{A*}$ = constant, equation (3) can be normalized and written as:

$$F_A(t) = \frac{R_A}{R_A + R_{A*}} = e^{-t/\tau} \qquad (4)$$

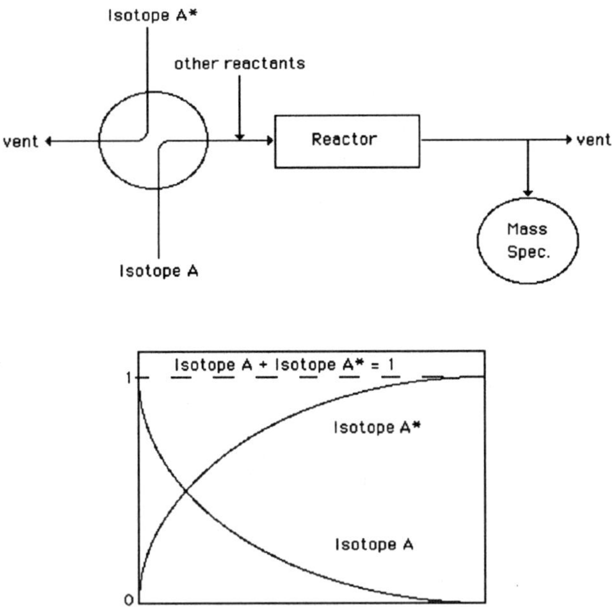

FIGURE 1 Switching the isotopic composition of the reactants does not disturb the steady-state of the reaction.

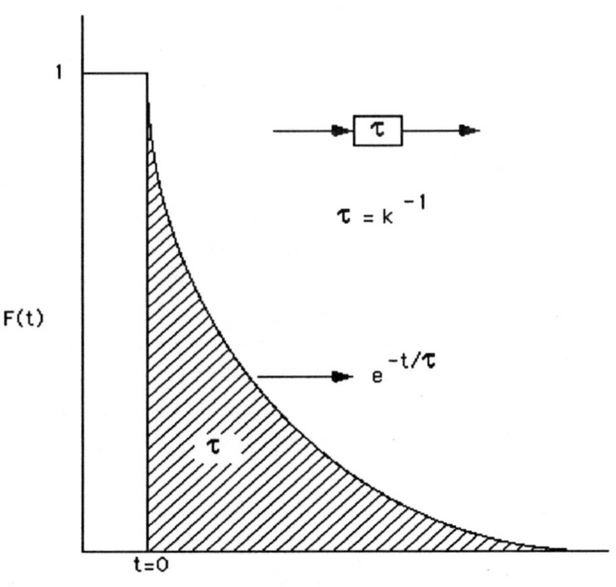

FIGURE 2 Response of a single, homogeneous surface pool to a step change in the isotopic concentration of one of the reactants.

Integration of equation (4) from $t=0$ to ∞ yields τ directly. This is represented by the shaded area in Figure 2.

Included in the measured τ are the τ's for the holdup in the gas phase and the additional holdup on the surface due to readsorption:

$$\tau_{measured} = \tau_{reaction} + \tau_{gas} + \sum_{i=1}^{N} \tau_{ads,i} \tag{5}$$

where N is the number of readsorptions each lasting for a time $\tau_{ads,i}$. $\tau_{measured}$ can be corrected for τ_{gas} by including an inert gas tracer in one of the isotopes to be switched. Since the inert gas does not interact with the catalyst surface the area under its transient is the hold-up in the gas phase, τ_{gas}. Values for τ_{ads} cannot be obtained directly, but readsorption can be minimized by 1) decreasing the bed length, 2) decreasing the holdup time in the reactor by increasing the flowrate, or 3) the addition of a gas which will compete for readsorption sites. It is therefore possible to approximate values for the true τ of surface reaction. This τ is an average value for the isotope in a particular product formed by all surface reaction pathways.

The surface concentration of reaction intermediates can be calculated by applying the formula:(17)

$$N_i = \tau * \text{(product yield)}$$

where product yield is defined as conversion multiplied by product selectivity. Values for Θ can be obtained by normalizing N_i by a monolayer coverage.

Catalysts and Gases

The cubic and monoclinic Sm_2O_3 were prepared by precipitation from a 100 ml H_2O solution of 5g $Sm(NO_3)_3*6H_2O$ (Aldrich 99.9%) by an excess of 10% oxalic acid solution (2.34 g $C_2H_2O_4*2H_2O$ (Prolabo 99.8%) in 50 ml H_2O) up to a pH of 2.2. The obtained product ($Sm_2(C_2O_4)_3*10H_2O$) was then washed by distilled H_2O, dried at 125°C for 15 hours and annealed up to 725°C for 15 hours to obtain Sm_2O_3 in cubic form and 850°C for 15 hours to form Sm_2O_3 in monoclinic form. Cubic Sm_2O_3 was also purchased from Alpha Products and found to be comparable in behavior to the synthesized cubic Sm_2O_3. Both Sm_2O_3 catalysts had a BET surface area = 1 m^2/gm and prior to reaction were pretreated by heating to reaction temperature in flowing oxygen then holding for 30 minutes prior to introduction of the reactants.

Argon (99.995% purity) and helium (99.995% purity) from Linde were used as diluents. The $^{12}CH_4$ (99% purity, purchased from Linde) contained 5.1% argon and the $^{16}O_2$ (99% purity, purchased from Linde) contained 5.1% helium which permitted correction of the transient results for gas-phase holdup. The $^{18}O_2$ (97.5 atom%) was purchased from Isotec and the $^{13}CH_4$ (99 atom%) from Monsanto.

Reaction System

The reaction system used in this study consisted of a 4 mm I.D. straight tube quartz reactor which narrowed to a 1 mm opening after the catalyst bed. Quartz wool was used to hold the catalyst in place. Catalyst loadings varied between 25 and 100 mg. The entrance and exit lines were capillary stainless-steel tubing and heated to 100°C. The above design was used to minimize holdup in the gas phase and gas phase reaction.

The product and reactant gases were separated with a 60/80 Carbosieve S-II column in a Varian G.C. and quantified with a thermal conductivity detector. Under the conditions used in this study, no oxygenates (CH_3OH, H_2CO) were detected.

An Extranuclear (Extrel) quadrupolar mass spectrometer coupled with a computerized data acquisition system was used to follow and record the evolution and decay of the different labeled products and reactants following an isotopic switch. Very low ionization energies (typically 15-24 eV) were used to minimize fragmentation of the molecules.

The oxygen exchange experiments were conducted using the following conditions: 43 cc/min Ar, 2 cc/min $^{16}O_2$ switched to 2 cc/min $^{18}O_2$, catalyst wt. = 100 mg, temperature = 600, pressure = 1 atm.

Separate experiments were conducted in which the isotopic composition of the CH_4 was switched from $^{12}CH_4$ to $^{13}CH_4$ and the O_2 from $^{16}O_2$ to $^{18}O_2$ during reaction. Reaction conditions were as follows: catalyst wt. = 25 - 100 mg., temperature = 600°C - 640°C, total pressure = 1 atm., CH_4 flow = 5 cc/min, and $He/CH_4/O_2$ = 94/10/1. For the monoclinic catalyst at 600°C, CH_4 conversion was 3.7%, CO_2/CO was 3.2, and C_2H_6 selectivity equaled 8%. For the cubic catalyst at 600°C, CH_4 conversion was 9.1%, CO_2/CO was 1.4, and C_2H_6 selectivity was 26%. At 640°C, CH_4 conversion was 7.8%, CO_2/CO equaled 1.1, and C_2H_6 and C_2H_4 selectivities were 37.5% and 6.9%, respectively.

RESULTS

Oxygen Exchange

Figures 3 and 4 depict the oxygen transients obtained during oxygen exchange over monoclinic and cubic Sm_2O_3, respectively. It is evident from a comparison of the relaxation profiles of the isotopes for the two catalysts that cubic Sm_2O_3 was capable of a more rapid exchange of oxygen. The final (pseudo-steady-state) offset of the $^{16}O^{18}O$ signal for cubic Sm_2O_3 was approximately twice that for the monoclinic Sm_2O_3 and 600 seconds was needed for the new isotopic pseudo-steady-state to be reached for cubic Sm_2O_3 compared to approximately 100 seconds for monoclinic Sm_2O_3.

The equivalent number of atomic layers readily available for exchange along with the lattice oxygen diffusivity were calculated according to the methods previously outlined(8) and are presented in Table 1. There was a drastic difference in the lattice oxygen mobility and the equivalent number of oxide layers readily available for exchange for the two Sm_2O_3 catalysts.

X-ray diffraction analysis following exchange indicated no change in crystal structure for the two Sm_2O_3 catalysts during oxygen exchange.

Oxygen Reaction Pathway

Figures 5-7 present the transients obtained when the flow of O_2 was switched from $^{16}O_2$ to $^{18}O_2$ during steady-state reaction over cubic Sm_2O_3. Under reaction, the total amount of ^{16}O leav-

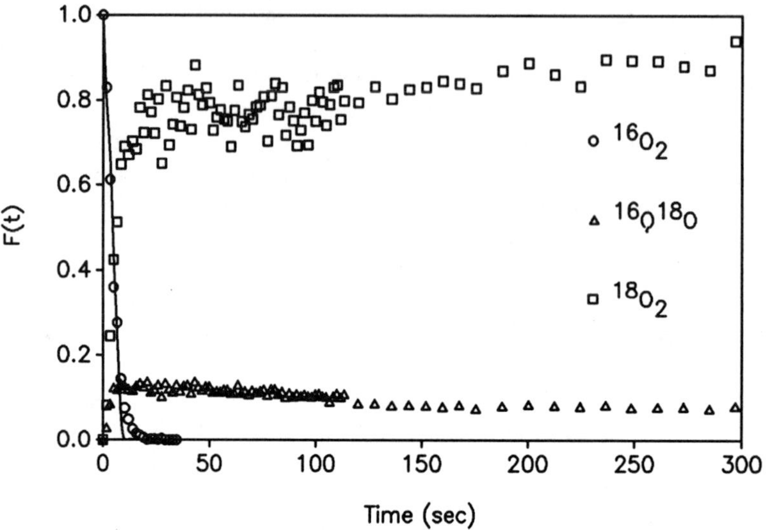

FIGURE 3 Oxygen exchange over monoclinic Sm_2O_3. Solid line indicates holdup in the gas phase.

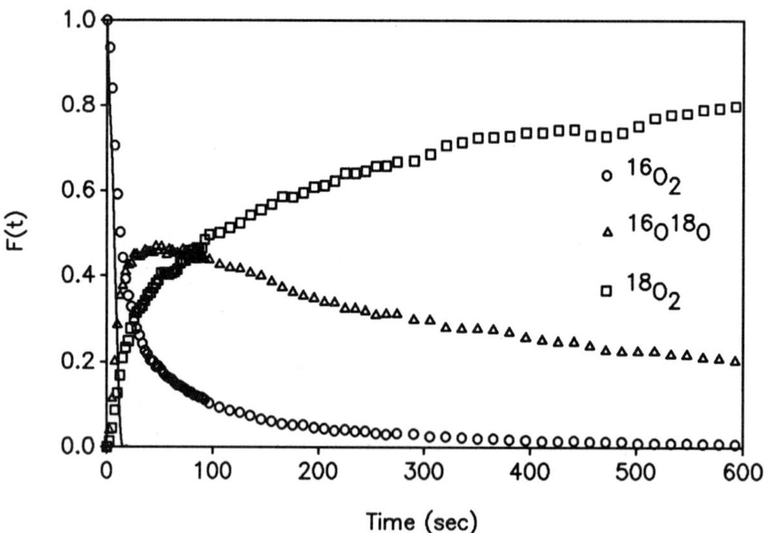

FIGURE 4 Oxygen exchange over cubic Sm_2O_3. Solid line indicates holdup in the gas phase.

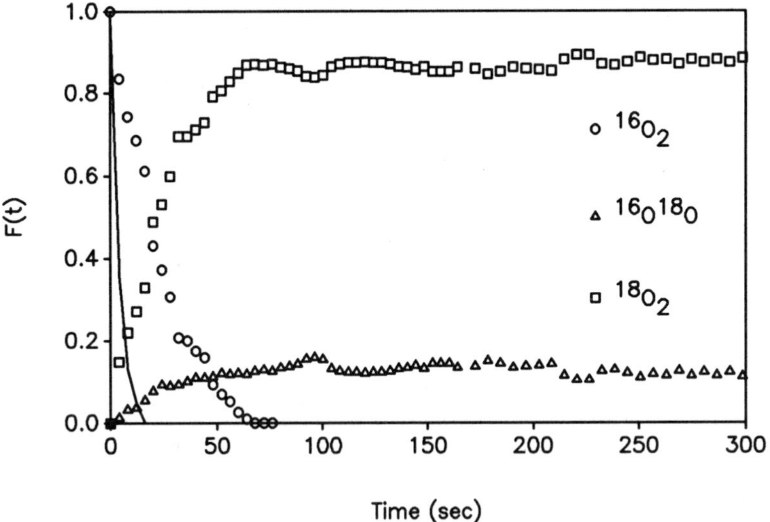

FIGURE 5 O_2 transients obtained when the flow of O_2 is switch from $^{16}O_2$ to $^{18}O_2$ during steady-state reaction over cubic Sm_2O_3. Solid line indicates inert holdup.

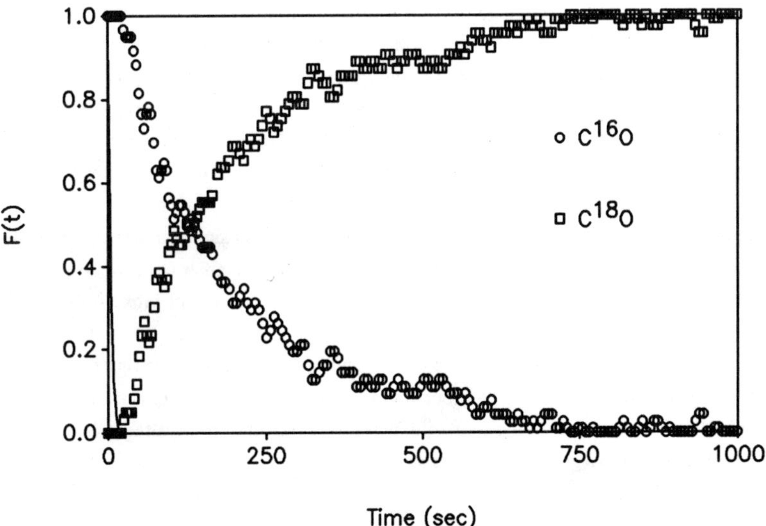

FIGURE 6 CO transients obtained when the flow of O_2 is switch from $^{16}O_2$ to $^{18}O_2$ during steady-state reaction over cubic Sm_2O_3. Solid line indicates inert holdup.

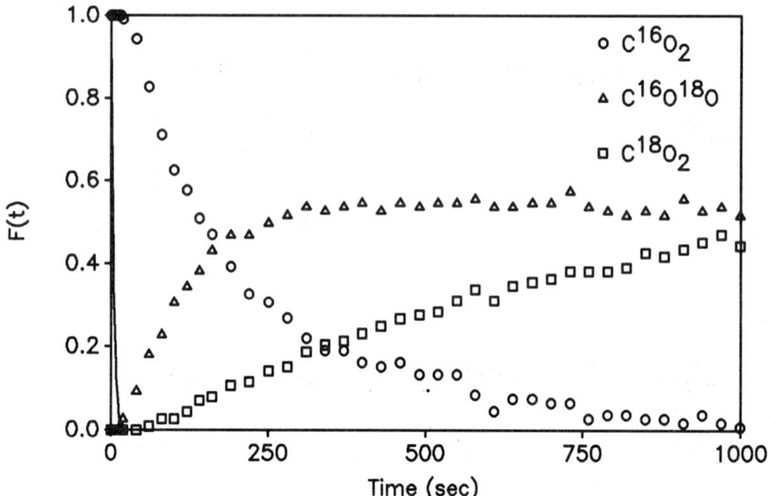

FIGURE 7 CO_2 transients obtained when the flow of O_2 is switch from $^{16}O_2$ to $^{18}O_2$ during steady-state reaction over cubic Sm_2O_3. Solid line indicates inert holdup.

ing the "surface" as $^{16}O_2$, $^{16}O^{18}O$, $C^{16}O$, $C^{16}O_2$, $C^{16}O^{18}O$, and $H_2{}^{16}O$ corresponded to 12.3x10^{20} atoms/gm. Calculations based on the oxygen exchange experiment (Table 1) indicate 13x10^{20} "surface" ^{16}O atoms/gm available for exchange for this particular catalyst.

The amount of "surface" ^{16}O atoms/gm leaving the catalyst as products during reaction was approximately equal to the amount of "surface" ^{16}O atoms/gm available for exchange as determined from the exchange experiments. However, the amount of "surface" ^{16}O desorbing during reaction as molecular oxygen was an order of magnitude less than the amount of such molecular ^{16}O desorbing during simple exchange. Continued contribution to the overall rate from the bulk oxygen can be noted in Figure 5 for the $^{16}O^{18}O$ transient and Figure 7 for the $C^{16}O^{18}O$ transient.

It is interesting to note the value for the final offset of the $C^{16}O^{18}O$ signal at 1000 seconds in Figure 7. The value of $F(t) = 0.5$ for the signal offset indicates that there is an equal probability of CO_2 containing an ^{16}O as an ^{18}O atom. Figure 6 does not indicate the equivalent offset for CO only for technical reasons inherent in the data collection and processing.

The effect of cubic Sm_2O_3 catalyst loading on the measured surface residence times along the oxygen reaction pathway is given in Table 2. The four-fold increase in surface oxygen residence times for CO, CO_2, and H_2O with a corresponding four-fold increase in the amount of catalyst is clear indication of re-exchange of the oxygen with the surface.

Carbon Reaction Pathway

Figure 8 presents the transients obtained when the isotopic composition of methane was switched from $^{12}CH_4$ to $^{13}CH_4$ during reaction at 600°C over monoclinic Sm_2O_3. Comparative

TABLE 1 Oxygen availability and lattice oxygen diffusivity as determined from oxygen exchange experiments for Sm_2O_3.

Sm_2O_3 Catalyst	Temp. (°C)	Total "surface" ^{16}O atoms desorbing/gm catalyst x10^{-20}	Equivalent # of layers	Diffusivity cm^2/sec x10^{15}
cubic	600	13	>100(a)	136
monoclinic	600	0.2	2(b)	9

(a) Based on a surface area = 1 m^2/gm and 17.2 $Å^2$/oxygen atom.

(b) Based on a surface area = 1 m^2/gm and 19.0 $Å^2$/oxygen atom.

TABLE 2 Residence times along the oxygen reaction pathway at 600°C for cubic Sm_2O_3 as a function of catalyst loading.

	τ_o (sec)	
Species	25 mg	100 mg
$^{16}O_2$	21	19
$H_2{}^{16}O$	81	294
$C^{16}O$	51	191
$C^{16}O_2$	44	226

transients for cubic Sm_2O_3 have been presented elsewhere(4) but the quantitative results from such isotopic transients are given in Table 3 for comparison with those of monoclinic Sm_2O_3.

The lack of strongly adsorbed CO_2 on cubic Sm_2O_3 is evident from the rapid relaxation of the CO_2 transient. This observation has been confirmed by a series of experiments in which the catalyst bed length was varied resulting in no change (within experimental error) in the CO_2 residence time as shown in Table 4.(4) This data clearly shows there is no readsorption of the carbon of the product molecules on Sm_2O_3.

The measured surface coverages of carbon intermediates leading to each product over cubic and monoclinc Sm_2O_3 are given in Table 3. On the order of 57% of the cubic Sm_2O_3 working surface was involved in the formation of products at any time while only 23% of the monoclinic

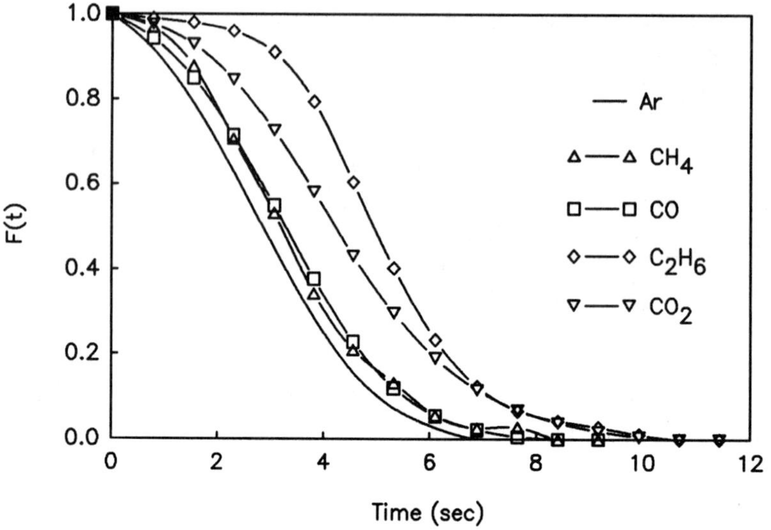

FIGURE 8 Transients obtained when the isotopic composition of methane is switched from $^{12}CH_4$ to $^{13}CH_4$ during reaction over monoclinic Sm_2O_3 at 600°C.

TABLE 3 Measured surface coverages of carbon intermediates and products for cubic and monoclinic Sm_2O_3 at 600°C.

	$\tau_c^{(a)}$ (sec)		$N_c^{(b)} x 10^{-18}$		$\theta_c^{(c)}$		Intrinsic Site TOF (sec^{-1})	
Species	cubic	mono	cubic	mono	cubic	mono	cubic	mono
C_2H_6	3.1	3.9	2.9	0.3	0.22	0.03	0.32	0.26
CO_2	4.0	3.2	3.2	1.9	0.24	0.17	0.25	0.31
CO	2.4	2.1	1.4	0.3	0.11	0.03	0.42	0.48
CH_4	0.3	0.4	5.6	8.5	0.43	0.77	----	----

(a) An error of ±0.2 sec was determined from the average of three separate experiments for the cubic Sm_2O_3 and ±0.3 sec for the monoclinic Sm_2O_3.

(b) Number of carbon atoms leading to products/gm of catalyst.

(c) Based on an assumption of $1x10^{19}$ sites/gm for a surface area of 1 m^2/gm and of 1 carbon atom intermediate/site. θ_c renormalized for a monolayer coverage.

TABLE 4 Surface residence times along the carbon reaction pathway at 600°C as a function of cubic Sm_2O_3 loading.

	τ_c (sec)		
Species	25 mg	45 mg	100 mg
C_2H_6	4.0	4.1	2.8
CO_2	2.7	3.4	2.7
CO	2.3	2.4	1.7

Sm_2O_3 working surface was involved in this formation. Since surface lifetime is inversely proportional to the intrinsic site TOF, as can be seen in Table 3 the rate of formation of C_2H_6 intermediates over cubic Sm_2O_3 occurred at a lower rate than the rate of formation of CO intermediates and over monoclinic Sm_2O_3 occurred at a lower rate than the rate of formation of CO and CO_2 intermediates. The rate of formation of C_2H_6 intermediates was greater on cubic than monoclinic Sm_2O_3 while the rate of formation of CO and CO_2 intermediates was greater on monoclinic than cubic Sm_2O_3.

In order to obtain a measurable C_2H_4 transient over the cubic Sm_2O_3 catalyst the temperature was raised to 640°C. All other reaction parameters remained the same. The transients obtained following a CH_4 isotope switch are shown in Figure 9. Measured surface lifetimes and coverages of carbon intermediates during the production of C_2H_4 over cubic Sm_2O_3 are given in Table 5.

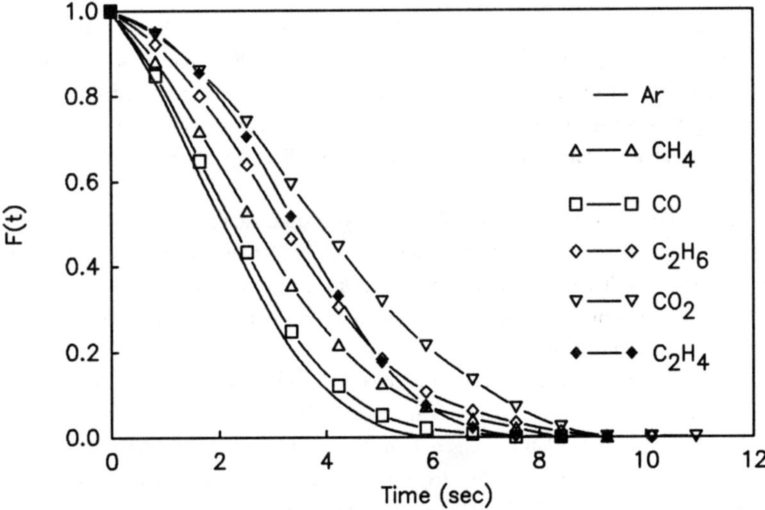

FIGURE 9 Transients obtained when the isotopic composition of methane is switched from $^{12}CH_4$ to $^{13}CH_4$ during reaction over cubic Sm_2O_3.

TABLE 5 Measured surface coverages of carbon intermediates and products for cubic Sm_2O_3 during the production of C_2H_4 at 640°C.

Species	$\tau_c^{(a)}$(sec)	$N_c^{(b)} x 10^{-18}$	$\theta_c^{(c)}$	Intrinsic Site TOF (sec^{-1})
C_2H_6	2.5	3.0	0.16	0.40
C_2H_4	2.6	0.6	0.03	0.38
CO_2	3.2	1.4	0.07	0.31
CO	1.5	0.7	0.04	0.67
CH_4	0.7	13	0.70	----

(a) Experimentally determined error of ±0.3 sec.

(b) Number of carbon atoms leading to products/gm of catalyst.

(c) Based on an assumption of $1x10^{19}$ sites/gm for a surface area of 1 m^2/gm and of 1 carbon atom intermediate/site. θ_c renormalized for a monolayer coverage.

DISCUSSION

The difference in lattice oxygen mobility and exchangeability between monoclinic and cubic Sm_2O_3 can be better understood by a close examination and comparison of the two crystalline structures.

The metal atom in the cubic structure is surrounded by 8 oxygen atoms at the corners of a cube with all oxygen distances the same.[18,19] However, one-fourth of the oxygen atoms are missing resulting in the metal atom being surrounded by six oxygen atoms and two lattice oxygen vacancies. These lattice oxygen vacancies are located in such a manner that they lie in straight lines through the body of the cubic crystal and essentially provide a "pipeline" for oxygen atoms to diffuse into and out of the bulk.

The monoclinic Sm_2O_3 structure has an additional oxygen atom along a threefold axis resulting in a distorted octahedron coordination about the Sm atoms.[20] Not all unit cells contain an "extra" oxygen atom so the monoclinic structure consists of both six- and seven-fold coordination about the metal atoms. This lack of symmetry about the metal atom results in a lower degree of lattice oxygen mobility for the monoclinic structure.

It is interesting to note the value for the offset of the $C^{16}O^{18}O$ signal in Figure 7. The value of F(t) = 0.5 for the signal offset indicates that there is an equal probability of CO_2 containing an ^{16}O as an ^{18}O atom. For this to occur, the lattice oxygen diffusivity must be great enough to directly compete with gas phase oxygen for vacant surface oxygen sites. In other words, when an oxygen atom desorbs from the Sm_2O_3 surface as O_2, CO, CO_2 or H_2O, there is an equal probability of the vacant surface oxygen site being filled with an oxygen from the gas phase or the subsurface lattice.

The fact that the CO_2 transients trail the CO transients in Figures 8 and 9 may indicate possible multistep surface oxidation pathways on monoclinic and cubic Sm_2O_3. However, the possibility that CO_2 is formed as a primary product and not from further oxidation of adsorbed CO cannot be ruled out from this data.

It is interesting to note in Table 3 that the rate of formation of C_2H_6 intermediates was greater on cubic than monoclinic Sm_2O_3 while the rate of formation of CO and CO_2 intermediates was greater on monoclinic than cubic Sm_2O_3. One possible explanation of this may be the difference in lattice oxygen mobility between the two catalysts. If the lattice oxygen served to activate adsorbed CH_4 selectively while adsorbed oxygen activated CH_4 nonselectively, increasing the mobility of the lattice oxygen would increase the ability of a catalyst to selectively activate CH_4. The selectivity of a catalyst would then be a matter of competition for adsorbed CH_4 between lattice oxygen and gas phase oxygen. The involvement of lattice oxygen in the selective coupling of methane over Sm_2O_3 has been reported.(21) As the flow of O_2 was interrupted, Sm_2O_3 preferentially formed C_2H_6.

It is interesting to note from Table 5 that the carbon intermediates leading to the formation of C_2H_4 remain on the surface for an amount of time equal to that of the carbon intermediates leading to the formation of C_2H_6. This may be interpreted either as gas phase oxidative dehydrogenation of C_2H_6 to form C_2H_4 or the dehydrogenation of a surface adsorbed C_2H_6 leading to a surface adsorbed C_2H_4 which then desorbs into the gas phase.

By comparing surface lifetimes along the carbon and oxygen reaction pathways the relative sizes of the active carbon and oxygen reservoirs on the catalysts can be determined. For CO and CO_2 the catalyst residence lifetimes along the oxygen reaction pathways are larger (at least by an order of magnitude) than the corresponding lifetimes along the carbon reaction pathways. The oxygen reservoir is larger than the carbon reservoir because it includes the bulk oxygen.

CONCLUSIONS

Isotopic switches under steady-state reaction conditions have provided a quantitative description of the reactive intermediates and reaction pathways over cubic and monoclinic Sm_2O_3. The isotopic transients presented here indicate that a working Sm_2O_3 surface is more than a methyl radical generator. On the order of 57% of the cubic Sm_2O_3 working surface was involved in the formation of products while only approximately 23% of the monoclinic Sm_2O_3 was involved. Both surfaces were active in the formation of CO, CO_2, and possibly even C_2H_6 and C_2H_4. Surface carbonates did not appear to be present in any large amount and exchange of the carbon of the product gases was not important. However, exchange of the oxygen of the product gases was important. The different quantitative results obtained for the carbon reaction pathways on cubic and monoclinic Sm_2O_3 suggest an important role for the lattice oxygen since the surface chemistry of these two crystal structures would not be significantly different.

ACKNOWLEDGMENTS

Financial support of this work by Amoco Corporation and The Gas Research Institute is gratefully acknowledged.

REFERENCES

1. Otsuka, K., Jinno, K., and Morikawa, A., Chemistry Letters, 499 (1985).
2. Otsuka, K., and Komatsu, T., Chemistry Letters, 483 (1987).
3. Otsuka, K., Liu, Q., Hatano, M., and Morikawa, A., Chemistry Letters, 467 (1986).
4. Peil, K.P., Goodwin, Jr., J.G., and Marcelin, G., J. Am. Chem. Soc., **112**, 6129 (1990).
5. Ekstrom, A., and Lapszewicz, J.A., J. Am. Chem. Soc., **110**, 5226 (1988).
6. Ekstrom, A., and Lapszewicz, J.A., J. Phys. Chem., **93**, 5230 (1989).
7. Peil, K.P., Goodwin, Jr., J.G., and Marcelin, G., J. Phys. Chem., **93**, 5977 (1989).
8. Peil, K.P., Goodwin, Jr., J.G., and Marcelin, G., J. Cat., in press.
9. Peil, K.P., Goodwin, Jr., J.G., and Marcelin, G., Proceedings of the Symposium on Natural Gas Conversion, Oslo, Norway, August 12-17, 1990.
10. Happel, J., "Isotopic Assessment of Heterogeneous Catalysis", Academic Press, Inc., New York, 1986.
11. Biloen, P., J. Mol. Cat., **21**, 17 (1983).
12. Soong, Y., Krishna, K., and Biloen, P., J. Cat., **97** 330 (1986).
13. Zhang, X., and Biloen, P., J. Cat., **98** 468 (1986).
14. Biloen, P., Helle, J.N., van den Berg, F.G.A., and Sachtler, W.M.H., J. Cat., **81** 450 (1983).
15. Tamaru, K., "Dynamic Heterogeneous Catalysis", Academic Press, Inc., New York, 1978.
16. Yang, C.-H., Soong, Y., and Biloen, P., J. Cat., **94**, 306 (1985).
17. Yang, C.-H., Soong, Y., and Biloen, P., Proceedings of the 8th International Congress of Catalysis, **2** 3 (1984).
18. Templeton, D.H., and Dauben, C.H., J. Chem. Soc., Chem. Comm., 5237 (1954).
19. Stone, G.D., Weber, G.R, and Eyring, L., in "Mass Transport in Oxides" Wacchtman, Jr., J.B., and Franklin, A.D., eds. National Bureau of Standards Special Publication 296, 1968.
20. Cromer, D.T., J. Chem. Soc., Chem. Comm., **61** 753 (1957).
21. Peil, K.P., Goodwin, Jr., J.G., and Marcelin, G., J. Cat., in press.

5

LANTHANA-BASED CATALYSTS FOR THE OXIDATIVE COUPLING OF METHANE. I. EVIDENCE FOR THE FORMATION OF SUPEROXIDE IONS IN THE La_2O_3/MgO SYSTEM

Zhicheng JIANG, Zhenqiang YU, Bin ZHANG, Shikong SHEN, Shuben LI and Hongli WANG*

Lanzhou Institute of Chemical Physics, Chinese Academy of Sciences Lanzhou 730000, China

ABSTRACT

Lanthana-based catalysts have been found to be good catalysts for the oxidative coupling of methane at 708°C and above with high stability. XPS study of La_2O_3/MgO catalyst gives evidence for the formation of O_2^- on the lanthanum oxide surface as the result of adsorption of oxygen on an oxygen-depleted surface of La_2O_3 at elevated temperatures. A preliminary model for the active surface of La_2O_3/MgO catalyst is presented.

INTRODUCTION

Rare earth metal oxides have been found to be good catalysts for the oxidative coupling of methane to produce C_2 compounds[1-4].

---

* *Also associated with the Dalian Institute of Chemical Physics, Chinese Academy of Sciences, Dalian, China 116032.*

Among which, lanthanum oxide has been found to be very active when used in two component systems together with an alkaline earth metal compound [3,4]. The lanthana based catalyst has the additional feature of being very stable. One such catalyst has been found to have no decline in catalytic activity in a run exceeding 500 h on stream in our laboratory. The nature of the active species of oxygen which is responsible for the abstraction of hydrogen atom from methane in these systems has aroused profound interest and much speculation.

The (M^+O^-) type center has been wisely believed to be active in the Li-promoted MgO system [5]. However, this type of center may not be present in lanthana-based catalysts. In fact, no O^- was found on La_2O_3 using ESR by which the existence of (M^+O^-) was elegantly discovered in Li^+/MgO [6]. Furthermore, the (M^+O^-) type center has been considered to be useful at temperatures lower than 700-720^{0}C [7], while many experiments on the oxidative coupling of methane are currently conducted at temperature of 750^{0}C and even higher. To look for other oxygen species which may be an important part of the active surface for the oxidative coupling of methane is therefore clearly desirable.

We report here the XPS study of the La_2O_3/MgO system and present evidence for the existence of a superoxide species O_2^- in such a system at temperatures up to 750^{0}C.

EXPERIMENTAL

La_2O_3/MgO catalyst was prepared by coprecipitation from a solution of $La(NO_3)_3$ and $Mg(NO_3)_2$ with $(NH_4)_2CO_3$ as the precipitating agent. The dried catalysts were calcined at 800^{0}C for 4 h before use. The sample studied by XPS contains 8% La_2O_3 and 92% MgO respectively.

XPS studies were conducted on a PHI 550 Photoelectron Spectrometer using Mg K_a radiation, and the pass energy of the cylindrical mirror analyzer was 25 eV.

The main difficulty in obtaining reliable XPS spectra for the sample of lanthanum oxide is the preponderance of impurities on the sample surface. As is well known, La_2O_3 is notoriously active toward water vapor and CO_2 abundant in the atmosphere. The hydroxyls and carbonate ions which are always present on the La_2O_3 surface are extremely difficult to be completely removed. We took great pains in cleaning the sample surface and produced a really clean La_2O_3/MgO surface with only lattice oxygen ions present besides La^{3+} as revealed by the XPS measurements. The oxygen adsorption as well as the subsequent XPS measurements were then undertaken at temperature from 400°C to 750°C, i.e., under conditions much closer to the reaction conditions than those reported in the literature to data. The binding energies of La $3d_{5/2}$, Mg 2p, and O 1s were taken in the usual manner.

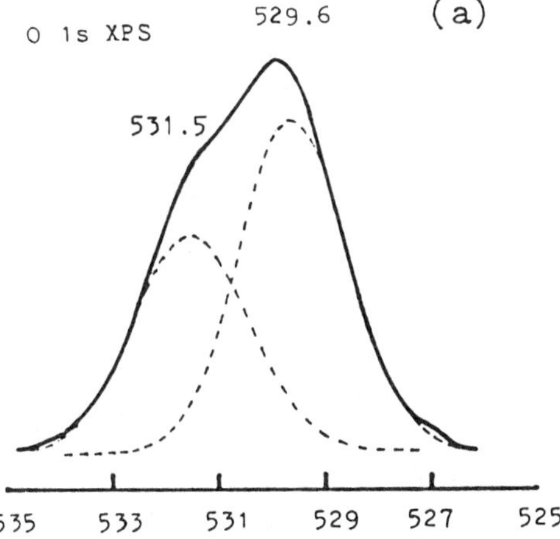

Fig. 1 *XPS O 1s spectra for La_2O_3/MgO catalyst.*
(a) Initial surface. (b) Clean surface. (c) Clean surface on exposure to oxygen at 773K.

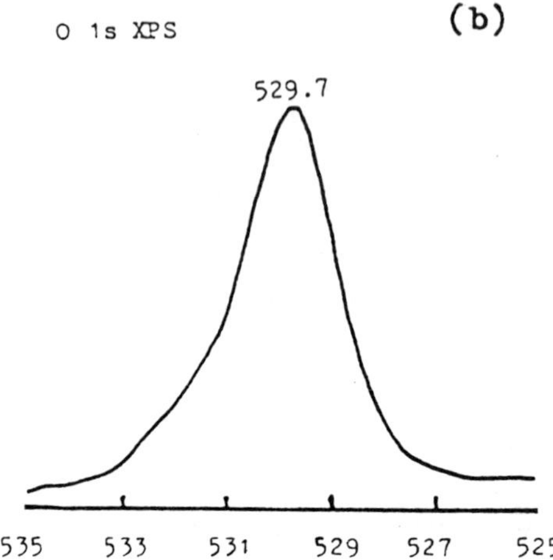

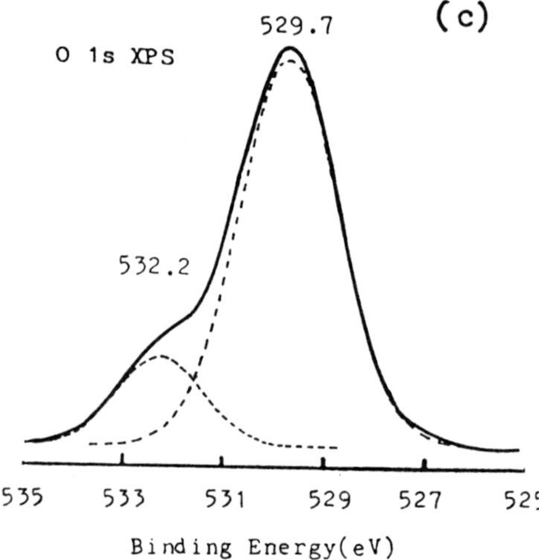

RESULTS AND DISCUSSION

The XPS O 1s and C 1s spectra of the initial catalyst sample are shown in Fig. 1(a) and Fig. 2(a) respectively. It is seen that for the O 1s spectrum, in addition to the peak of lattice oxygen O^{2-} of both oxides which is centered at 529.6 eV, a shoulder at 531.5

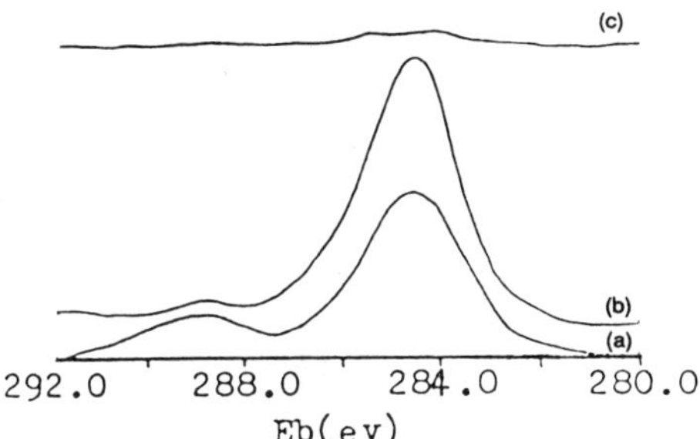

Fig. 2 *XPS C 1s spectra for La_2O_3/MgO catalyst*
(a) Initial surface. (b) Initial surface after high temperature treatment. (c) Clean surface.

eV also appears which is resulted from the presence of OH^- and CO_3^{2-} both of which have a binding energy of about 531 eV. The presence of these impurities in obviously large amount broadens the O 1s peak considerably. While for the C 1s spectrum, there is another peak appeared at about 289 eV as the result of the presence of carbonate ions on the surface in addition to 284.6 eV peak of the adventitious carbon which was taken as a reference.

When the La_2O_3/MgO sample was subjected to repeated cycles of dehydrating at 673-873 K, exposure to oxygen at 873 K and degassing at 1000K first in prechamber(10^{-5}torr) and subsequently in UHV(10^{-9}torr) for a long period of time, a clean surface of the sample was attained. This is clearly shown by the O 1s spectrum in which only a narrow and smooth peak of lattice oxygen ions with a binding energy of 529.7 eV is observed as shown in Fig. 1(b). The shoulder due to hydroxyls and carbonates have been totally eliminated. It is to be noted that this is believed to be the first O 1s spectrum of a uncontaminated La_2O_3 surface ever obtained to our knowledge. Similarly, from the C 1s spectrum it is seen in Fig. 2(b) that the carbonate peak was reduced in intensity after high temperature treatment and essentially all the contaminated carbon was removed at the end of the cleaning procedure as shown in Fig. 2(c).

On a clean sample of La_2O_3/MgO maintained at high temperature, O 1s spectrum was again recorded after exposure to pure oxygen. The resulted O 1s spectrum presented in Fig. 1(c) shows a shoulder at 532.2 eV besides the original peak of 529.7 eV ascribed to the lattice oxygen of La_2O_3 as well as that of MgO.

What is the nature of this new species of oxygen? Is it a charged particle or a neutral one? Is it in atomic form or in molecular form? For a system like La_2O_3/MgO at elevated temperatures, it would be expected that an ample supply of electrons and holes will be available on the surface [6,8]. It is thus reasonable to expect that charged particles will be formed under our experimental conditions. An adsorbed neutral molecule of oxygen would either desorb or transform into ions before or after it splits into atoms. Therefore, it appears that the oxygen species detected in our system can only be O_2^- or O^-.

A clue as to its identity comes from the value of its binding energy. It is known from the literature that the binding energy of O 1s for the molecular form of oxygen O_2 is 532.4 eV [9]. The binding energy of O 1s for O_2^- is expected to be close to this value but somewhat lower. While for the species of O^-, it was reported to be about 530.1 eV [10], a value much less than the one observed in our experiment, 532.2 eV. We are therefore led to the conclusion that the oxygen species formed upon the exposure of pure oxygen to a clean surface of La_2O_3/MgO is O_2^-. This species of oxygen was also detected by Lunsford and his associates under different conditions on La_2O_3 using ESR technique [6]. The possibility of such a peroxide ion being involved in the catalytic partial oxidation of methane has been raised by different authors [1,7].

Another interesting point to be described here is the shift of the binding energy for both La $3d_{5/2}$ and Mg 2p, but in opposite directions, upon the introduction of oxygen to the clean surface of the catalyst La_2O_3/MgO. The XPS data for both the catalyst and its components La_2O_3 and MgO after different treatments are given in Table 1.

It is seen from Table 1 that the binding energy of La $3d_{5/2}$ decreases from 834.4 eV to 833.5 eV, and that of Mg 2p increases from 49.05 eV to 49.6 eV when the sample was thermally treated in UHV at 700°C. This may be rationalized by the migration of lattice

Table 1 *XPS data for the La_2O_3/MgO catalyst and its components after different treatments.*

Sample	State of the sample	Binding Energy, eV		
		La $3d_{5/2}$	Mg 2p	O 1s
La_2O_3	*Initial sample*	*834.4*		*528.9*
				530.8
	Clean sample	*833.7*		*528.9*
				531.7
MgO	*Initial sample*		*49.0*	*529.4*
				531.3
	Clean sample		*49.0*	*529.4*
La_2O_3/MgO	*Initial sample*	*834.4*	*49.05*	*529.6*
				531.5
	Clean sample	*833.5*	*49.6*	*529.7*
	Clean sample on exposure to oxygen	*834.0*	*49.6*	*529.7*
				532.2

oxygen from La_2O_3 to MgO, forming La_2O_{3-x} and MgO_{1+x} surface species respectively, with oxygen vacancies present on the La_2O_3 surface. Upon adsorption of oxygen, concomitant with the appearance of a shoulder on the O 1s peak, the binding energy of La $3d_{5/2}$ rises to 834.0 eV. That is to say, together with the formation of O_2^-, the La_2O_3 is reoxidized. We are thus inclined to believe that this O_2^- is generated by the adsorption of oxygen from the gas phase and is associated with the oxygen vacancy on the La_2O_{3-x} surface. Furthermore, this O_2^- is apparently stabilized by the presence of oxygen-rich MgO nearby. It is to be noted here that the O_2^- species found on pure La_2O_3 by Lunsford et.al was thermally stable only under 200^0C.

The active surface for the La_2O_3/MgO catalyst may thus be schematically represented by

$$(MgO_{1+x})(La_2O_{3-x})\square O_2^-$$

Where $\square$ stands for the oxygen vacancy created on the surface of La_2O_3 at elevated temperatures, and x is related to the amount of oxygen migrated from La_2O_3 to MgO. The stability of one of the lanthana based catalysts observed experimentally presumably reflects the stability of an active surface of similar nature under reaction conditions.

REFERENCES

1. K. Otsuka and K. Jinno, *Inorg. Chemica Acta* 121:237(1986).
2. K. Otsuka, K. Jinno and A. Morikawa, *J. Catal.* 100: 353(1986).
3. Z.-Q. Yu, S.-K. Shen, S.-B. Li and H.-L. Wang, *J. Mol. Catal.* (China) 3: 304(1989).

4. Z.-Q. Yu, S.-K. Shen, S.-B. Li and H.-L. Wang, Preprints of 3B Symposium on Methane Activation, Conversion and Utilization, PACIFICHEM' 89, p.126.

5. T. Ito, J.-X. Wang, C.-H. Lin and J.H. Lunsford, J. Am. Chem. Soc. 107: 5062(1985).

6. J.-X. Wang and J.H. Lunsford, J. Phys. Chem. 90: 3890(1986).

7. J.H. Lunsford, Catalysis Today 5: 235(1990).

8. A. Kooh, J.L. Dubois, H. Mimoun and C.J. Cameron, Preprints of 3B Symposium on Methane Activation, Conversion and Utilization, PACIFICHEM' 89, p.60.

9. B. Grzybowska, J. Haber, W. Marczewski and L. Ungierp J. Catal. 42: 327(1976).

10. Karl C.C. Kharas and J.H. Lunsford, J. Am. Chem. Soc. 111: 2336(1989). The O 1s binding energy of O^- is assumed to be close to that of O_2^{2-} reported in this article.

6

ADDITIVE EFFECT ON THE PERFORMANCE OF PbO/Al_2O_3 CATALYST FOR THE OXIDATIVE COUPLING OF METHANE

Y. Xu, W. Liu, J. Huang and Z. Lin

Laboratory for Catalysis,
Dalian Institute of Chemical Physics, Academia Sinica,
P. O. BOX 110, Dalian, China, 116023

ABSTRACT

The effect of various additives on the performance of alumina supported lead oxide catalyst for the titled reaction was evaluated in a fixed bed reactor at atmospheric pressure. Li, Ca and La oxides were used as additives. All additives are effective for promoting the performance of oxidative coupling of methane. XRD and XPS measurements show that part of the additive covers and disperses on the surface of PbO species and part covers the surface of the Al_2O_3 support and forms a complex oxide such as $LiAlO_2$ and $LaAlO_3$. It is found that the rate of the production of ethane and ethylene, R, is correlated to the atomic ratio of additive/Pb and also linearly correlated to the atomic ratio of Pb/Al on the surface. The promotional effect is attributed to the enrichment of PbO species on the surface, the suppression of the acidity of the alumina support, and the interaction between the PbO species and the additive species which cover the PbO species.

INTRODUCTION

The production of C_2 intermediates for chemicals by the oxidative coupling of methane over metallic oxides catalysts has recently become a highlight in research on chemical catalysis. It is of tremendous significance for the transfer of the raw materials for the chemical and fuel industries in the world and useful utilization of natural gas. In addition, to activate methane, the most stable paraffin, at moderate conditions is a big challenge to catalyst scientists.

Many researchers have reported that various catalyst systems are effective for the oxidative coupling of methane. Also, several mechanisms have been proposed to explain the nature of the catalytic process [1-3]. Based on the intrinsic features of the radical mechanism, which is now generally accepted, it is claimed that development of a catalyst with high activity and selectivity, to obtain yields of ethane and ethylene of about 25%, will be very difficult if not impossible. Meanwhile, greatest efforts are still in process in the world to try to create new processes for the direct use of natural gas in the chemical and fuel industries.

It is well known that catalysts effective for the oxidative coupling can be classified into four kinds of oxides:

(1) reducible metal oxides;
(2) irreducible metal oxides;
(3) rare earth metal oxides;
(4) complex metal oxides.

Lead oxide is an important reducible metal oxide that has been investigated to a considerable extent.

The effect of various lead-containing compounds and different supports and preparation methods on the performance of the oxidative coupling of methane have been reported [4-15]. PbO/Al_2O_3 catalysts as a rule result in a high productivity [16, 17].

In our previous report [18], the performance of PbO/Al_2O_3 catalyst for the oxidative coupling of methane was studied from the point of view of metal-support interaction and the effect of various additives on the reaction was also preliminarily reported. It was found that various oxides play a positive role for the performance of the oxidative coupling of methane over PbO/Al_2O_3 catalyst. To further improve their activity and selectivity, the effect of various additives were investigated in considerably more detail.

EXPERIMENTAL

1. Catalyst Preparation

Catalysts used in this work were prepared by impregnation. Al_2O_3 support, with the surface area of 258 m^2/gcat (BET), was calcined at 873 K for 4h and sieved to obtain a 20-40 mesh product. Lead nitrate was then used as impregnating solution. The mixture of Al_2O_3 support and lead nitrate solution was kept overnight. The mixture was next evaporated and dried in a steam bath, and was then calcined at 973 K for 3h. The PbO nominal content of the sample was 10 mol% and coded as Pb10A.

An additive oxide was incorporated into the Pb10A catalyst by successive impregnation; various nitrates were used to produce these oxides. Samples were dried and then calcined again at 973 K for 3h. In this work, three oxides were individually incorporated as an additive into the Pb10A catalyst. These are oxides of Li, Ca, and La. All the additives are tentatively thought to be in their normal form of oxide, i. e. Li_2O, CaO and La_2O_3. The additive content is expressed in mole percent and coded as, e. g., Ca4Pb10A to mean 4 mol% CaO and 10 mol% PbO supported on Al_2O_3. Chemical composition of the samples was analyzed by a ICP/5000 atomic absorption spectrometer and the surface area of the sample was measured in a Digisorb 2500 apparatus by N_2 adsorption.

2. Catalyst Evaluation

Catalyst evaluation was performed in a fixed bed reactor at atmospheric pressure. A quartz reactor with 8 mm i. d. was used. Usually the catalyst charge in the reactor was 200 mg. A small thermocouple well was located at the side of the reactor in such a way it just contacts the top of catalyst sample. Methane, air and helium flowrates were each measured by mass controllers (Brooks Co.,); these gases were mixed before entering the reactor. Usually the total flowrate was 50 ml/min. For a low partial pressure of O_2, the volumetric ratio of CH_4/O_2 was about 10:1 and these gases were supplemented with He (F_{CH_4} = 20 ml/min, F_{air} = 10 ml/min and F_{He} = 20 ml/min). The ratio of CH_4/O_2 was about 3.33:1 (F_{CH_4} = 20 ml/min and F_{air} = 30 ml/min). No He stream was used when the partial pressure of O_2 was high.

The reaction products consist of H_2O, CO_2, CO, C_2H_6 C_2H_4 with a small amounts of C_3H_8, and sometimes C_3H_6. Two GC units were used for on-line separating and analyzing the effluent. A column of Carbon sieve of 3 mm X 3000 mm and TCD was used for CH_4 and the permanent gases. A column of Porapak R of 3 mm X 2000 mm and FID was employed to separate and analyze the hydrocarbons. Using N_2 as an internal standard, the carbon balance was calculated and its value usually ranged from 0.95 to 1.05.

Performance of the oxidative coupling of methane over a catalyst was expressed as follows:

Conversion of methane, X_{CH_4};

Conversion of oxygen, X_{O_2};

Selectivities for various products:

S_{C_2} for ethane and ethylene;

S_{CO_2} for carbon dioxide;

S_{CO} for carbon monoxide;

Yields of ethane and ethylene, Y_{C_2};

Ratio of ethylene and ethane in the product, r;

Rate of the production of ethane and ethylene, R, was calculated in terms of X_{CH_4} and S_{C_2} and used as a measure of the activity of various catalysts in ml/m^2h.

3. X-Ray Diffraction (XRD) Measurements

XRD measurements were made with a Rigaku X-ray diffractometer using a Cu rotary target at room temperature. The operating conditions were selected as 40 KV X 100 mA at the scanning rate of 8 degree/min. All XRD spectra were recorded, stored, processed and assigned using a computer system attached to the instrument.

4. XPS Measurements

XPS measurements were performed using a VG MICRO LAB II electron spectrometer. Al K_α excitation was used as the X-ray source (1253.6 eV). Ground samples were first evacuated in a pretreatment chamber and then transferred into a UHV chamber for XPS measurements. The background pressure was 3 X 10^{-9} mbar during measurements. All

XPS data were recorded, stored and processed in a computer system. The binding energies were corrected using C_{1s} (284.6 eV) as the internal standard. The atomic ratio of elements concerned on the catalyst surface was calculated using corrected integrated peak areas:

$$N_1/N_2 = (I_1/I_2) \times (a_2/a_1) \times (E_{k2}/E_{k1})^{1/2} \quad (1)$$

where N_i, the number of ith atom on the surface, I_i, the intensity of a peak in XP spectra as calculated from the integrated peak area, E_{ki}, the kinetic energy of the XP spectrum and a_i, the cross-section for ionization.

RESULTS AND DISCUSSION

1. Relationship between the Atomic Ratios of Additive Promoted Pb10A Catalysts in the Bulk and on the Surface

The chemical composition of additive promoted Pb10A catalysts and their surface areas are listed in Table 1. Fig. 1 shows the variation of the surface area with the atomic ratio of additive/Pb in the bulk of the sample. When more additive is incorporated into the Pb10A catalyst, the surface area of the sample becomes less. Li_2O additive has a more severe effect on the surface area reduction while CaO and La_2O_3 additives follow almost the same trend as shown in Fig. 1. Because of the difference in the surface areas of the samples, the rate of the production of ethane and ethylene, R, is expressed in terms of ml/m^{2}h in this context.

Fig. 2 shows the relationship between the atomic ratios of Li/Pb, Ca/Pb and La/Pb in the bulk and on the surface of the additive promoted Pb10A catalysts. All the results follow a linear relationship between the atomic ratios of additive/Pb on the surface and in the

TABLE 1 The chemical composition and surface area of additive promoted Pb10A catalysts

Catalyst	Chem. Compos., mol% PbO	Add.	Surf. Area m2/g	Atomic Ratios in the Bulk Pb/Al	Add./Pb
Pb10A	11.2	--	110.9	0.063	--
Li4Pb10A	9.7	2.0	94.8	0.055	0.41
Li8Pb10A	12.9	4.4	81.0	0.078	0.68
Li14Pb10A	18.1	9.5	67.4	0.125	1.05
Li20Pb10A	14.7	13.0	58.0	0.108	1.77
Li25Pb10A	13.0	20.5	53.6	0.099	3.15
Ca2Pb10A	8.6	2.2	95.6	0.048	0.26
Ca4Pb10A	9.2	4.2	94.9	0.053	0.46
Ca6Pb10A	9.5	6.8	91.1	0.057	0.72
Ca8Pb10A	9.9	7.9	84.8	0.060	0.88
Ca10Pb10A	8.5	11.7	79.0	0.053	1.38
La2Pb10A	10.2	1.5	99.2	0.058	0.29
La4Pb10A	9.9	3.2	90.7	0.057	0.65
La6Pb10A	9.9	5.9	81.6	0.059	1.19
La8Pb10A	9.8	8.5	73.6	0.060	1.74
La10Pb10A	9.7	11.3	67.6	0.061	2.33

bulk at low loadings and then level off at higher loadings of the additive. This is typical of a so-called highly dispersed state (monolayer state) of supported oxide sample as proposed by Kerkhof and Moulijn [19]. This suggests that part of additive oxide covers and disperses on the surface of PbO species and so leads to higher atomic ratios of Li/Pb, Ca/Pb and La/Pb on the surface than in the bulk. The three additives show almost the same trends at low loadings,

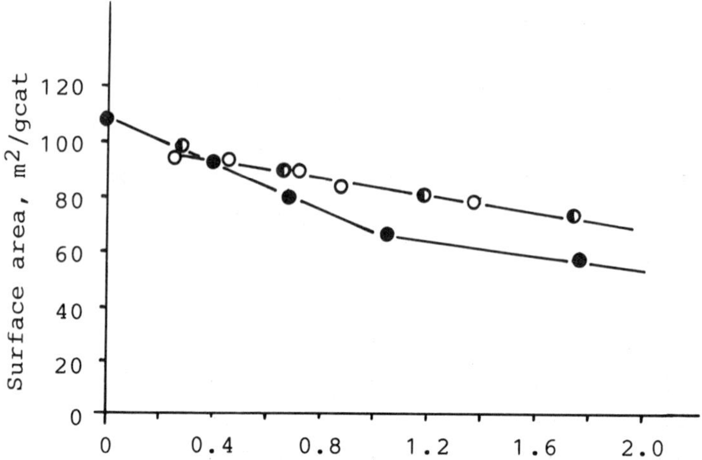

Fig. 1 Variation of the surface area with the atomic ratio of additive/Pb in the bulk.
●, for Li, ○, for Ca, and ◐, for La promoted Pb10A.

but have quite different value of the saturated capacity of the monolayer state as shown in Fig. 2. The atomic ratio corresponding to the saturated capacity is 1.0 for a Li_2O promoted Pb10A catalyst, but is 0.6 for La_2O_3 and CaO promoted samples. With the increase in the amount of additive beyond the saturated capacity, the dispersion degree of the additives becomes smaller and smaller.

XPS measurement also shows a slight chemical shift on the binding energies of $Pb_{(4f)}$ in the additive promoted Pb10A catalyst and implies that there is some type of interaction between the additive and the PbO species on the surface as shown in Table 2. Low loadings of the Li additive lead to a slight increase, but high loadings lead to a decrease in the binding energies of $Pb_{(4f)}$. Ca and La additives apparently have

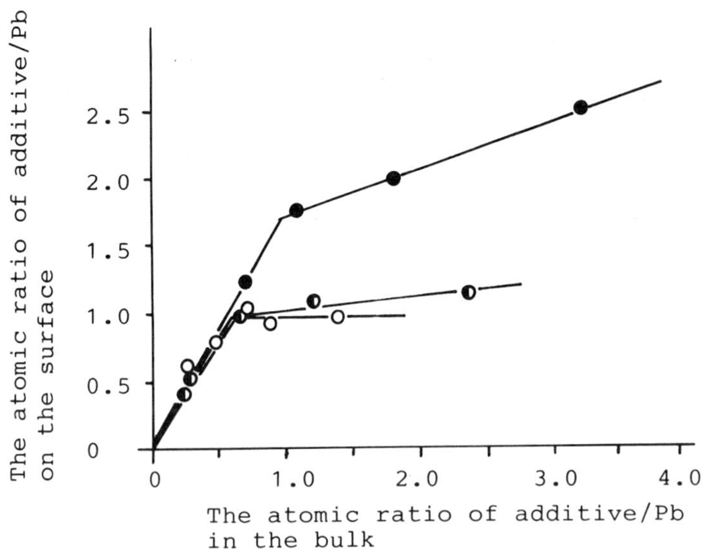

Fig. 2 Relationship between the atomic ratios of additive/Pb on the surface and in the bulk. ●, for Li, ○, for Ca, and ◐, for La promoted Pb10A.

no effect on the binding energies of $Pb_{(4f)}$ at high loadings, while at low loadings, La additives decrease the binding energies of $Pb_{(4f)}$. Ca additives follow the same trend as Li additives.

XRD measurements on the fresh catalysts show that part of Li_2O and La_2O_3 forms complex oxides such as $LiAlO_2$ and $LaAlO_3$ at medium ranges of the additive loading; this conclusion is based on the results obtained during the preparation. The peak intensity characteristic of these oxides increases with increases in the additive amount. No similar complex oxides were detected for CaO promoted Pb10A sample.

2. Effect of Various Additives

Fig. 3-5 show how the performance of promoted Pb10A catalyst was affected by the additive amount, expressed

TABLE 2 The binding energies of $Pb_{(4f)}$ and atomic ratios of additive promoted Pb10A catalysts

Catalyst	Binding Energies, eV, $Pb_{(4f)}$		Atomic Ratios on the Surface Pb/Al	Add./Pb
Pb10A	143.6	138.8	0.049	--
Li4Pb10A	143.9	139.1	0.066	--
Li8Pb10A	143.9	139.1	0.076	1.23
Li14Pb10A	143.3	138.4	0.088	1.75
Li20Pb10A	143.2	138.7	0.084	1.99
Li25Pb10A	143.3	138.4	0.105	2.51
Ca2Pb10A	143.8	138.9	0.038	0.60
Ca4Pb10A	143.5	138.8	0.041	0.77
Ca6Pb10A	143.4	138.9	0.046	1.02
Ca8Pb10A	143.6	138.8	0.056	0.92
Ca10Pb10A	143.6	138.9	0.058	0.96
La2Pb10A	143.4	138.7	0.072	0.50
La4Pb10A	143.6	138.8	0.075	0.96
La6Pb10A	143.6	138.7	0.082	1.08
La10Pb10A	143.7	138.9	0.087	1.15

in the atomic ratio of Li/Pb, Ca/Pb and La/Pb on the surface. The rate of the production of ethane and ethylene, R, increases linearly with an increase in the atomic ratio of Li/Pb on the surface at a reaction temperature of 973 K for either high or low partial pressure of O_2. In the former case, the slope is even higher than that in the latter case.

CaO as an alkaline earth oxide additive displays a positive effect by promoting the performance of Pb10A catalyst for the oxidative coupling of methane as shown

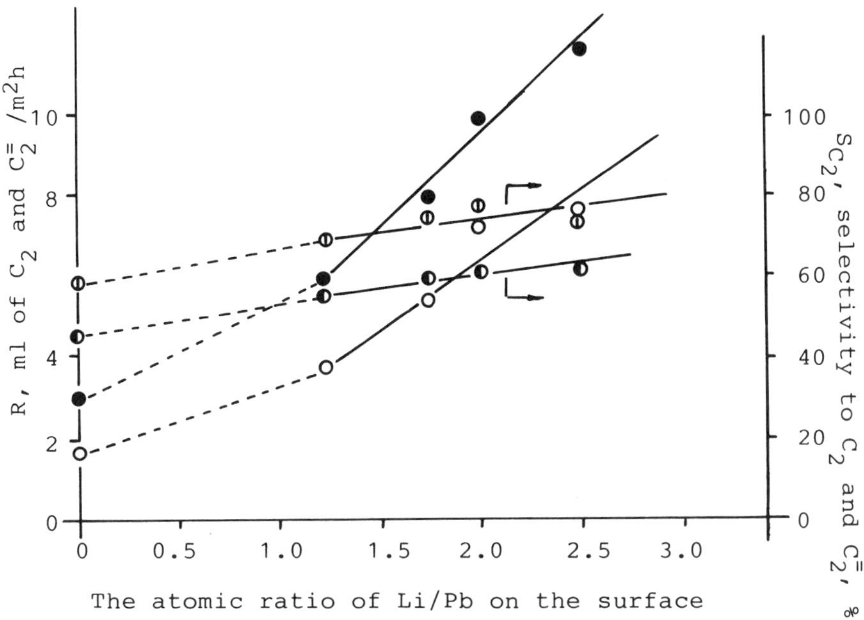

Fig. 3 Dependence of R and S_{C_2} on the atomic ratio of Li/Pb on the surface over Li_2O promoted Pb10A.
○, for the value of R at the low partial pressure of O_2;
●, for the value of R at the high partial pressure of O_2;
⦶, for the value of S_{C_2} at the low partial pressure of O_2;
◐, for the value of S_{C_2} at the high partial pressure of O_2.

in Fig. 4. Higher atomic ratios of Ca/Pb on the surface resulted in higher R values for the catalyst at both high and low partial pressures of O_2.

La_2O_3 as a rare earth oxide additive also exhibits a positive effect to the performance of Pb10A for the oxidative coupling of methane as shown in Fig. 5. The rate of the production of ethane and ethylene, R, increases with the increase in the atomic ratio of La/Pb on the surface in general, but the slope is less than those of Li_2O and CaO promoted Pb10A catalysts.

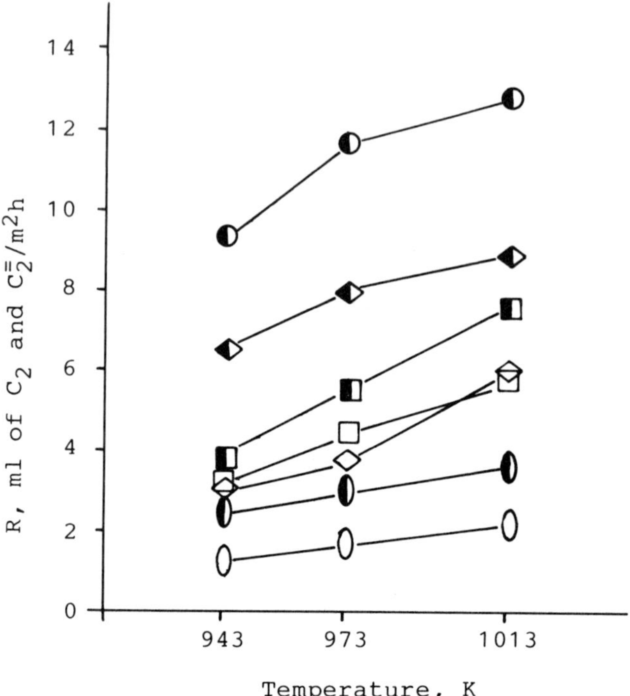

Fig. 6 Dependence of the value of R on the reaction temperature over various additives promoted Pb10A. ○, for Li, ◇, for Ca, and □, for La promoted Pb10A (0). □, at the low and ◧, at the high partial pressure of O_2.

coupling of methane of the Pb10A catalyst. For comparison, the temperature dependence of R and S_{C_2} over Pb10A catalyst are also shown in Fig. 6 and 7. The catalysts with the additive of Li and Ca oxide exhibit the same temperature dependence on R, i.e. the temperature effect will be high if it is lower than 973 K, otherwise it will be low if the reaction temperature is higher than 973 K. The temperature effect of S_{C_2} also displays a similar trend as shown in Fig. 7. In contrast to Li_2O and CaO promoted Pb10A catalysts, the promotional effect of the La_2O_3 additive

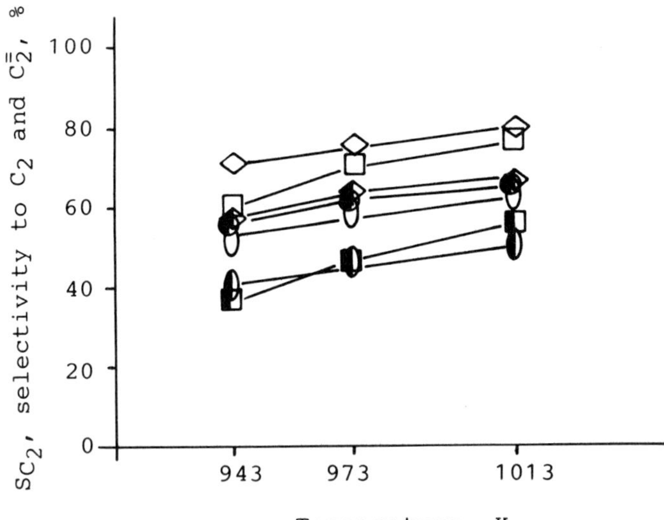

Fig. 7 Dependence of the value of S_{C_2} on the reaction temperature over various additives promoted Pb10A. ○, for Li, ◇, for Ca, and □, for La promoted Pb10A(○). □, at the low and ◧, at the high partial pressure of O_2.

on R becomes more obvious at a higher temperature (1023 K) than it does at a low temperature (943 K), while the temperature effect on S_{C_2} is more severe at low temperatures than at high temperatures for La_2O_3 promoted Pb10A catalysts. This implies that the mechanism of the promotional effect is different with Li_2O and CaO and with La_2O_3 as additives incorporated in Pb10A catalyst.

A detailed examination of the temperature dependence on the performance of Li25Pb10A catalyst for the oxidative coupling of methane was performed and the results are shown in Fig. 8. The ratio of ethylene to ethane, r, almost linearly increases as the reaction temperature increases from 843-1013 K. However, the temperature dependence on R and S_{C_2} are more complicated

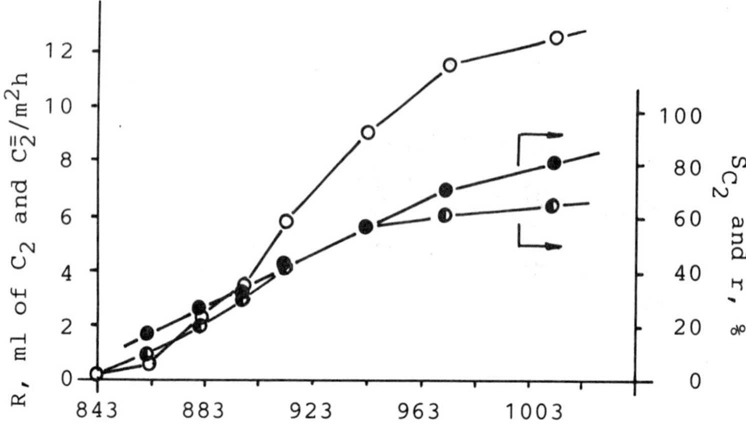

Fig. 8 Temperature dependence of the performance of the oxidative coupling of methane over Li25Pb10A at the high partial pressure of O_2.
○, ◐, and ●, for the values of R, S_{C_2}, and r respectively.

and both display a S-shape curve as shown in Fig. 8. R and S_{C_2} increase with the reaction temperature very rapidly in the temperature range of 883-973 K, but both have almost constant values at temperatures beyond 973 K or lower than 883 K. The value of R for Li25Pb10A catalyst shows very strong temperature dependence in the range of 863-973 K. This suggests that for Li25Pb10A catalyst, the best reaction temperature should be between 913-973 K.

4. Origin of the Promotional Effect Caused by the Additive

In our previous study on the performance of the oxidative coupling of methane over alumina supported lead oxide catalysts [18], it was shown that R, the rate of the production of ethane and ethylene, is linearly

17. U. Preuss and M. Baerns, Chem. Eng. Technol., 10: 297 (1987).

18. Y. Xu, W. Liu, J. Huang and Z. Lin, Proc. 1st Tokyo Conf. Adv. Catal. Sci. Technol., Tokyo, Japan, (1990), in press.

19. F. P. J. M. Kerkhof and J. A. Moulijn, J. Phys. Chem., 83: 1612 (1979).

I. INTRODUCTION

Oxidative coupling of methane(OCM) promises to be an effective method for the chemical utilization of the abundant natural gas resources, and becomes a subject of intensive research interests(1). Since the pioneering work of Keller and Bhasin(2), a wide variety of catalysts have been found to be effective for the OCM to obtain ethane and ethylene. Recently, special attention has been given to enhancing the ethylene selectivity(3—5).

Alkali metal chloride doped oxides show higher ethylene selectivity than other catalysts(1,3). Fundamental studies are rarely reported on such a system(6). In this work, a series of $NaCl/B_2O_3/Fe_2O_3$ catalysts were prepared for the OCM. The importance of the third additive B_2O_3, and the performance of the catalysts are addressed. The activation mechanism and the complicated deactivation process are discussed with the aid of XPS and XRD data.

II. EXPERIMENTAL

A. *CATALYSTS*

The catalysts used were prepared by two—step impregnations: (a) Known quantities of Fe_2O_3 were soaked in H_3BO_3 solutions of different concentrations for 24 hr., dried at 393K and calcined at 673K and 973k for 2 hr. respectively, to obtain the B_2O_3/Fe_2O_3 samples with the B/Fe atomic ratio from 0 to 0.3 (labeled as FB00 to FB30 or 0 — 30 (at)% B_2O_3/Fe_2O_3). (b) Then the B_2O_3/Fe_2O_3 were impregnated with NaCl solution and dried at 393K, to obtain the $NaCl/B_2O_3/Fe_2O_3$ catalysts with the Na/Fe atomic ratio of 0.25 (labeled as FBN0025 — FBN3025). These samples were pelletized and sieved to 50—70 mesh for reaction, and ground to powder for XPS and XRD measurements.

B. *METHANE CONVERSION*

The catalytic reaction was performed in a laboratory scale fixed—bed reactor operated in a flow mode at atmospheric pressure. The reactor was constructed from a 8 mm(i. d.) quartz tube connected to a 1 mm(i. d.) quartz capillary. A 4 mm(i. d.) quartz tube with one end sealed was provided for a thermocouple to the catalyst bed. The general reaction conditons were as follows: T=1023 K, t=0. 24 g. s. cm^{-3}, P_{CH_4}/P_{O_2}= 4, P_{CH_4}=32 kPa, and balanced to atmospheric pressure by helium.

The reactants and products were analysed by a gas chromatograph with a thermal conductivity detector and carbon zeolite columns. The main products of reaction were CO, CO_2, C_2(=C_2H_4+ C_2H_6), H_2O and HCl. H_2O and HCl were trapped in a water—ice condenser downstream of the reactor. C_3 hydrocarbons were produced at most in trace amount and were neglected. The conversion of methane was calculated on a carbon atom basis. The selectivities were calculated on the basis of the conversion of methane (expressed as mole percentage) to each product.

C. *XPS AND XRD MEASURMENTS*

XPS data were obtained on a VG ESCALAB MKII Photoelectron Spectrometer using Al Ka X—ray excitation (hv=1486. 6 eV). The experimental resolution was better than 0. 1 eV. The charge effect was corrected for by referring to the 1s level of the background carbon at a binding energy of 284. 6 eV.

XRD data were obtained on a Rigaku D/MAX—RB X—ray diffractiometer using a copper target at 40KVx100mA and scanning speed of 8 Deg. /min.

IV. RESULTS and DISCUSSION

A. *EFFECTS OF BORON*

Addition of the third additive B_2O_3 improves the desired catalytic properties of NaCl/Fe_2O_3 significantly. Even 1(at)% B_2O_3 additive increases the C_2 yield 5 times. When the addition approches 15(at)% B_2O_3, the C_2 yield and selectivity reach their maxima, 17% and 75% respectively(Fig. 1). The addition of B_2O_3 also greatly suppresses the complete oxidation to CO_2, but increases the amount of CO formed.

XPS data show that there are two well distincted peaks in the O_{1S} XPS spectrum. One peak with B. E. 529 eV is assigned to the lattice oxygen of Fe_2O_3, and the other peak with B. E. 531. 1 eV is assigned to the so—called additive oxygen of OH, B_2O_3, or both of them, The addition of B_2O_3 increases the ratio of the additive oxygen to the lattice oxygen of Fe_2O_3(Fig. 2). The improved catalytic properties may result from the modification of the lattice oxygen by the additive oxygen.

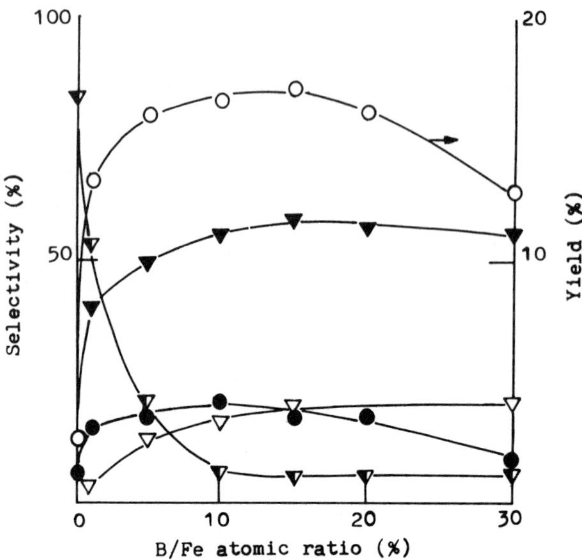

Fig. 1 *Optimum amount of boron*

○ : C_2 *yield*, ▼ : C_2H_4 *sel.*, ● : C_2H_6 *sel.*, ▽ : *CO sel.*, ▼ : CO_2 *sel.*; *NaCl*/B_2O_3/Fe_2O_3 *Catalysts with* 25(*at*)% *Na/Fe*, $T = 1023K$, $t = 0.24$ *g. s. cm*$^{-3}$, $P_{CH4}/P_{O2} = 4$, $P_{CH4} = 32kPa$.

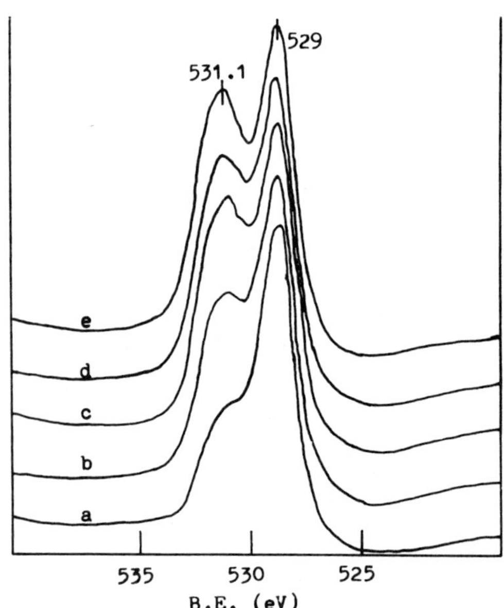

Fig. 2 O_{1s} *XPS spectra of the FB—series*

a: 00, *b*: 01, *c*: 05, *d*: 10, *e*: 20 (*at*)% B_2O_3/Fe_2O_3.

B. *EFFECTS OF REACTION CONDITIONS*

The effects of reaction conditions including reaction temperature, contact time and P_{CH4}/P_{O2} ratio, are summarized in Table 1. The conversion of methane increases with increasing temperature, but not the C_2 selectivity because the C_2 formation and the complete oxidation both increase with temperature. The C_2 yield increase is simply due to the methane conversion increasing with temperature.

Long contact time evidently increases the C_2H_4/C_2 ratio; the C_2H_4/C_2 is as high as 95% when t=2.40 g. s. cm^{-3}, but the total C_2 selectivity decreases. Although the

Table 1 *Effects of reaction conditions on the catalyst performance*

No.	1	2	3	4	5	6
Temp. (*K*)	973	1023	1073	1023	1023	1023
$t(g.s.cm^{-3})$	0.24	0.24	0.24	1.20	2.40	0.24
P_{CH4}/P_{O2}	4	4	4	4	4	10
Conv. (%)						
CH_4	10.5	22.8	29.8	29.8	30.4	14.8
O_2	~34	~64	~99	~93	~99	~96
Sel. (%)						
C_2H_4	33.6	58.5	60.8	59.1	56.6	57.0
CO	31.2	16.4	7.4	5.5	3.2	26.5
CO_2	9.9	5.4	12.2	24.7	30.5	7.0
C_2 *Sel.* (%)	64.8	74.9	68.2	64.6	59.8	83.5
C_2 *yield* (%)	6.8	17.1	20.4	19.2	18.2	12.4
C_2H_4/C_2 (%)	52	78	89	91	95	68

Cat.: *FBN*1525

methane conversion drops with increasing P_{CH4}/P_{O2} ratio, the C_2 selectivity increases significantly; the C_2 selectivity reaches 84% when $P_{CH4}/P_{O2}=10$.

C. *DEACTIVATION*

The most serious problem of the alkali metal chloride doped oxide catalysts is their deactivation. In the published literatures(4,5) this is attributed to a loss of chloride, i. e., a decrease of chloride concentration. The present results reveal that the deactivation of the $NaCl/B_2O_3/Fe_2O_3$ catalysts is a complicated process.

1. The XPS data show that the surface concentration of chloride does not change substantially while the activity decreases rapidly in the initial period of reaction (Fig. 3).

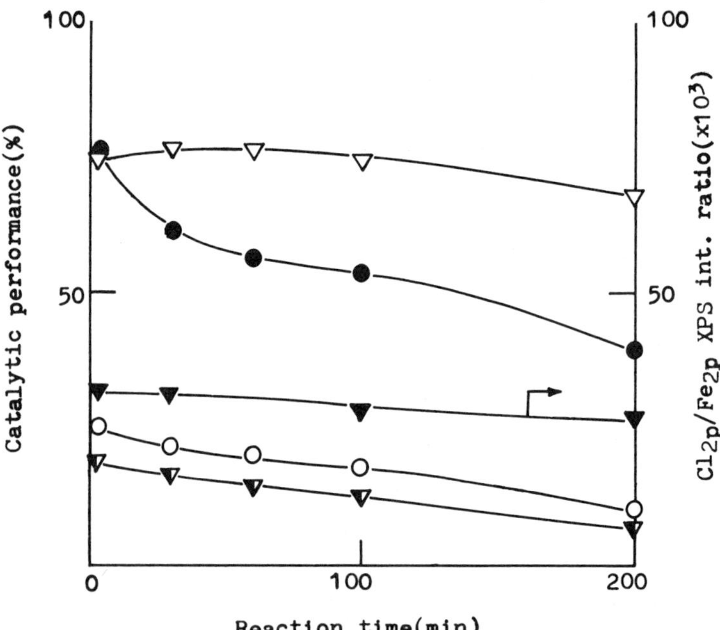

Fig. 3 *Change of catalytic performance and Cl_{2p}/Fe_{2p} ratio with reaction time*
● : O_2 *conv.*, ▽ : C_2 *sel.*, ○ : CH_4 *conv.*, ▼ : C_2 *yield*, ▼ : Cl_{2P}/Fe_{2P} *ratio*; *FBN*1025 *catalyst*, $T=1023K$, $t=0.24$ $g.s.cm^{-3}$, $P_{CH4}/P_{O2}=4$, $P_{CH4}=32$ *kPa*.

The deactivation of the catalysts is hence not simply due to a chloride loss. The loss of chloride is, nevertheless, one of the important deactivation factors; e. g. , about 20% of chloride on FBN1025 was lost as HCl after 40 hr. of reaction.

2. The amount of B_2O_3 has important effect on the deactivation rate When the addition of B_2O_3 is less than 5(at)% , the total activity is high and does not change appreciably with time, but the C_2 selectivity decreases to the level of $NaCl/Fe_2O_3$. When the addition exceeds 20(at)% B_2O_3, the total activity decreases rapidly, e. g. the activity of FBN3025 decreases to the level of FB30 after 100 min. of reaction. 15(at)% B_2O_3 addition can effectively restrain the deactivation (Fig. 4).

This effect can be understood from the XPS data(Fig. 5). After 200 min. of reaction, the lattice oxygen(B. E. 529 eV) is the major oxygen species on the catalyst sur-

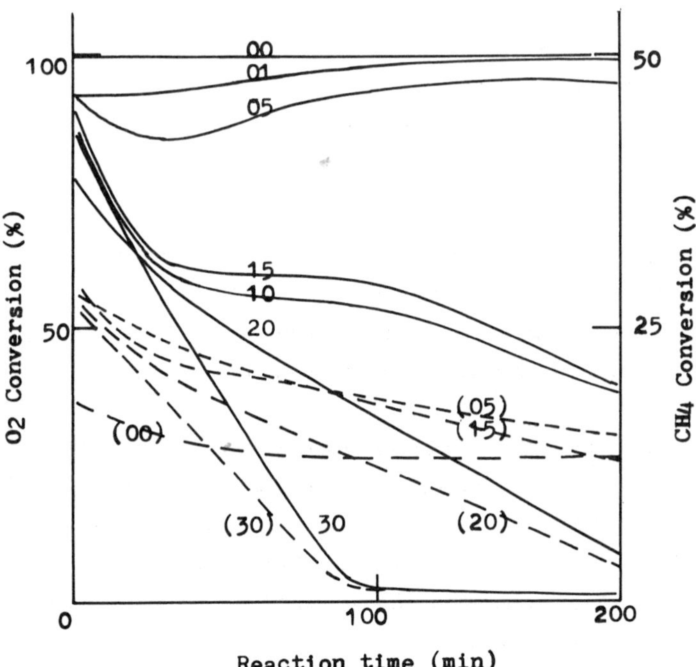

Fig. 4 *Effect of boron on deactivation*

Solid line: O_2 conv. , dash line: CH_4 conv.. Numbers on curves are the B/Fe atomic ratio in different catalysts with 25(at)% Na/Fe.

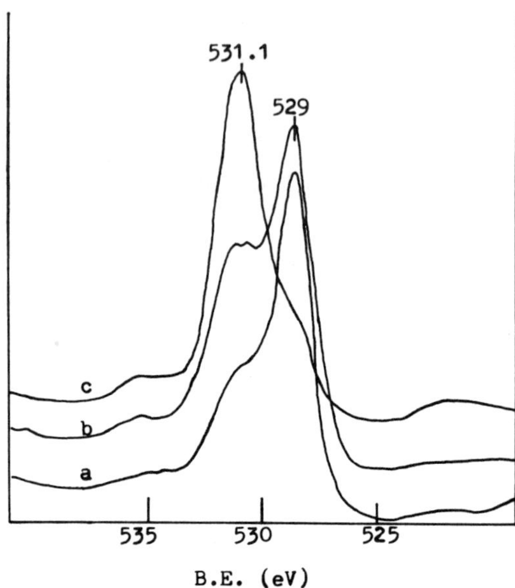

Fig. 5 *O_{1S} XPS spectra of FBN—series*

*a: FBN*0125, *b: FBN*1025, *c: FBN*3025 *after* 200 *min of reaction.*

face of FBN0125 , which provides high activity but poor C_2 selectivity. On the surface of FBN1025, another surface oxygen species (B. E. 531. 1 eV) modifies the property of the lattice oxygen. This results in better catalytic performance. When the additive oxygen becomes the dominant oxygen species, as on the surface of FBN3025, the total activity is quite low.

3. Serious corrosion of the quartz reactor was observed. By employing XPS to follow the reaction and deactivation process (Table 2), it can be seen that (a) the chloride loss is serious as reported in published works(4,5); (b) boron loss is also quite serious; and (c) silica from the corrosion of the reactor has deposited on the catalyst surface causing deactivation. XRD data show that a new compound, $NaFeSi_2O_6$, is formed on FBN1025 after 40 hr. of reaction (Fig. 6) .

By grinding the FBN1525 catalyst together with fine quartz powder, and calcining at 1023K for 2 hr. in air, a silica containing catalyst was obtained (labeled as FBN1525Q). Its catalytic performance is compared with FBN1525 (Fig. 7). During the first hour of reaction, the activity of FBN1525Q is as high as that of FBN1525,

Table 2 *Changes of the catalyst performance and XPS intensity ratio with time*

Catalyst	reac. time(hr)	Conv. (%) CH_4	Sel. (%) C_2	Yield(%) C_2	XPS int. ratio(%)[a] O_S/O_l	Cl/Fe	Na/Fe	B/Fe	Si/Fe
	0.5	29.1	65.3	19.0	85	9	41	3	0
*FBN*1025 *	10	27.0	61.8	16.7	94	8	40	2	1
	40	12.4	53.8	6.7	150	6	33	1	4
*FBN*0125	3.3	15.0	10.7	1.6	38	3	15	0	
*FBN*1025	3.3	10.9	68	7.4	59	3	18	1	
*FBN*3025	3.3	0.2	—	—	360	8	60	4	

a O_S/O_l: *the XPS intensity ratio of surface oxygen to lattice oxygen. FBN*1025 * $\sim t = 1.2$ *g. s. cm*$^{-3}$; *others* $\sim t = 0.24$ *g. s. cm*$^{-3}$; $T = 1023K$ $P_{CH_4}/P_{O_2} = 4$, $P_{CH_4} = 32$ *kPa.*

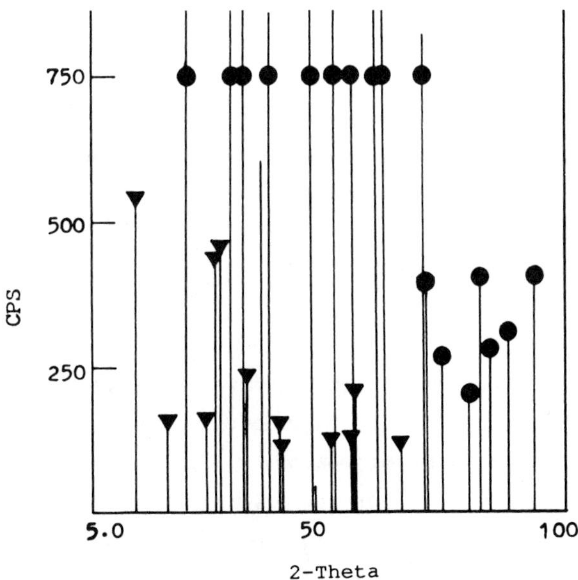

Fig. 6 *A new compound detected by XRD*

▼: *$NaFeSi_2O_6$*, ●: *Fe_2O_3*; *FBN*1025 *catalyst after* 40 *hr. of reaction.*

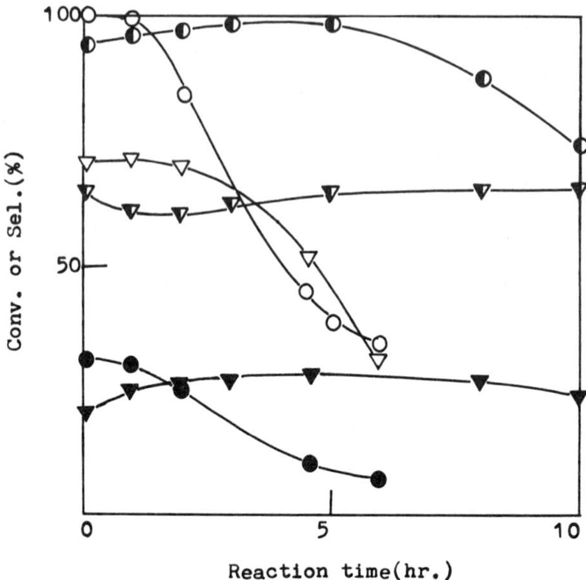

Fig. 7 *Effect of silica on FBN1525 catalyst*
FBN1525Q catalyst, ○ : O_2 *conv.*, ▽ : C_2 *sel.*, ● : CH_4 *conv.*; *FBN1525 catalyst*, ◐ : O_2 *conv.*, ▼ : C_2 *sel.*, ▼ : CH_4 *conv.*; $T=1023K$, $t=0.24$ *g. s. cm*$^{-3}$, $P_{CH4}/P_{O2}=4$, $P_{CH4}=32kPa$.

but thereafter it decreases rapidly. This also suggests that silica deposition on the catalyst surface is a significant factor causing deactivation.

D. *MECHANISMS OF ACTIVATION AND DEACTIVATION*

The match of the lattice oxygen (B. E. 529 eV) with the additive oxygen (B. E. 531. 1eV), and the presence of alkali metal chloride are the necessary conditions for a good catalyst. As mentioned above (Fig. 5 and Table 2), the lattice oxygen provides high activity but low selectivity; the additive oxygen has poor activity. Only when the lattice oxygen and additive oxygen are well matched, can high activity and selectivity be obtained. By employing XPS to follow the reaction, it is also revealed that well matched additive oxygen with lattice oxygen of Fe_2O_3 provides high activity and selectivity, and the activity decreases substantially when the additive oxygen becomes dominant on the surface of the catalyst(Fig. 8 and Table 3).

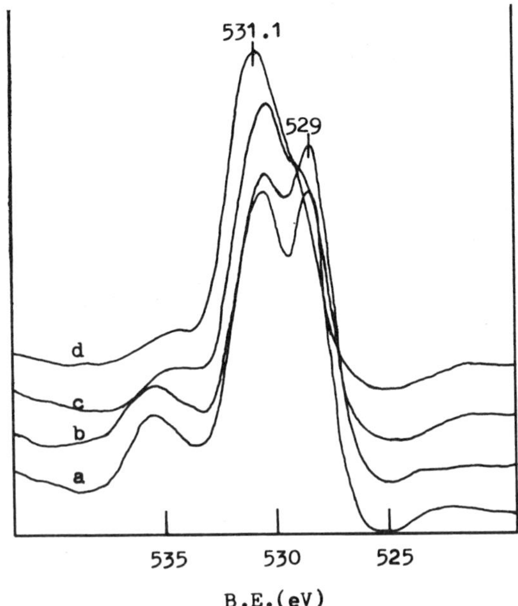

Fig. 8 *Change in O_{1s} XPS spectra with reaction time*
*FBN*1025 *catalyst after* 1/2(*a*), 10(*b*), 40(*c*) *hr. of reaction*; *d*: *FBN*1525*Q* *catalyst after* 6 *hr. of reaction.*

Table 3 *Synergistic effect of NaCl and B_2O_3*

Catalyst	Conversion (%)		Selectivity (%)					Yield (%)
	CH_4	O_2	CO	CO_2	C_2H_4	C_2H_6	C_2	C_2
BN *	1.2	2	—	2.1	23	75	98	1.2
*FB*20	5.9	24	49.8	10.0	4.8	35.4	40.2	2.4
*FN*25	18.2	100	0.5	70	15.7	13.7	29.4	5.4
*FBN*2025	21.4	61	18.8	5.8	56.9	18.6	75.5	16.1

$t=0.24$ *g. s. cm*$^{-3}$, $P_{CH4}/P_{O2}=4$, $P_{CH4}=32$ *kPa*, $T=1023K$ *except BN* * ; *BN* * *denotes* 8 (*wt*)% $B_2O_3/NaCl$, *ground thoroughly, pelletized and sieved to* 50—70 *mesh.* $t=1.20$ *g. s. cm*$^{-3}$, $P_{CH4}/P_{O2}=4$, $P_{CH4}=32$ *kPa*, $T=1023$ *K*.

It is noted that NaCl and B_2O_3 show a synergistic effect: $NaCl/Fe_2O_3$(FN25) has high activity but produces mainly CO_2, B_2O_3/Fe_2O_3 (FB20) has poor activity and $B_2O_3/NaCl$(BN *) is not active, but the $NaCl/B_2O_3/Fe_2O_3$(FBN2025) has high activity and selectivity for the OCM (Table 3).

Based on above results, it is suggested that the addition of B_2O_3 adjusts the balance of the additive oxygen and lattice oxygen of Fe_2O_3. The lattice oxygen of Fe_2O_3 is the methane activation oxygen species, and the additive oxygen inhibits the complete oxidation. The alkali metal chloride participates in methane activation and dehydrogenation of C_2H_6 to C_2H_4, therefore high ethylene selectivity results with $NaCl/B_2O_3/Fe_2O_3$. The mechanisms of the methane activation and the catalyst deactivation are proposed as follows:

$$CH_4 + NaCl + FeO_X(O) \longrightarrow CH_3Cl + NaOH + FeO_X \quad (1)$$
$$CH_3Cl \longrightarrow CH_3\cdot + Cl\cdot \quad (2)$$
$$2CH_4 + FeO_X(O) \longrightarrow 2CH_3\cdot + H_2O + FeO_X \quad (3)$$
$$CH_3\cdot + CH_3\cdot \longrightarrow C_2H_6 \quad (4)$$
$$C_2H_6 + Cl\cdot \longrightarrow C_2H_5\cdot + HCl \quad (5)$$
$$C_2H_5\cdot + Cl\cdot \longrightarrow C_2H_4 + HCl \quad (6)$$
$$HCl + NaOH \longrightarrow NaCl + H_2O \quad (7)$$
$$FeO_X + 1/2O_2 \longrightarrow FeO_X(O) \quad (8)$$
$$NaOH + SiO_2 \longrightarrow (Na,Si)O_X \quad (9)$$
$$(Na,Si)O_X + Fe_2O_3 \longrightarrow NaFeSi_2O_6 \quad (10)$$

Eq. (1) and Eq. (3) are the methane activation processes. Eq. (2) and Eq. (3) show the methyl radical generation steps. Eq. (4) shows the formation of C_2H_6 through methyl radical coupling. Eq. (5) and Eq. (6) explain why the alkali metal chloride doped oxides shows higher ethylene selectivity than other kind of catalysts. Eq. (7) is the chloride recycling process and Eq. (8) is the catalytic redox process for catalyst. Eq. (9) and Eq. (10) show the deactivation processes caused by silica.

V. CONCLUSION

Addition of boron to $NaCl/Fe_2O_3$ adjusts the balance of the additive oxygen and lattice oxygen of Fe_2O_3, which results in high activity and selectivity.

NaCl and B_2O_3 show a synergestic effect: B_2O_3 provides additive oxygen which inhibits complete oxidation. NaCl participates in the methane activation. The lattice oxygen of Fe_2O_3 is the methane activation oxygen species.

The catalyst deactivation is a complicated processes related to the chloride and boron loss of the catalysts and silica deposition on the catalyst surface. The latter results in the formation of a new compound $NaFeSi_2O_6$.

VI. REFERENCES

1. J. S. Lee and S. T. Oyama; *Catal. Rev. Sci. and Eng.* 30(1988)249.
2. G. E. Keller and M. M. Bhasin; *J. Catal.* 73(1982)9.
3. Liu Qin, Shao yun, et al ; *Cui Hua Xue Bao (CN)*, 9(1988)214.
4. K. Otsuka, Qin Liu, et al ; *Chem. Lett.* 903(1986).
5. Kiyoshi and K. Otsuka; *Methane Conv. Symposium*, New Zealand, 1987.
6. A. G. Anshits, N. P. Kirik, et al; *Catal. Today* 4(1989)399.

8

CONVERSION OF METHANE TO ETHYLENE AND ACETYLENE BY THE CHLORINE-CATALYZED OXIDATIVE PYROLYSIS PROCESS

S.M. Senkan* and R. Yildirim
Chemical Engineering Department
University of California
Los Angeles, CA 90024

and

D. Palke and G.E. Lewis
Dow Chemical Company
Hydrocarbons Research
Freeport, TX 77541

ABSTRACT

Chlorine-catalyzed oxidative pyrolysis (CCOP) process was recently developed to convert methane, the major component in natural gas, into higher molecular weight hydrocarbons such as C_2H_4 and C_2H_2. According to the CCOP process, methane is chlorinated first to chlorinated methanes (CM), i.e. CH_3Cl and others, using any one of the well established technologies. Chlorinated methanes are then oxidatively pyrolysed in a flow reactor at about 900 °C, forming primarily C_2H_4, C_2H_2 and synthesis gas. The presence of oxygen in the system inhibits the formation of carbonaceous deposits, such as soot, tar, and coke thereby decreasing the down time of the pyrolysis reactors. In this paper, we report on the results of the oxidative pyrolysis of CH_3Cl conducted in the presence of different levels of CH_4.

---

* To whom correspondence should be addressed.

INTRODUCTION

The production of olefins and aromatics from methane is of considerable practical significance because of the availability of large quantities of natural gas. At present, less than 10% of the natural gas produced is used as a chemical feedstock according to the estimates by the Gas Research Institute [1], [2]. This figure includes natural gas used in the manufacture of synthesis gas (i.e., mixtures of hydrogen and carbon monoxide), carbon black, as well as the direct production of higher molecular weight hydrocarbon products. One of the major reasons for such a low utilization rate is related to the stability of the methane molecule. Because methane is stable and unreactive, high temperatures must be employed to effect its conversion. Commercial processes utilize electric and plasma arcs, flames, and furnaces to achieve the necessary high temperatures [3]. However, at the high temperatures associated with these processes, the selectivities of valuable hydrocarbon products are too low unless reaction times are kept extremely short [4]. On the other hand, short reaction times decrease the extent of conversion of methane. In addition, carbonaceous deposits, which include coke, soot, and tars almost always form in current commercial processes. The use of oxygen reduces carbon formation significantly; however, this further decreases the selectivities of valuable products because of the formation of CO, CO_2, and H_2O. Consequently, a significant fraction of current research on "methane activation chemistry" is focused on low-temperature heterogeneously-catalyzed conversion processes [5]. However, highly active and selective catalysts with adequate life spans are yet to be developed.

Over forty years ago, Gorin [6] proposed a gas-phase, chlorine-catalyzed process in which methane conversion is achieved by CH_4 chlorination first, followed by the pyrolysis of chlorinated methanes (CM) and formation of C_2+ products and HCl. The HCl produced could either be converted into chlorine via the well-known Deacon reaction and recycled, or used to oxychlorinate methane, thereby completing the catalytic cycle for chlorine. More recently Benson [7] patented a related process involving the simultaneous chlorination of methane and pyrolysis of CM in a single chamber by igniting a mixture of CH_4 and Cl_2. In a follow up study, Weissman and Benson [8] examined the kinetics of the pyrolysis of CH_3Cl in a flow reactor. As discussed in both of these investigations, the pyrolysis of CM leads to the formation of significant levels of carbonaceous deposits, rendering these processes less attractive from commercialization point of view. The formation of excessive carbonaceous deposits, however, is not surprising because of the well known soot (carbon) promoting effects of chlorine and chlorinated hydrocarbons in hydrocarbon combustion [9].

In the Chlorine-Catalyzed Oxidative-Pyrolysis (CCOP) process, developed recently in our laboratories, the formation of carbonaceous deposits has been minimized by the use of oxygen during CM pyrolysis [10]. As also discussed in Karra and Senkan [11], the primary effect of O_2 in the CCOP process is to intercept and convert C_2H_3 to CH_2O and CHO, thus inhibit further molecular weight growth, without significantly altering the levels of stable C_2 hydrocarbons produced by the chemically activated recombination of C_1 radicals [12]. In an earlier investigation the co- pyrolysis of CH_3Cl and CH_4 was studied by Weissman and Benson [8], in which the selectivity for C_2H_4 was noted to increase with increased CH_4 concentration, again with the formation of substantial levels of carbonaceous deposits. The conditions in the CCOP process are different because of the presence of O_2 and flammability issues. The potential for flame formation in the CCOP process imposes operational complications in which the reaction temperature, residence time and pressure must be matched with each $CH_3Cl/CH_4/O_2$/Diluent mixture composition to avoid combustion reactions. Flame formation clearly is undesirable because it results in the significant reduction in C_2 product selectivity accompanied by the simultaneous increase in the formation of carbon oxides, as well as soot and coke because of the fuel rich nature of the mixtures [13].

EXPERIMENTAL

The experimental facility used to study the oxidative pyrolysis of CH_3Cl/CH_4 mixtures is shown in Figure 1. The reactions were carried out at atmospheric pressure in a 2.1 cm ID quartz reactor which was about 100 cm long. At this diameter and at the temperatures and pressures (1 atm) involved, the oxidative pyrolysis processes are expected to be dominated by gas-phase kinetics, with minor contributions from surface induced reactions. The reactor was placed inside a 3-zone Linbergh furnace. The reactants (CH_3Cl/O_2 mixtures) were introduced into the reactor by using a 6 mm OD quartz probe that was placed centrally at the upstream end of the reactor through radially directed injection holes. The injector was air cooled to prevent premature reactions. The first zone of the furnace, which was about 15 cm long, was used to preheat the carrier gas (N_2) and CH_4.

Gas flows were established by the combined use of rotameters and needle valves. The needle valves were operated under critical flow conditions in order to establish uncoupled flow rates. Mean gas flow velocities were in the range 1-10 m/s during the experiments, suggesting laminar flow conditions were present. However, the expected deviation from ideal plug flow behavior would be in the range 10-15%, about the same order of magnitude as the other experimental errors [14]. This was due to high dilutions of the reactants, and high temperatures present in the system, as well as due to the sufficiently high aspect ratio of the reactor, which was in excess of 15 in our studies.

Species sampling were accomplished by withdrawing gases through a quartz microprobe having an orifice diameter of about 0.10 mm at its tip. With the exception of its tip, the sampling probe was cooled by hot-oil (140 °C) to prevent the condensation of heavy molecular weight products. The sampling probe pressure was maintained at about 100 Torr (13.3 kN/m^2) to provide rapid removal of gases from the reaction zone and to minimize the condensation of high molecular weight species.

The identification and quantification of species were accomplished by using a computer-controlled gas chromatograph (GC) system. The GC was equipped with

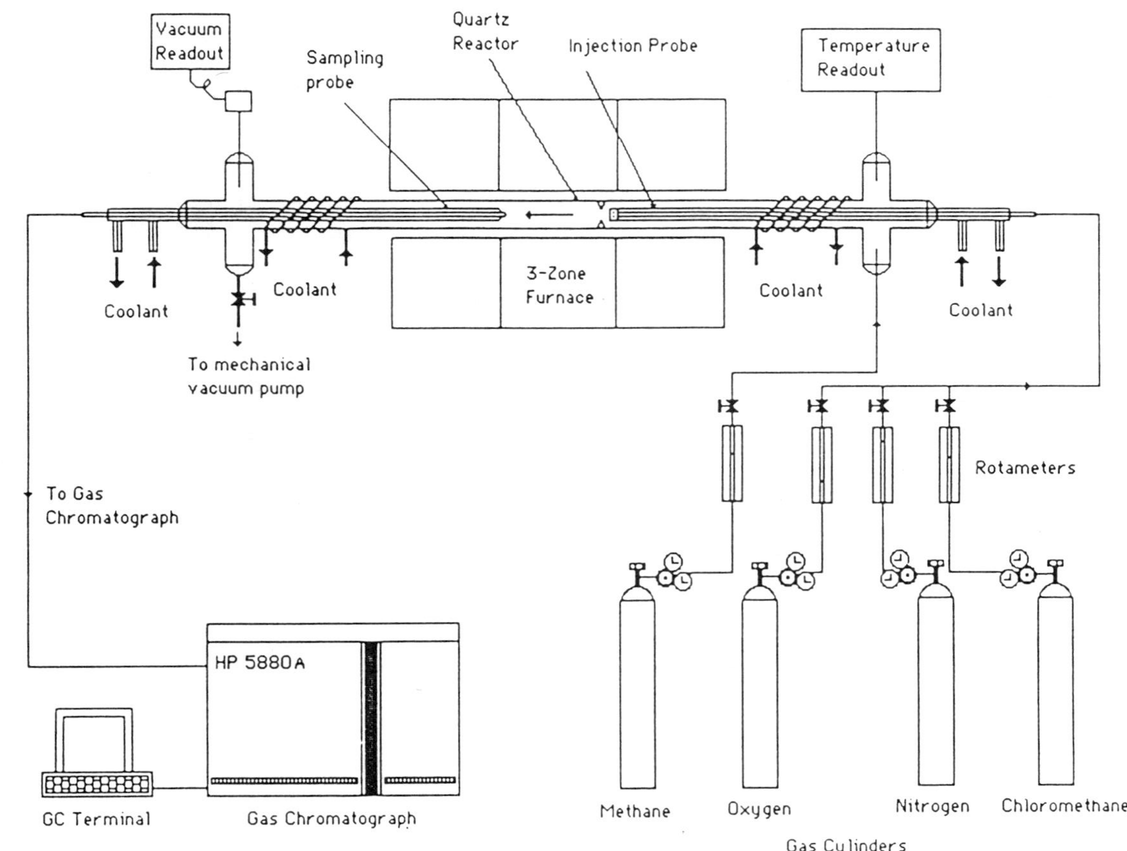

Figure 1. Schematics of the experimental system.

two packed columns (20 ft Porapak QS and 6 ft molecular sieve 13A) which were connected to a thermal conductivity detector that was operated with helium carrier gas. The samples were introduced into the gas chromatograph by computer controlled valve injection, and analyzed using standard GC techniques. Calibration of reactants was made by withdrawing unreacted gases directly through the sampling probe. For all gaseous intermediates and products, certified gas mixtures acquired from Matheson Co. were used. Species profiles were generated by moving the injector probe relative to the stationary sampling probe.

Temperature profiles were measured by axially moving a 0.075mm chromel-alumel thermocouple placed into the reactor. The temperature profiles were sufficiently uniform within the reaction zone of interest, and remained essentially isothermal (within ± 10 °C) throughout the experiments.

RESULTS AND DISCUSSION

In order to better ascertain the effects of CH_4 on the oxidative pyrolysis of CH_3Cl a series of experiments were conducted with different CH_4/CH_3Cl ratios and at different temperatures and total hydrocarbon concentrations. In all the experiments, the O_2 concentration was kept at about 3.7% and N_2 was used as the diluent. In Table 1, the specific compositions investigated are shown. In this paper we report on experimental results involving only the major species, which were the reactants (CH_3Cl, O_2), N_2 carrier gas, and products C_2H_2, C_2H_4, C_2H_6, CH_4, HCl, CO and CO_2. The formation of minor species and the process of coke formation and suppression by O_2 will be the subject of a future communication.

In Figure 2, the product distributions for the oxidative pyrolysis of pure CH_3Cl (17.32 % in the pre-reaction mixture, i.e. Mixture D) are presented as a function of temperature at a nominal residence time of about 0.43 s. In this and subsequent figures lines have been drawn through experimental data points (indicated by symbols) to show trends. The data in this figure were acquired by sampling gases

Table 1. Pre-Reaction Compositions of the Mixtures Studied (Mole %)

Mixture	CH_4	CH_3Cl	O_2	N_2	CH_4/CH_3Cl
A	73.14	8.02	3.89	14.95	9.12
B	63.80	17.54	3.72	14.95	3.64
C	57.48	23.64	3.76	15.12	2.43
D	00.00	17.32	3.68	80.11	0.00
E	17.34	17.56	3.72	61.38	0.99
F	22.78	22.80	3.72	50.69	1.00
G	10.31	10.03	3.70	75.95	1.03
H	29.81	22.84	3.73	43.62	1.31
I	15.78	22.77	3.71	57.52	0.69
J	36.69	11.83	3.71	47.84	3.11

by fixing the spatial positions of both the injection and sampling probes and by changing the temperature of the furnace. In addition, the feed composition and feed gas flow rates were maintained the same. Consequently, the data points shown in Figure 2 do not precisely correspond to identical residence times because of differences in gas density, caused by differences in temperature, and the total number of moles, caused by different extents of reaction. However, the effects of both of these variations on residence times would be small. For example, since differences in temperatures between data points were at most 140 K (990-850° C), reaction times would be different by at most 12.5 % because of changes in gas density, i.e. (140 K/1123 K)x100 = 12.5 %). In addition, the increase in gas flow rate caused by increases in the number of moles would also be less than 10% based on the complete conversion of CH_3Cl into C_2H_4, C_2H_2, H_2 and HCl. Both these

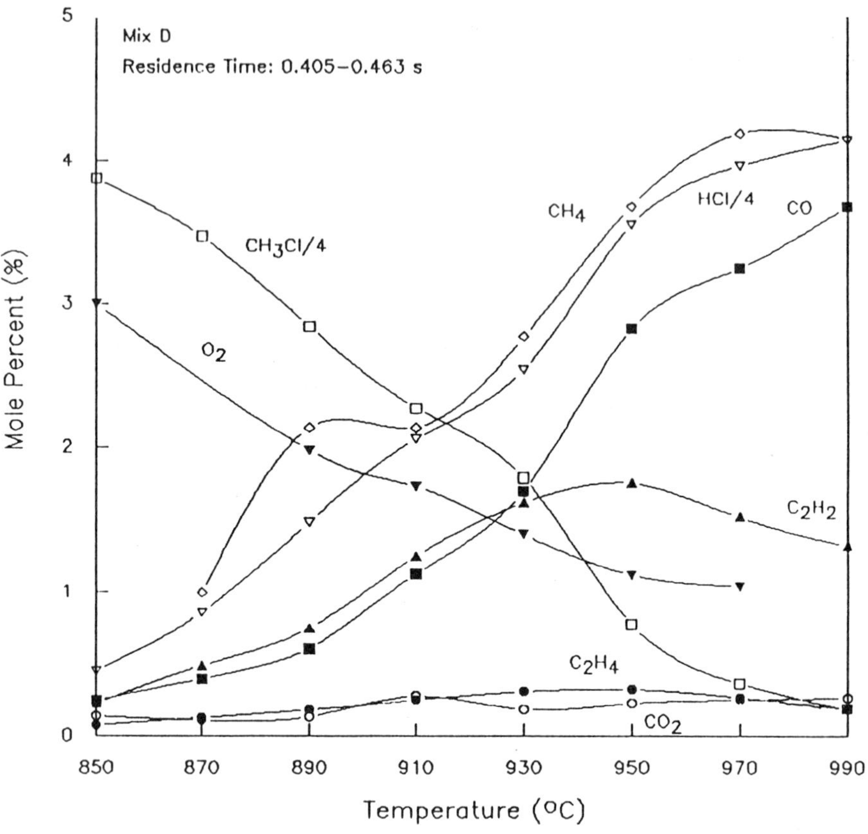

Figure 2. The effects of temperature on conversion and product distributions for Mixture D.

errors are well within the normal experimental limits of $\pm 15\%$ expected in these types of studies.

As evident from Figure 2, C_2H_2 is the major C_2 product in CH_3Cl oxidative pyrolysis, reaching levels as high as 1.8 % in the reaction mixture at about 950°C, with the corresponding selectivity of about 60 % based on the amount of CH_3Cl reacted. The selectivity for C_2H_4 formation is considerably less, with concentrations below 0.5%. CH_4 also forms as a significant by- product in the oxidative pyrolysis of CH_3Cl reaching levels as high as 4%. Although some C_2H_3Cl were also present in the product mixture, it was not quantified because of low concentrations. As expected, CO production increases with increasing temperature due to the oxidation of C_2H_4 and C_2H_2, and ultimately becomes the major product at high temperatures. It is important, however, to note that CO_2 production remains low, indicative of the absence of flame reactions.

In Figure 3, the product distributions are presented for Mixture A, which had the highest CH_4/CH_3Cl ratio of 9.12. A comparison of this figure with Figure 2 shows that the presence of CH_4 alters the product distribution significantly and

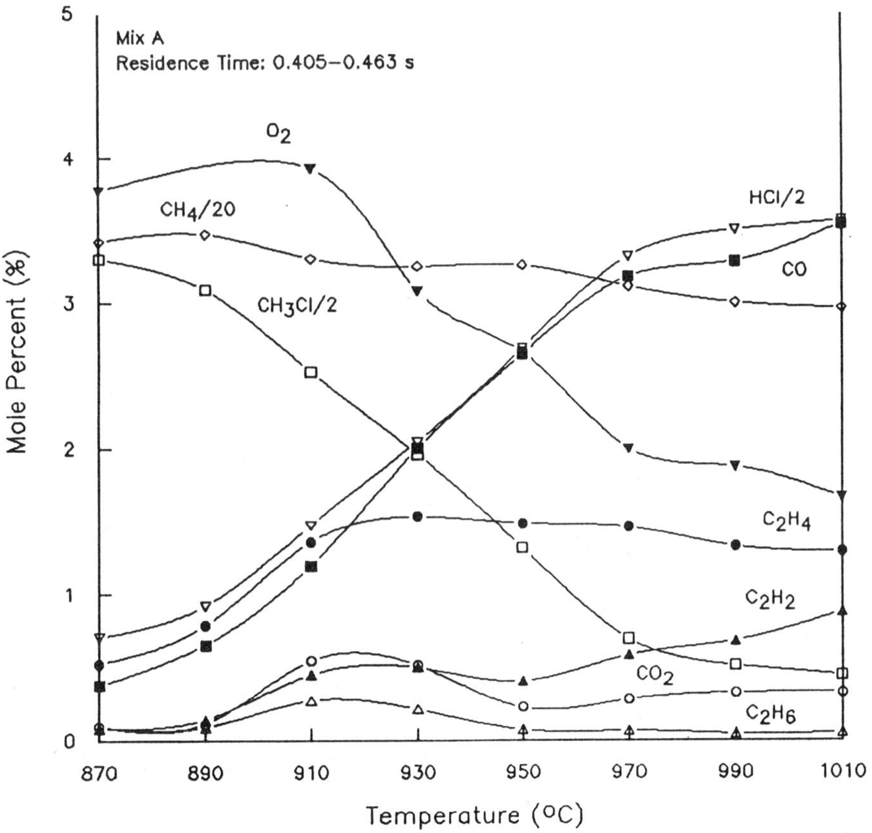

Figure 3. The effects of temperature on conversion and product distributions for Mixture A.

promotes the formation of C_2H_4 over C_2H_2. Some C_2H_6 also did form in these experiments. The enhanced C_2H_4 formation follows the earlier observations [8], in spite of the presence of O_2 in the system. This is a surprising and gratifying result in view of the high H/Cl ratio of the mixture and the potential for combustion reactions due to the presence of O_2. Apparently the amount of CH_3Cl present and the HCL produced in the system is sufficient to suppress flame formation and to preserve the yields of C_2 products.

The effects of temperature and pre-reaction mixture composition on C_2H_4 and C_2H_2 production are shown in Figures 4 and 5, respectively for residence times in

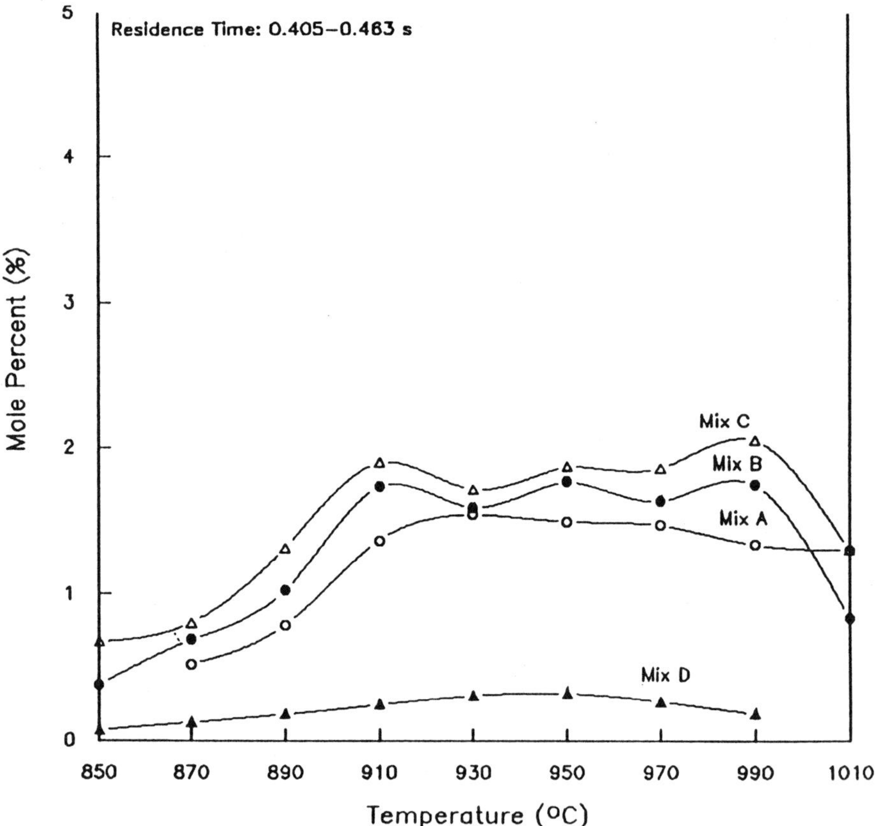

Figure 4. The effects of temperature and mixture composition on C_2H_4 production. Mixtures A-H.

the range 0.405-0.463 s. It is particularly interesting to note from Figure 4 that maximum C_2H_4 concentrations in the product mixtures were achieved for Mixtures B and C with CH_4/CH_3Cl ratios 2.43 and 3.64, respectively. It is particularly significant to note that Mixture A, in spite of its high CH_4/CH_3Cl ratio leads to lower C_2H_4 concentrations, primarily because of high CH_4 dilution. The lowest C_2H_4 concentrations were obtained for the case of pure CH_3Cl. At high temperatures the selectivities for C_2H_4 decrease because of its thermal decomposition.

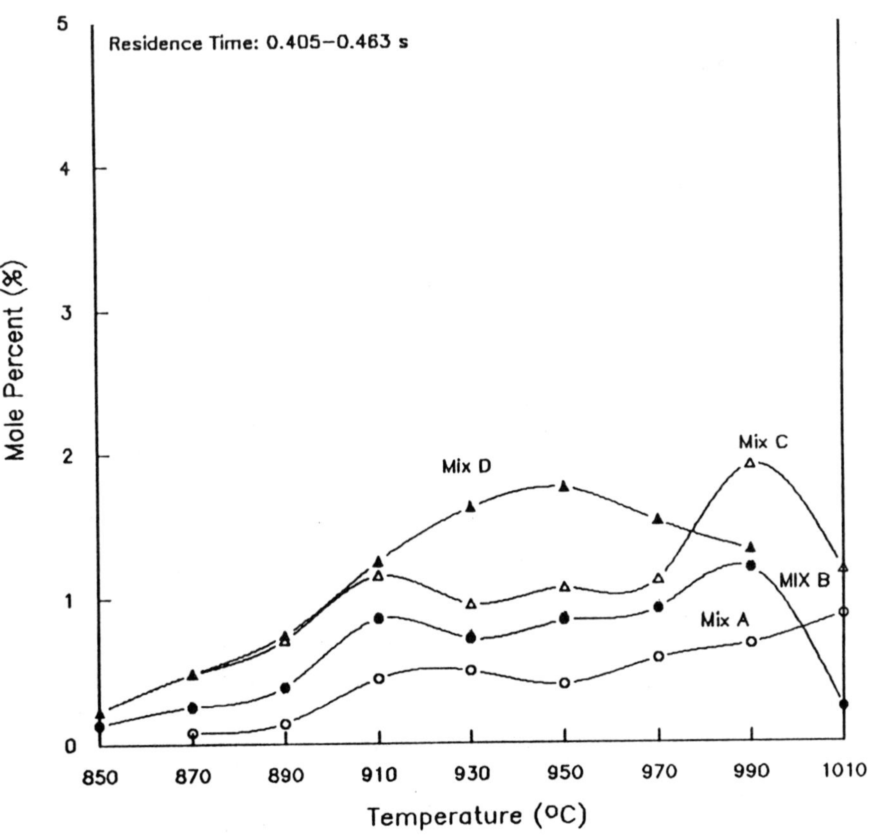

Figure 5. The effects of temperature and mixture composition on C_2H_2 production. Mixtures A-H.

From Figure 5 it is clear that, on the overall, the highest C_2H_2 concentration levels were reached in the absence of CH_4 (Mixture D). Mixtures B and C also lead to significant C_2H_2 production, with concentrations reaching levels as high as 1 %. Again at higher temperatures C_2H_2 yields decrease due to thermal decomposition.

In Figure 6 the effects of total hydrocarbon concentration on product distributions are presented at a reaction temperature of 910°C, a nominal residence time of 0.44 s, and for a CH_4/CH_3Cl ratio of 1.0. As apparent from this figure the

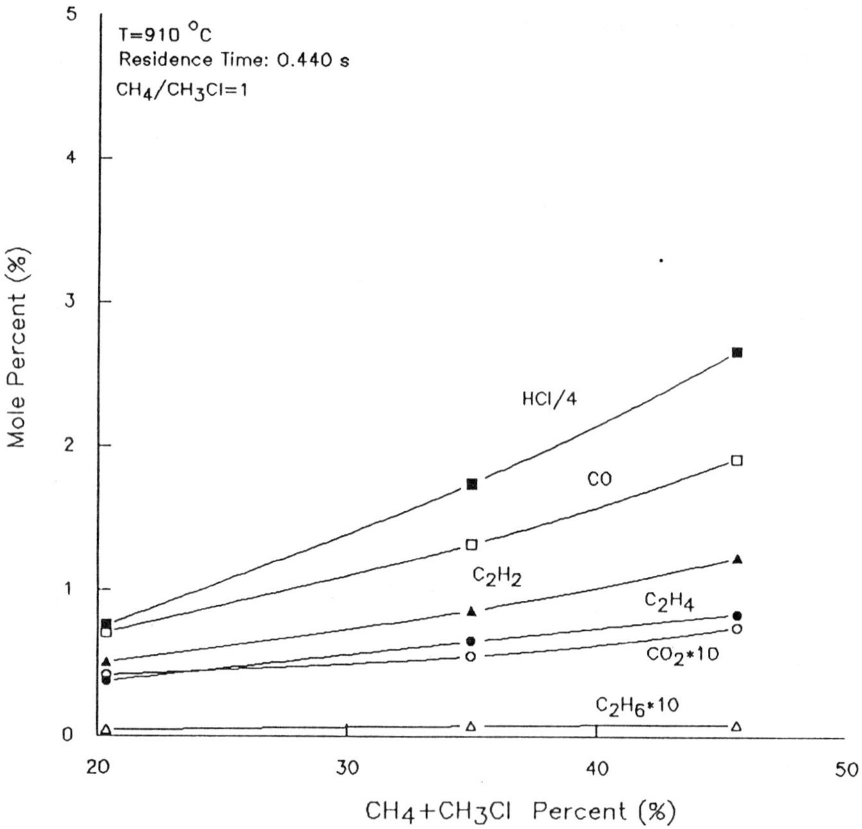

Figure 6. The effects of total hydrocarbon concentration on product distribution. Mixtures A-H.

As evident from the above discussion, increasing CH_4 increases the production of CH_3, and this favors reaction pathways leading to the formation of C_2H_4. It is also important to note that under the conditions studied, all of the unimolecular decomposition and radical combination reactions of interest would be in the fall-off regime. This necessitates that rate parameters must be evaluated as a function of the density of the system for the development of quantitative models [12].

C_2H_4 formed then undergoes the following destruction processes:

$$C_2H_4 + Cl \rightarrow C_2H_3 + HCl \quad (17)$$

$$C_2H_4 + (CH_2Cl, CH_3) \rightarrow C_2H_3 + (CH_3Cl, CH_4) \quad (18)$$

leading to the formation of one of the most important C_2 radicals in the system, namely C_2H_3. In the absence of oxygen, the primary reaction pathways available for C_2H_3 are its polymerization:

$$C_2H_3 + C_2H_2 \rightarrow CH_2CHCHCH \quad (19)$$

and to a lesser extent:

$$C_2H_3 + C_2H_4 \rightarrow CH_2CHCH_2CH_2 \quad (20)$$

or its decomposition to acetylene by reaction (16). The $CH_2CHCH_2CH_2$ and $CH_2CHCHCH$ radicals subsequently undergo dehydrogenation, hydrogenation, further addition reactions with C_2H_3 and C_2H_2, cyclize and ultimately result in the formation of carbonaceous deposits such as tar, soot and coke. Although the subsequent chemical kinetic steps leading to the formation of solid deposits are not fully known at present, the process nevertheless is well known to be extremely rapid [15], and reactions such as (19) and (20) are believed to play a pivotal role [8], [16], [17].

In the presence of oxygen, however, the C_2H_3 radicals are rapidly oxidized by O_2 via [18]:

$$C_2H_3 + O_2 \rightarrow HCOH + HCO \qquad (21)$$

Reaction (21) effectively compete with reactions (19) and (20) for the C_2H_3 radicals, thereby suppressing the processes that lead to higher molecular weight products. The HCOH and HCO formed by reaction
(18) subsequently are converted into CO via the following steps:

$$HCOH + Cl \rightarrow HCO + HCl \qquad (22)$$

$$HCOH + (CH_2Cl, CH_3) \rightarrow HCO + (CH_3Cl, CH_4) \qquad (23)$$

$$HCO + M \rightarrow CO + H + M \qquad (24)$$

Since the presence of methane increases C_2H_4 production, which in return leads to more C_2H_3 formation, CO production is also enhanced when CH_4 is present in the system. It must be recognized, however, that CO is a gaseous product which can be handled much more easily than carbonaceous deposits. Furthermore, CO is an essential constituent of the "synthesis gas", thus can itself be used to manufacture higher molecular weight hydrocarbons.

ACKNOWLEDGMENTS

This research was supported by funds from the Dow Chemical Company and the U.S. Environmental Protection Agency, Grant No: R-815136-01.

REFERENCES

1. Samsa, M.E., Hedman, B.A., and Solomon, I.J., "Status and Outlook for Natural Gas as a Chemical Feedstock", The Gas Research Institute Report, August 1984.

2. Solomon, I.J., "R & D Opportunities in the Use of Natural Gas as a Chemical Feedstock", The Gas Research Institute Report, September 1985.

3. Kirk-Othmer, "Encyclopedia of Chemical Technology", 1, p. 211, John Wiley, New York, 1978.

4. Back, M.H., and Back, R.A., "Thermal Decomposition and Reactions of Methane", in *Pyrolysis: Theory and Industrial Practice,* Albright, L.F., Crynes, B.L., and Corcoran, W.H., Eds., Academic Press, New York, NY 1983.

5. Pitchai, R., and Klier, K., *Cat. Rev. Sci. Eng.,* 28, 13 (1986).

6. Gorin, E., US Patent 2,320,274, 1943.

7. Benson, S.W., US Patent 4,199,533, 1980.

8. Weissman, M., and Benson, S., *Int. J. Chem. Kinetics.,* 16, p. 307 (1984).

9. Senkan, S.M., Robinson, J.M., and Gupta, A.K., *Combust. Flame,* 49, 305 (1983).

10. Senkan, S.M. US Patent 4,714,796, 1987.

11. Karra, S.B., and Senkan, S.M., *Ind. Eng. Chem Research,* 27, 45 (1988).

12. Karra, S.B. and Senkan, S.M., *Ind. Eng. Chem. Research,* 27, 447 (1988).

13. Granada, A., Karra, S.B., and Senkan, S.M., *Ind. Eng. Chem. Research,* 26, 1901 (1987)

14. Cathonnet, M., Boettner, J.C., James, H., Eighteenth *Symposium (International) on Combustion,* p. 903, The Combustion Institute, Pittsburgh (1981).

15. Trimm, D.L., "Fundamental Aspects of the Formation and Gasification of Coke", in *Pyrolysis: Theory and Industrial Practice,* Albright, L.F., Crynes, B.L., and Corcoran, W.H., Eds., Academic Press, New York, NY 1983

16. Frenklach, M., Clary, D.W., Gardiner, W.C., and Stein, S.E., *Twentieth Symposium (International) on Combustion,* p. 887, The Combustion Institute, Pittsburgh (1984)

17. Colket, M.B., *Twenty First Symposium (International) on Combustion,* The Combustion Institute, Pittsburgh, p. 851 1988.

18. Park, J.Y., Heaven, M.C., and Gutman, D., *Chem. Phys. Lett.,* 104, p. 469 (1984).

9

ETHYLENE AND PROPYLENE FROM METHANE: THE OXYPYROLYSIS OF NATURAL GAS

A. Robine and C. J. Cameron

Division de Recherche Cinétique et Catalyse,
Institut Français du Pétrole,
1 avenue de Bois-Préau, BP 311, 92506 Rueil Malmaison, France.

I. INTRODUCTION

The direct, selective oxidation of methane to higher hydrocarbons, known as oxidative coupling of methane (OCM), has attracted significant academic and industrial interest since its conception at Union Carbide. An article by Keller and Bhasin, which was submitted in Oct. 1980 and published in Jan. 1982 [1], contains the first clear evidence for this reaction. Both the 'redox' and 'co-feed' modes of operation were studied; however, the redox mode was shown to be superior under the conditions and with the particular reactor used in the study. Catalysts based on reducible manganese, lead, thallium and cadmium oxides were found to be particularly effective catalysts in the redox mode. This article clearly precedes other submission or filing dates on record.

Several months later (Aug. 1982), Jones, Leonard and Sofranko from Arco filed for the first OCM redox patents [2], and Baerns (Bochum) filed for the first co-feed patent using reducible metal oxide catalysts (Oct. 1982) [3].

After this initial activity, this research field remained dormant until August 1983, when Arco filed for several continuation-in-part patents based on the applications filed in 1982 [4].

Although 1982 was the period of discovery, international interest in OCM did not emerge until 1984. In late 1983, an article by Baerns appeared in the German literature [5]. This was followed, in early 1984, by the publication of his patent application [3] and the issue of the first OCM patent to Arco [6].

Processes based on OCM are considered to be potential competitors to the conventional syngas processes for the production of synthetic gasoline. Whereas most of the industrial research efforts in the field of OCM have been primarily directed to gasoline as the end product, our research efforts have targeted ethylene as the desired end product. By changing the primary objective of our approach to methane upgrading, we were able to pursue hypotheses which would not have been otherwise considered, such as separating the C_2+ fraction out of natural gas before carrying out the selective oxidation, then reinjecting the same fraction into the oxygen-free OCM effluent [7-11].

By effecting the selective oxidation (OCM) step at significantly higher temperatures ($> 850°C$) and substantially higher linear space velocities (> 0.45 m/s) than previously reported [12,13], the heat released by the highly exothermic OCM reaction can be utilized *in situ* for the endothermic pyrolysis of ethane to ethylene and hydrogen. It should be noted that the high temperature of the initial OCM effluent is substantially in excess of that required to convert ethane, the primary selective OCM product, into ethylene and hydrogen. Thus the previously removed C_2+ fraction is reinjected into the OCM effluent stream, thereby providing a thermochemical quench.

Combining two high temperature reactions (OCM and C_2+ pyrolysis) in the same reactor significantly decreases the required investment for a process based on OCM, due to the elimination of a costly steam cracker and its unique separation train [14]. The oxypyrolysis of natural gas enables a more efficient heat recovery and the production of an useful secondary product - hydrogen.

Hydrogenation of some of the non-selective oxidation products, to regenerate methane, increases the overall carbon selectivity of the reaction, albeit at the expense of hydrogen which is converted to water. Processes based on the oxypyrolysis of natural gas can be employed for both the synthesis of olefins and synthetic fuels. This process has been investigated more recently by researchers at CSIRO in Australia [15,16].

II. EXPERIMENTAL

The catalysts referred to as supported catalysts B and C have been described in a recent article and references therein [11]. Catalysts B and C contain barium-promoted lanthanum and strontium-promoted lanthanum respectively. The catalysts referred to as 'previous' and 'new' generation supported catalysts are composed of lanthanum and strontium active components.

The sintered alumina reactor assembly used in these experiments has been detailed elsewhere [13,17]. The catalyst bed was composed of 3.2 ml of supported catalyst without dilution. The pre-catalyst zone of the reactor was filled with calcined quartz grains which served as a preheating contact material. The reactor assembly was modified in order to simulate conditions comparable to those which might be found in an ethane steam cracker, that is: a total pressure of up to 2.4 bar (absolute) and a charge composed of a molar ratio of water to hydrocarbon (i.e. H_2O/CH_4) of up to 0.25. A motor driven syringe (Braun) was used for water injection into the methane/oxygen stream at the top of the reactor. The addition of a micro-valve between the reactor exit and the water condensor served as a back-pressure regulator. Both the effects of total pressure and initial water vapor concentration were examined using fixed conditions found to be suitable for the catalysts and reactor assembly in this study. The standard test conditions utilized for these studies were as follows: effective reactor cross sectional area = 1.48 cm^2, methane and oxygen flow rates of 1000 and 99 ml/min (NTP) respectively, and catalyst bed hot-spot (hs) temperature fixed at 880°C.

A. CONTOUR AND 3D GRAPHICS

The variation in C_2+ yield has been studied as a function of total flow rate and hs temperature using catalyst C. Seventy three approximately equally spaced data regions between 545 and 1945 ml/min total flow rate (NTP) and 795 and 925°C were used in this study. These data were interpolated into a 20 by 25 data matrix then transformed into both 2D contour (Fig. 1) and 3D (Fig. 2) maps, as previously described [13]. The measured oxygen conversion was less than 90 % only for experiments below 810°C with total flow rates above 1100 ml/min and below 830°C with total flow rates above 1600 ml/min (i.e. the bottom right hand corner of Fig. 1). A maximum C_2+ yield of 13.0-13.2 % (% yield = [CH_4 conv. $\times$ C_2+ sel. $\times$ 100]) occurred at about 1350-1450 ml/min total flow rate and 900-910°C hs temperature. This corresponds to methane and oxygen conversions of 16.1-16.3 % and > 99 % respectively, and a C_2+ selectivity of 81.0 %.

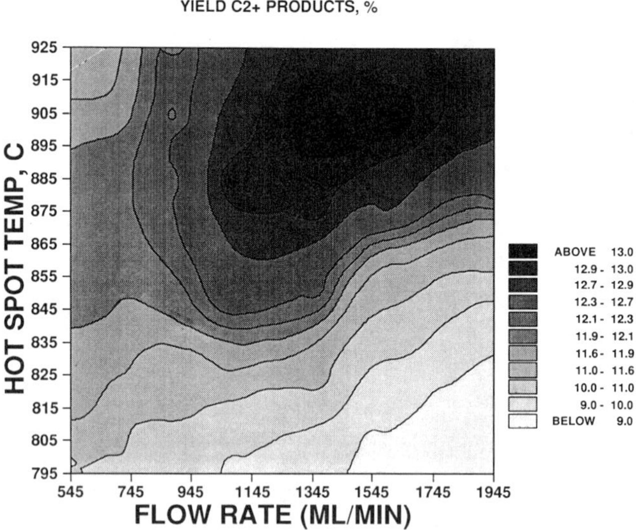

Fig. 1 *Two dimensional contour map of the C_2+ yield as a function of the catalyst bed hot-spot temperature and the total flow rate using supported catalyst C and a charge composed of 9 vol% O_2.*

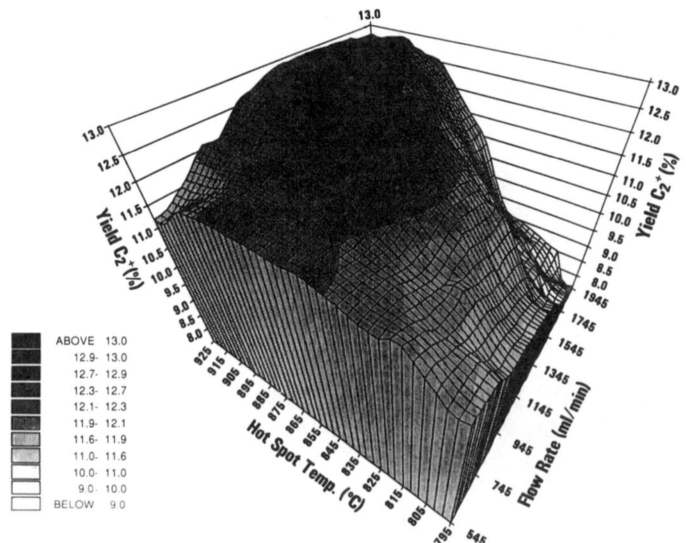

Fig. 2 *Three dimensional map of the C_2+ yield as a function of the catalyst bed hot-spot temperature and the total flow rate using supported catalyst C and a charge composed of 9 vol% O_2.*

The variation in C_2+ selectivity, using catalyst C, has been studied as a function of total flow rate and volume per cent oxygen in the charge (vol% $O_2 = 100 \times [\text{vol } O_2]/[\text{vol } O_2 + \text{vol } CH_4]$) at a fixed hs temperature of 880°C. The 89 data regions used for this study were collected in the region between 545 and 1945 ml/min total flow rate (NTP) and 8 to 13 vol% O_2. These data were also interpolated into a 20 × 25 data matrix for use in the contour map, Fig. 3. The oxygen conversion was 100 % over the entire range studied in this series of experiments. The maximum selectivity (82.3 %) occurs at the lowest oxygen concentration (8 vol%) and at a total flow rate of 1100 ml/min. The corresponding CH_4 conversion and C_2+ yield were 14.6 and 12.0 % respectively. The maximum yield (15.0 %) was attained with 13 vol% O_2 at 1100 ml/min total flow rate, corresponding to 20.3 % CH_4 conversion and 73.8 % C_2+ selectivity.

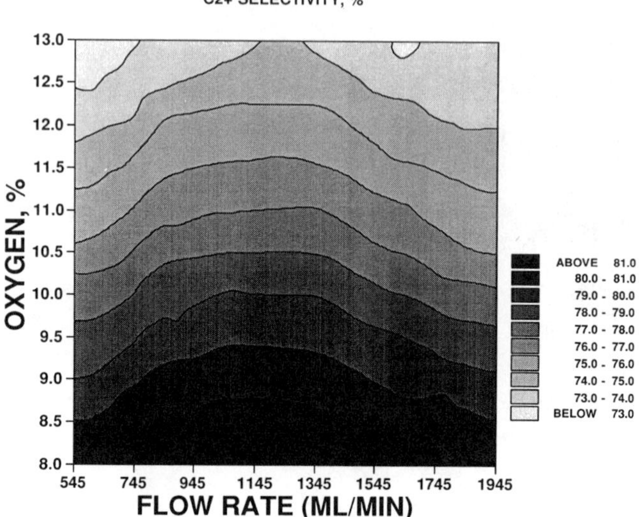

Fig. 3 *Two dimensional contour map of the C_2+ selectivity as a function of the initial oxygen concentration in pure methane and the total flow rate using supported catalyst C and at a catalyst bed hot-spot temperature of 880° C.*

III. RESULTS AND DISCUSSION

The oxidative coupling of methane is not a typical catalytic reaction. It is perhaps better described as a heterogeneous/homogeneous catalytic process. The OCM reaction is generally accepted to be initiated at the catalyst surface *via* the abstraction of a proton from methane by a non-fully reduced oxygen species (i.e. not $O^{=}$) to yield gas phase methyl radicals. The C_2+ selectivity is therefore highly dependent on the types of oxygen on the catalyst surface and the subsequent gas phase reactions undergone by the initial selective product, the methyl radical, $CH_3^{\bullet}$. The engineering aspects of the reaction are also important. The yield of coupling products can be maximized only when the appropriate parameters are selectioned for the particular reactor design under examination [12,13].

A. OCM CATALYSTS

Some of the more important requirements for an industrially interesting OCM catalyst are (1) high activity and (2) selectivity, as well as (3) stability, under (4) the operating conditions required to obtain 100 % O_2 conversion while producing (5) some olefins from (6) natural gas (as opposed to methane). Finally, once the catalyst requirements are met, an OCM process should either (7) be economically or geopolitically interesting. This last target is, unfortunately, the most difficult to attain.

B. SECONDARY PRODUCTS

Two secondary products which should not be ignored are hydrogen and heat. The hydrogen can eventually be used either to generate heat, which can in turn be converted to electrical energy in the proposed process, or to recover potentially lost carbon by converting CO and/or CO_2 back to methane. Very little results exist regarding the quantity of hydrogen in the OCM effluent [7-11,18-20]. The high temperature effluent stream can be utilized for both the thermal cracking of C_2+ alkanes [7-11,14-16] and the generation of high pressure steam.

C. OPERATING CONDITIONS

We have recently reported that OCM catalysts can function at relatively high catalyst bed hs temperatures ($> 850°C$) and at high linear space velocities ($>$ 0.45 m/s) [10-13]. We have purposely chosen to examine the higher temperature domain of catalyst activity for two reasons. First, higher temperatures typically lead to increased reaction rates and thus permit a much lower residence time in the reactor. Second, by operating at higher temperatures and flow rates, the O_2 and CH_4 conversions, C_2+ selectivity, H_2 concentration and effluent temperature all significantly increase. This is unusual for an oxidation reaction. Higher temperatures, in selective oxidation chemistry, are generally associated with a decrease in selectivity. The lower stability of the methylperoxyradical, at higher temperature, is thought to disfavor non-selective, gas phase reactions. This may be an important factor in determining the selectivity of the reaction; however, operating under plug-flow conditions at high linear space velocity and at quantitative oxygen conversion are certainly important. The limitation of back-mixing of the oxygen deficient product gases with the oxygen-containing charge ensures that there is little molecular oxygen in contact with the effluent.

The activity/selectivity of an OCM catalyst can be visualized most clearly by presenting the results in the form of contour and 3D maps. The yield of C_2+ products as a function of both catalyst bed hs temperature and total flow rate (9 vol% O_2), using supported catalyst C, are shown in Fig. 1 and 2. The steam reforming reaction begins to become important at about 935°C. At this temperature, CO formation increases with concomitant loss of ethylene and ethane. Therefore, the temperatures used in this study do not exceed 925°C. The non-smooth contour line transitions result from not submitting the initial data to a massage algorithm. The data matrix shows only minor imperfections and is considered to be representative for this particular catalyst. A sufficiently large number of experimental regions, required for the generation of these maps, necessitates over 30 working hours of catalyst testing. This is only possible with highly stable catalysts.

The oxygen conversion is quantitative above a line which is roughly between 800°C/545 ml/min and 880°C/1945 ml/min. The lower four contour lines approximately follow the oxygen conversion. The region of non-complete oxygen conversion is readily visible on the 3D map. There is an identifiable trench in the region of high flow rate and low hs temperature. The region of maximum activity for this catalyst is an oval shaped plateau centered at about 905°C and a flow rate of 1400 ml/min. This represents a linear space velocity at the hot spot of about 0.63 m/s. Note that this calculation does not take into account the volume occupied by the catalysts itself. The actual non-occuppied cross-section is less than 10 % of the total. Therefore the true linear space velocity in this region of the catalyts bed is in excess of 6 m/s.

The variation in C_2+ selectivity as a function of initial O_2 concentration and flow rate, at a fixed hs temperature of 880°C, is shown in Fig. 3. This contour map clearly implies that the selectivity is directly proportional to the initial quantity of oxygen and the linear space velocity (flow rate). The contour lines in this figure change at every per cent selectivity suggesting that the selectivity is linearly proportional to the initial oxygen concentration at a fixed flow rate. The distance between contour lines, centered at about 1145 ml/min, is similar, indicating that the initial quantity of oxygen in the charge is an important factor in determining the overall selectivity. Thus for a catalyst bed hs temperature of 880°C in the chosen reactor, this catalyst exhibits a maximum in selectivity at a flow rate of 1150 ml/min.

We have also undertaken catalytic testing in the presence of water vapor, in order to examine its effect on the C_2+ selectivity. The results obtained, using catalyst C and the chosen standard conditions, when the water concentration in the reactant gases is increased from 0 to 0.25 mol% H_2O per mol CH_4 are shown in Table 1. The H_2O/hydrocarbon mole ratio of 0.25 is close to that used in ethane steam cracking.

TABLE 1 *The effect of water vapor on the oxidative coupling of methane**

H_2O/CH_4	0	0.09	0.11	0.18	0.25
CH_4 Conversion (%)	15.5	16.0	16.1	16.3	16.4
C2+ Selectivity (%)	80.0	80.9	81.1	81.6	82.0
C2+ Yield (%)	12.4	12.9	13.1	13.3	13.5
C_2H_4/C_2H_6	0.90	0.90	0.90	0.91	0.95

**These results were obtained at a catalyst bed hs temperature of 880°C, methane and oxygen flow rates of 1000 and 99 ml/min (NTP) respectively, and an effective cross sectional area of 1.48 cm^2, with 3.5 g of supported catalyst C.*

The increase in yield from 12.4 to 13.5 % upon the addition of water is significant. Water vapor is known to serve as a passivator for the interior of reactor walls; however, the effect in these experiments is too important to be consistent with either a passivation or a simple dilution factor. The addition of water vapor may shifts the surface oxygen equilibrium toward surface peroxide, which has been proposed previously as the active and selective OCM surface species [21-24]. The beneficial effect of added water vapor is not unusual, and has been previously reported for lithium-promoted MgO and CaO catalysts [25,26] and other alkali promoted catalysts [27-30].

The effect of total pressure on the performance of catalyst B under standard conditions indicates that pressure has a detrimental effect on C_2+ yield, Table 2. A drop in yield from 12.3 to about 11.5 % was obtained when the absolute pressure was increased from 1.28 to 2.4 bar. The higher pressure represents that typically found in an ethane steam cracker. A much more detailed study on the effect of pressure is necessary in order to ascertain its effects on the ideal operating temperature of the catalyst and the linear space velocity for the reaction. The variation in yield as a function of pressure, hs temperature and total flow rate at 9 vol% O_2 would be of great interest.

TABLE 2 *The effect of pressure on the oxidative coupling of methane**

Relative Pressure, bar	0.28	0.50	0.75	1.00	1.40
CH_4 Conversion (%)	15.6	15.0	15.0	14.8	15.0
C2+ Selectivity (%)	78.9	78.7	78.4	77.6	77.3
C2+ Yield (%)	12.3	11.8	11.8	11.5	11.5
C_2H_4/C_2H_6	0.75	0.8	0.87	0.96	1.06

**These results were obtained with 3.5 g of supported catalyst B see Table 1 for conditions.*

D. CATALYST LIFETIME

Supported rare earth catalysts have been found to be active and selective under the operating conditions required to obtain 100 % oxygen conversion. Catalyst lifetime studies have been performed with three different supported catalysts, containing lanthanum and strontium, in order to determine catalyst stability, Fig. 4. These tests were performed using the previously defined standard conditions in the absence of added water vapor. All three tests employed a constant catalyst bed hs temperature of 880°C and led to quantitative oxygen conversion. All attempts to synthesize selective supported catalysts, prior to the discovery of catalyst C, resulted in major decreases in the C_2+ selectivity with respect to a non-supported powdered catalyst (catalyst A, see reference [11]). Catalyst C exhibits a much higher stability over 600 working hours and a significantly higher initial C_2+ selectivity, 80 versus 73.7 %, from the previous generation supported catalyst. The new generation supported

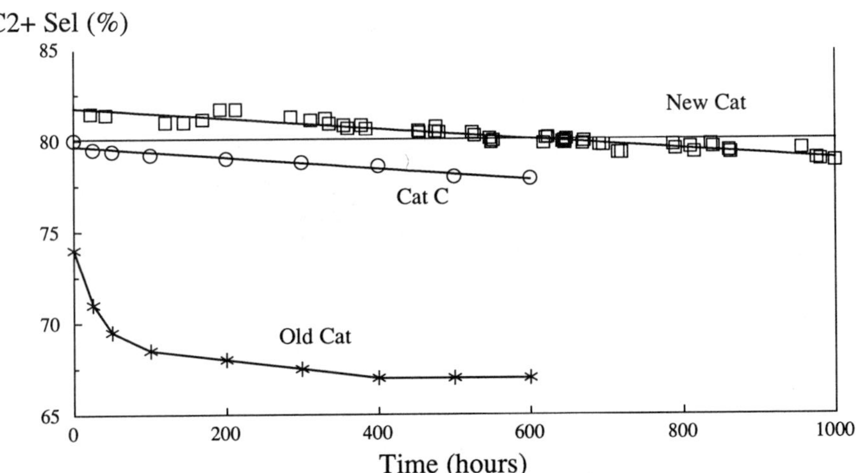

Fig. 4 *Catalyst lifetimes of three supported strontium promoted lanthanum catalysts. Experiments were performed at a constant catalyst bed hot-spot temperature of 880° C and a fixed charge of 1000 ml/min CH_4 and 99 ml/min (NTP) O_2 using a reactor with an effective cross sectional area of 1.48 cm^2.*

catalyst (see New Cat, Fig. 4) has been tested for 1000 h and exhibits a slow but constant deactivation. The catalyst activity drops from an initial 16.2 % CH_4 conversion, 81.4 % C_2+ selectivity (13.2 % C_2+ yield) to 15.6 % CH_4 conversion 78.8 % C_2+ selectivity (12.3 % C_2+ yield) after 1000 h of continuous operation.

IV. PROCESS IMPLICATIONS

Prior to the discovery of the IFP natural gas oxypyrolysis process [7] the adverse effects of the presence of C_2+ alkanes on the OCM reaction were not known. Before 1989, only one patent specifically referred to the C_2+ alkanes normally present in natural gas. One of the process schemes proposed, by Maffia at Arco, was to separate the C_2+ alkane components from natural gas before oxidatively coupling methane by the 'redox' mode. The redox effluent would then be oligomerized and the unreacted C_2+ alkanes combined with those previously separated. The combined higher alkanes could then be oxidatively dehydrogenated and oligomerized, Fig. 5 [31].

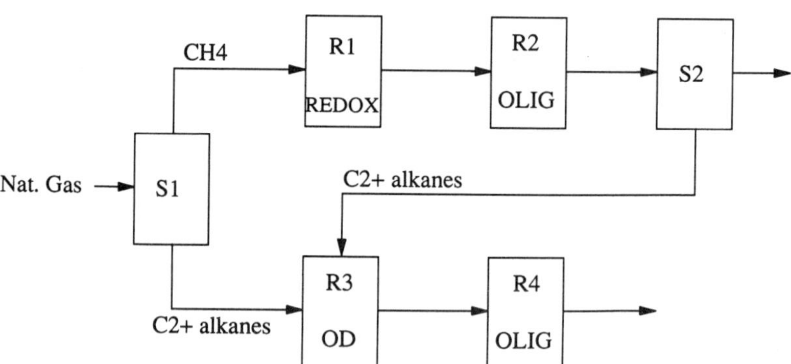

Fig. 5 *A box diagram for a process based on the oxidative coupling of methane in the redox mode using a separate treatment for the C_2+ alkanes in natural gas.*

A. OXIDATION AND PYROLYSIS

A particularly interesting concept patented by Brazdil, Hattenberger, Hilderbrand and Bartek of Standard Oil of Ohio (BP America), which was published subsequent to the application of the IFP patents [7-9], suggested using the heat generated by the OCM reaction to drive the thermal cracking of the formed alkanes to produce unsaturated hydrocarbons and hydrogen, Fig. 6 [20]. The experimental results presented in this patent were obtained from runs at temperatures between 900°C and 1084°C with a $SrO\text{-}La_2O_3$ catalyst. The ratio of C_2 unsaturates (acetylene and ethylene) to ethane were between 6.88 and 32. Significant quantities of acetylene were formed at these temperatures and ratios of unsaturates to saturates. Important quantities of acetylene, in the ethylene product stream, are particularly undesirable in ethylene synthesis.

The IFP natural gas oxypyrolysis process also incorporates the utilization of the exothermic heat, produced by the selective oxidation reaction between methane and oxygen, for the endothermic conversion of ethane to ethylene and hydrogen. The Sohio and IFP processes differ, however, in the application of this process to natural gas upgrading. The reactivity of ethane

SOHIO: Brazdil, Hattenberger, Hilderbrand, Bartek

US4822944, 18 Apr 1989

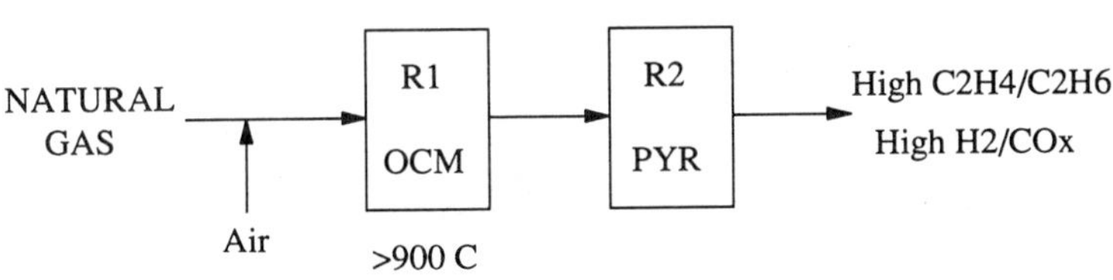

Exothermic + Endothermic

Fig. 6 *A box diagram for a process based on the oxidative coupling of methane (or natural gas) with a subsequent pyrolysis of the gaseous effluent.*

has been shown to significantly interfere with the OCM reaction [7-11]. The presence of C_2+ hydrocarbons in methane, at concentrations above about 1 vol%, are expected to have a substantially detrimental effect on both methane conversion and C_2+ selectivity. The lower overall C_2+ selectivity and methane conversion in the presence of higher hydrocarbons is probably due to the significantly lower C-C bond strength of ethane, H_3C-CH_3 88 kcal/mol, with respect to H-C_2H_5 (98 kcal/mol) and H-CH_3 (104 kcal/mol). The production of $CH_3^\bullet$ radicals before the catalyst bed is thought to lead to the non-selective, gas phase formation of carbon monoxide [11].

The higher reactivity of ethane may also pose problems for the upgrading of natural gas by other non-conventional means, such as: (1) the high temperature pyrolysis of natural gas, (2) the direct synthesis of methanol from natural gas/oxygen mixtures and (3) the synthesis of gasoline (*via* $CH_{4-x}Cl_x$ oligomerization on zeolite catalysts) from natural gas/chlorine mixtures. These processes could be seriously affected by the presence of appreciable quantities of C_2+ hydrocarbons. Ethane would undoubtedly be more reactive than methane in these processes, leading either to the formation of coke (pyrolysis), a low conversion of methane or a low selectivity (methanol) or the formation of coke from polyhalide decomposition (gasoline *via* $CH_{4-x}Cl_x$ oligomerization).

B. INTEGRATED NATURAL GAS OXYPYROLYSIS

A conceptual representation of an adiabatic, fixed bed, oxypyrolysis reactor is shown in Fig. 7. The reactor is divided into four zones: mixing, catalyst bed, C_2+ injection and pyrolysis zones. The rapid temperature rise at the beginning of the catalyst bed is an important factor in obtaining a rapid and complete oxygen conversion as well as a high C_2+ selectivity. The pyrolysis zone can be considered to be a completely independent reactor, in that it has a negligible effect on the oxygenate (CO + CO_2) to C_2+ hydrocarbon distribution [11]. The pyrolysis zone almost exclusively serves to crack the primary OCM coupling product (ethane) and added C_2+ hydrocarbons to olefins

9. C. Cameron, H. Mimoun, A. Robine, S. Bonnaudet, P. Chaumette and D.V. Quang, *Fr. Pat. Appl. 89/00188* (1989).

10. H. Mimoun, A. Robine, S. Bonnaudet and C.J. Cameron, *Chem. Lett.*, *12:* 2185 (1989).

11. H. Mimoun, A. Robine, S. Bonnaudet and C.J. Cameron, *Appl. Catal.*, *58:* 269 (1990).

12. A. Kooh, J.-L. Dubois, H. Mimoun and C.J. Cameron, *Preprints of 3B Symp., Pacifichem 89*, Honolulu, 60 (1989).

13. A. Kooh, J.-L. Dubois, H. Mimoun and C.J. Cameron, *Catal. Today, 6:* 453 (1990).

14. C. Raimbault and C.J. Cameron, *Natural Gas Conversion Symposium, 9th Internat. Symposium on C_1 Reactions*, Oslo, Norway, (1990).

15. J.H. Edwards and R.J. Tyler, *Chem. Eng. Aust., 14:* 5 (1989).

16. J.H. Edwards, K.T. Do and R.J. Tyler, *Preprints of 3B Symp., Pacifichem 89*, Honolulu, 160 (1989).

17. A. Kooh, H. Mimoun and C.J. Cameron, *Catal. Today, 4:* 333 (1989).

18. R.F. Hicks, *U.S. Pat. 4780449* (1988).

19. J.M. DeBoy and R.F. Hicks, *J. Chem. Soc. Chem. Commun.*, 982 (1988); ibid., *Ind. Eng. Chem. Res., 27:* 1577 (1988); ibid., *J. Catal., 113:* 517 (1988).

20. J.F. Brazdil, J.S. Hattenberger, R.E. Hildebrand and J.P. Bartek, *U.S. Pat. 4822944* (1989).

21. K. Otsuka and K. Jinno, *Inorg. Chim. Acta, 121:* 237 (1986).

22. M.Yu. Sinev, V.N. Korshak and O.V. Krylov, *Kinet. Catal. 28:* 1188 (1987); ibid., *Proceedings, All-Union Conference on Mechanisms of Catalytic Reactions*, Moscow, *2:* 27 (1986), reference taken from: *Proc. VI Int. Symp. Heterogeneous Catal.*, Sofia, *1:* 450 (1987).

23. K. Otsuka, A.A. Said, K. Jinno and T. Komatsu, *Chem. Lett.*, 77 (1987).

24. A.M. Gaffney, C.A. Jones, J.J. Leonard and J.A. Sofranko, *J. Catal., 114:* 422 (1988).

25. J.B. Kimble and J.H. Kolts, *Amer. Inst. of Chem. Eng.*, New Orleans (1986).

26. J.B. Kimble and J.H. Kolts, *Energy Progress, 6:* 226 (1986).

27. A.M. Gaffney, *U.S. Pat. 4788372* (1988).

28. D.W. Leyshon, *U.S. Pat. 4886932* (1989).

29. A.M. Gaffney, *U.S. Pat. 4848571* (1989).

30. D.W. Leyshon, *U.S. Pat. 4886932* (1989).

31. G.J. Maffia, *U.S. Pat. 4533780* (1985).

32. C. Moneuse, M. Cassir, C. Piolet and J. Devynck, *Appl. Catal., 63:* 67 (1990).

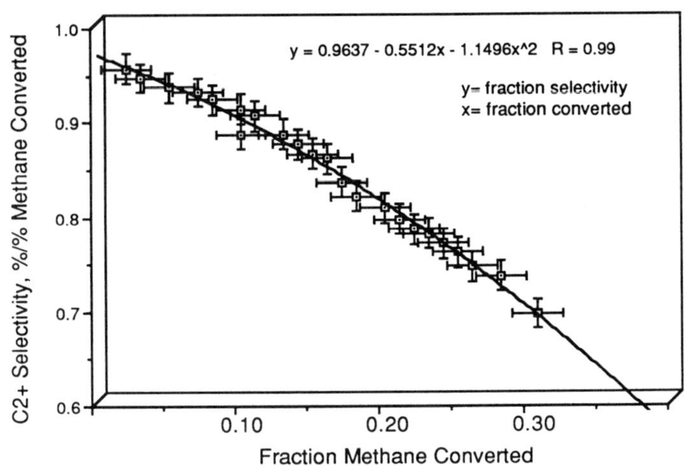

Figure 6. Conversion/Selectivity Plot for Methane Conversion. Error bars based on Standard Error. Data for 15% Mn, 5% Na4 P207/Si02 catalyst.

Rctr Fd/# C2+

60
50
40
30
20
10
0

0.0 0.1 0.2 0.3 0.4

methane conv

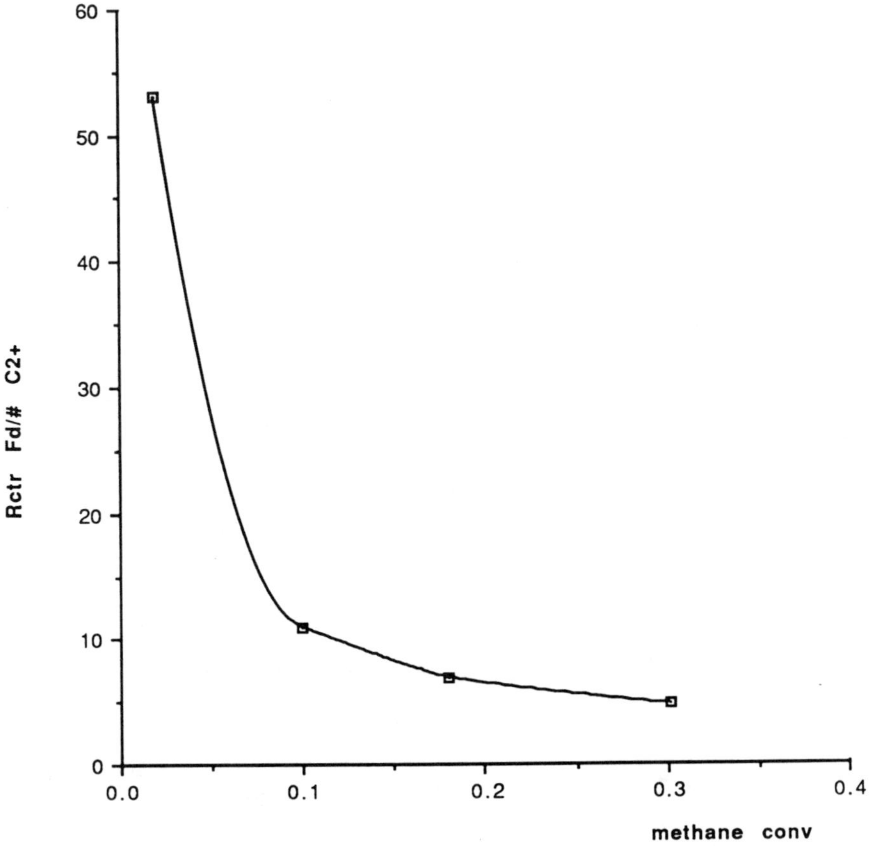

Figure 7. Hydraulic Loading on the Plant Versus Methane Conversion.

a function of feed rate is a good indication of how difficult the temperature control will be in an adiabatic reactor. This relationship is presented in Figures 7 and 8. Note, the hydraulic loading on the plant decreases as the conversion increases but simultaneously the heat released per pound of feed increases making greater demands on the heat rejection system.

The discovery of this technology provided the possibility of the first two-step route from natural gas to gasoline range products. In the first step, methane is coupled to ethane and via secondary reactions to ethylene and other light olefins. In the

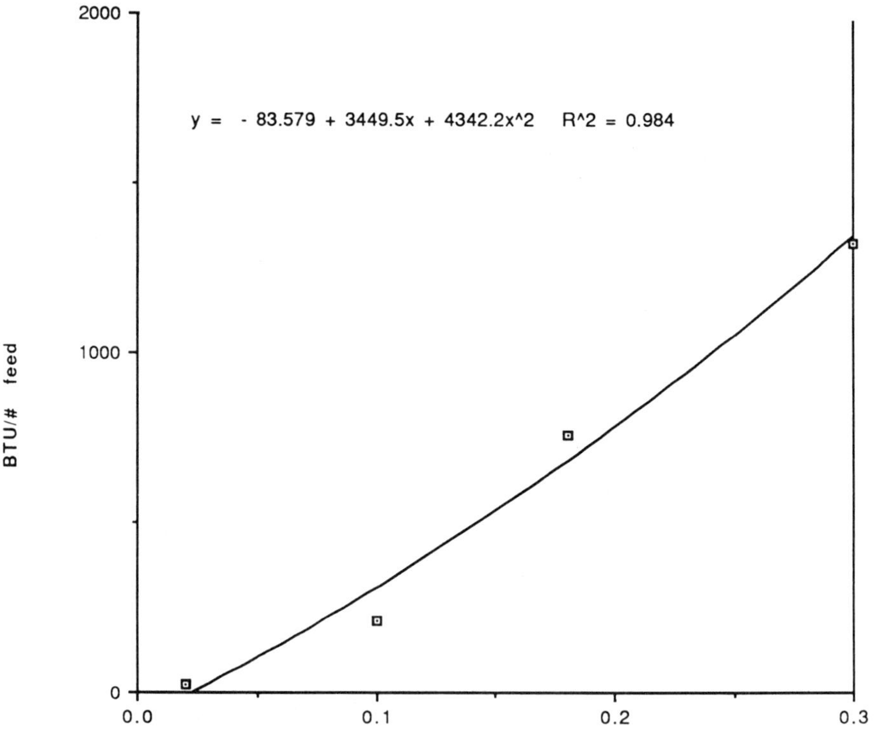

Figure 8. Heat Release as a Function of Conversion.

second step, several conventional catalysts will oligomerize C_2-C_4 olefins to gasoline.

ACC PROCESS

Typical natural gas composition and pretreatment processing are indicated in Figure 9. Condensable liquids are recovered and depending on the economics, a separation between methane and ethane may be performed. Any pretreatment for contaminent removal is not shown.

Ethane, propane and butanes can be processed in a conventional steam pyrolysis plant to make light olefins. The material balance shows the pretreatment yield of methane for oxidative coupling processing.

In conjunction with developing a process around the methane conversion catalyst, demethanization technology has been explored extensively by ACC (82, 83 for example) and others (84, 85). In summary, considerations in demethanization, range from energy conservation to the prevention of carbon dioxide freeze-out. Since subsequent oxidative coupling processing will operate with modest methane conversions there is potential to integrate the front end demethanization with the in plant function. This depends on the value of the gas, plant capacity and if once through or recycle processing is the most economical.

In the next step, demethanizer overheads are preheated and fed to the methane conversion reactor system.

ACC REACTOR DESIGNS

Two basic reactor options exist for the conversion of methane using the ACC process; reduction/oxidation (redox) swing process and co-feed (oxygen and methane simultaneously over a fixed bed of

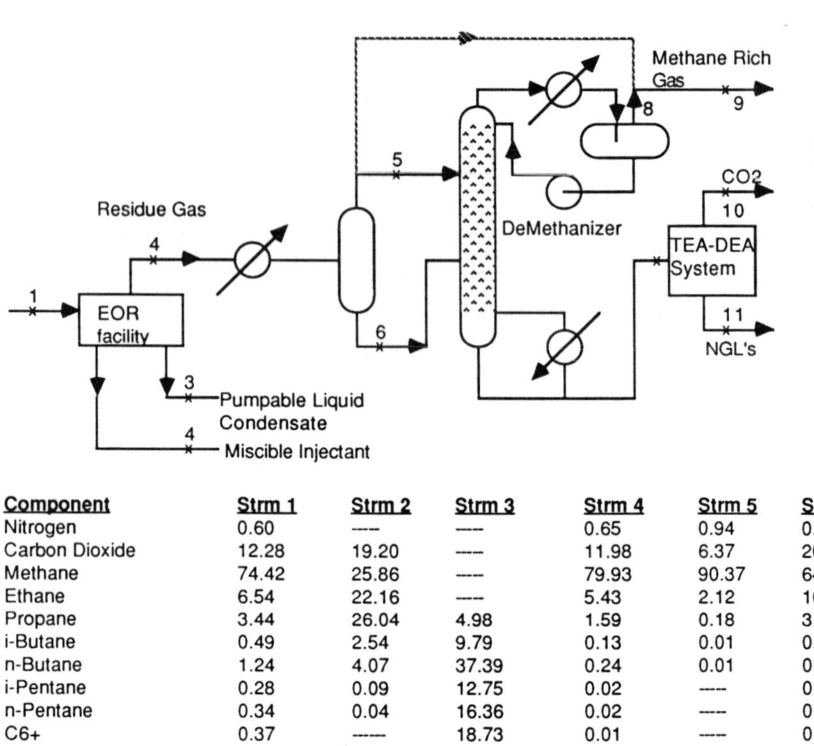

Component	Strm 1	Strm 2	Strm 3	Strm 4	Strm 5	Strm 6
Nitrogen	0.60	----	----	0.65	0.94	0.23
Carbon Dioxide	12.28	19.20	----	11.98	6.37	20.36
Methane	74.42	25.86	----	79.93	90.37	64.30
Ethane	6.54	22.16	----	5.43	2.12	10.39
Propane	3.44	26.04	4.98	1.59	0.18	3.69
i-Butane	0.49	2.54	9.79	0.13	0.01	0.31
n-Butane	1.24	4.07	37.39	0.24	0.01	0.58
i-Pentane	0.28	0.09	12.75	0.02	----	0.06
n-Pentane	0.34	0.04	16.36	0.02	----	0.06
C6+	0.37	-----	18.73	0.01	----	0.02
FLOW per 100 MOLS of STRM 1	***100***	***7.29***	***1.96***	***90.74***	***54.37***	***36.37***

Component	Strm 7	Strm 8	Strm 9	Strm 10	Strm 11
Nitrogen	0.35	----	0.76	----	----
Carbon Dioxide	1.99	55.84	5.03	100	5.03
Methane	96.65	1.88	92.31	----	4.03
Ethane	1.01	28.48	1.78	----	61.26
Propane	----	10.82	0.12	----	23.26
i-Butane	----	0.90	trace	----	1.93
n-Butane	----	1.71	trace	----	3.68
i-Pentane	----	0.16	----	----	0.35
n-Pentane	----	0.16	----	----	0.35
C6+	----	0.05	----	----	0.11
FLOW per 100 MOLS of STRM 1	***23.96***	***12.41***	***78.33***	***6.64***	***5.77***

Figure 9. Front End Processing and Material Balance.

catalyst). In terms of unit operations, fixed or fluidized bed may be used in either option.

Depending on the option chosen for the reactor, auxiliaries and remaining processing will be quite different. When operating in a co-feed manner, the catalyst maintains a constant oxidation state, and from an engineering standpoint, functions in a familiar heterogeneous catalysis mode. There are a number of configurations or reactor designs to remove the heat of reaction. In the redox version, the metal oxide is exposed to methane and then alternatively air as part of a cyclical process. The metal oxide in this case is functioning as a reagent. Evaluation studies have indicated the redox process to be preferred, mostly due to the elimination of the requirement of an oxygen plant.

Mechanistically, the conversion of methane to ethylene and higher hydrocarbons proceeds via the partial oxidation of methane on or near the surface of the catalyst. As a first step, methyl radicals are formed and combine to form ethane. Residual hydrogen is oxidized to water. Unsaturates are formed via an uncatalyzed pyrolysis mechanism of ethane and other products; C-H bonds in ethane and higher hydrocarbons are also activated via the metal oxide.

Within the context of a redox process for the conversion of methane, there are basically two different reaction design options: fixed bed operating in a switching bed arrangement and fluid bed with catalyst transfer. The fluid bed design option is shown for comparative purposes in Figures 10 and 11. The fixed bed operating in a switching redox mode is discussed further in this paper.

Downstream of the reactor, processing will depend on the desired products. If the goal is to make ethylene and other light olefins, processing will be similar to conventional steam prolysis.

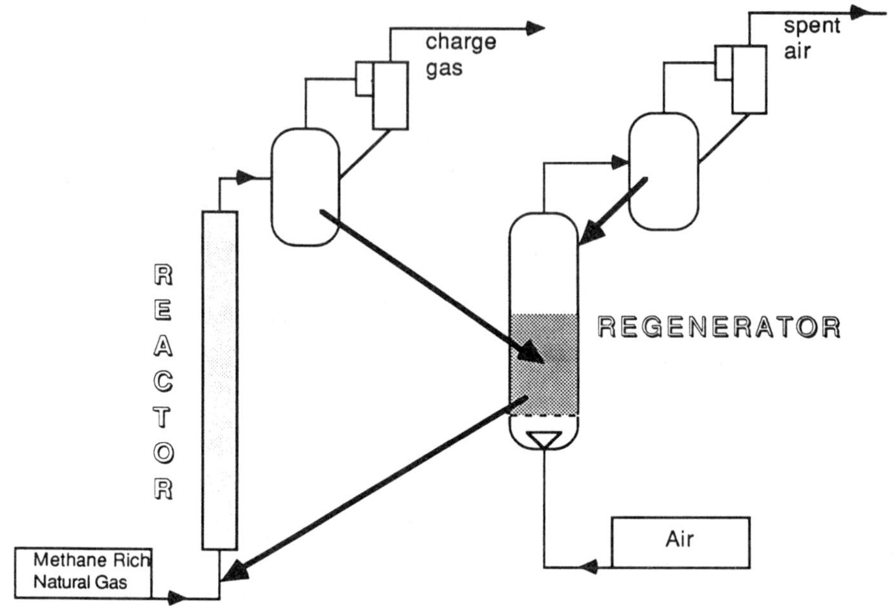

Figure 10. Remote gas conversion in a fluid bed reactor operating in a redox mode.

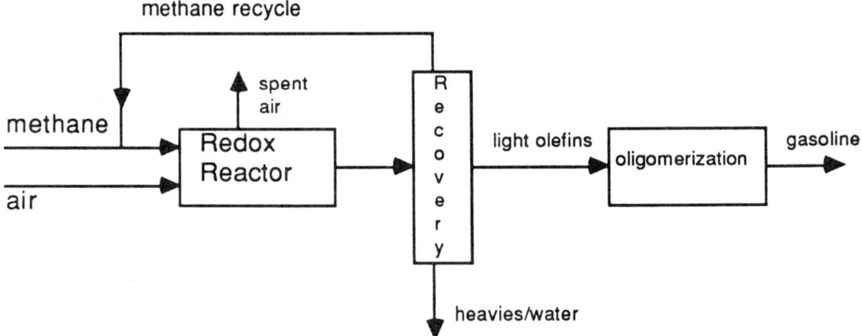

Figure 11. Remote gas conversion to gasoline in a two step process with methane coupling via redox reactor.

If gasoline is the desired product, then an oligomerization step will be required.

A process flow diagram for the ethylene from methane option is presented in Figure 12. A conventional gas (C_2-C_4) fed steam pyrolysis plant is shown in the process flow diagram in Figure 13. The redox option operating on ethane feed is shown in Figure 14.

The bed reduction (methane oxidation step) is approximately thermoneutral and the bed reoxidation (air reduction) is exothermic. The heat of reaction, which depends on the carbon dioxide selectivity, is released during the bed reoxidation. Typical selectivities, conversions, reactor loadings and associated heat release are indicated in Figures 4-8.

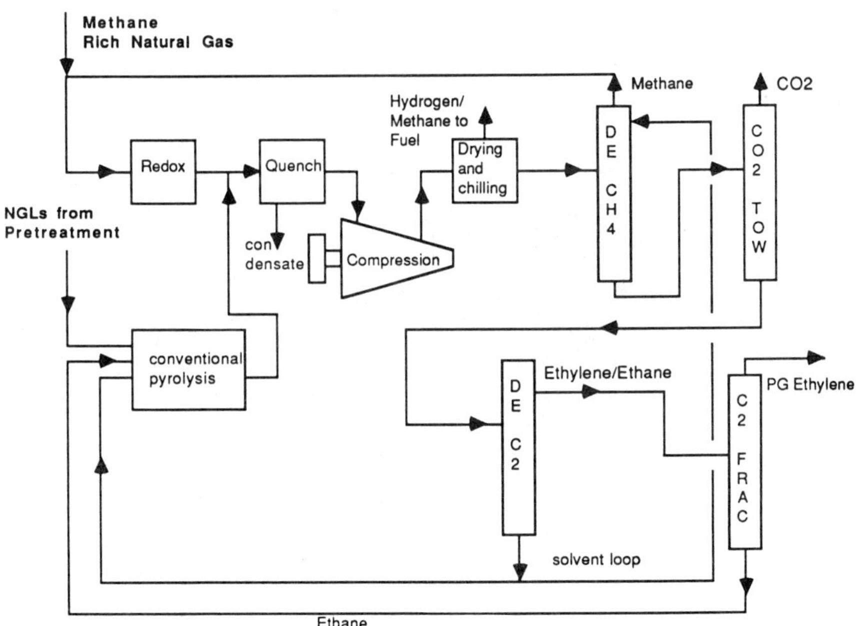

Figure 12. Olefins from Natural Gas via Fixed Bed Redox.

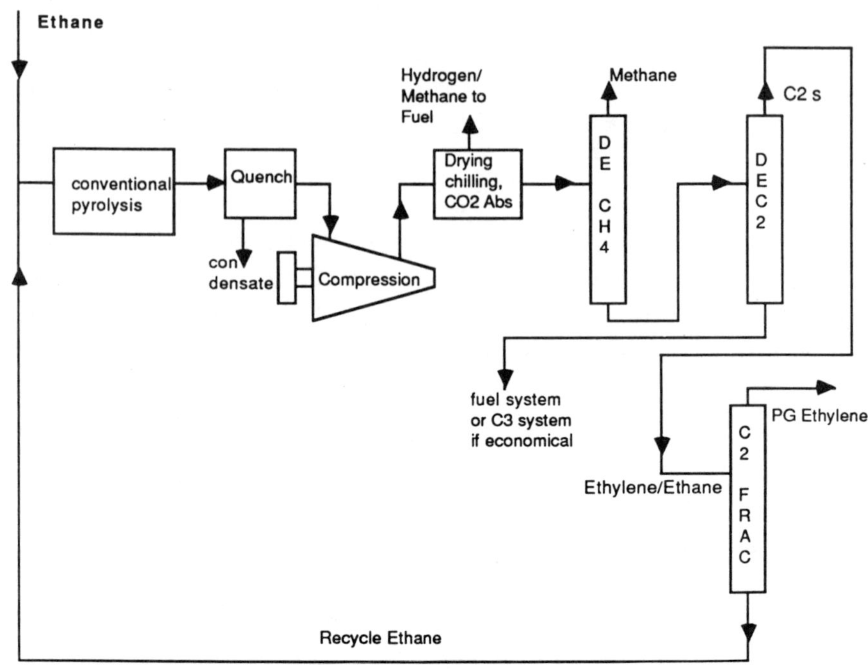

Figure 13. Olefins from Ethane via Conventional Pyrolysis.

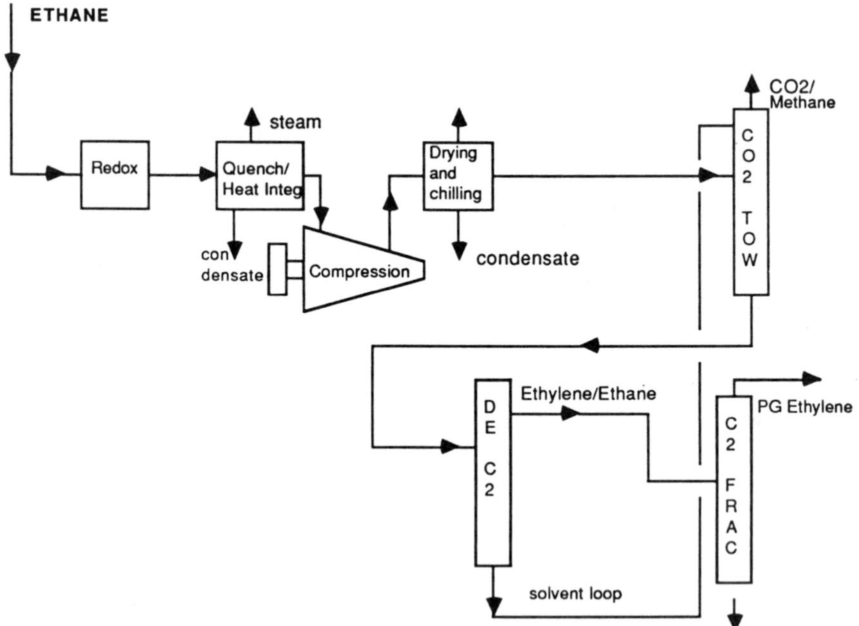

Figure 14. Olefins from Natural Gas via Fixed Bed Redox.

It is important to maintain a sufficient oxygen content of the catalyst in order to maintain high rates and catalyst productivity. Oxygen transfer is indicated by the following equation in which the oxidation is first order in oxygen concentration and reduction is zero order in oxygen concentration:

$$\frac{df_{ox}}{dt} = k_{ox}\,(1 - f_{ox})\,PO_2 - k_{reduction}\,f_{ox}$$

where k_{ox} is the oxidation rate constant of the catalyst.
and $k_{reduction}$ is the reduction rate constant of the catalyst.

The problem at hand then becomes one of maintaining good levels of conversion to keep the plant loadings down, keep selectivities high for good raw material utilization and modest heat release, and to design a reaction flow system and sequence to balance the exothermic heat of reaction. To accomplish this, a processing arrangement consisting of methane conversion and catalyst oxidation separated by a variety of purging, smoothing and cooling steps was tested experimentally and in modelling studies.

It was envisioned that full scale versions of a switching fixed bed redox system would operate adiabatically in which the large heat of catalyst reoxidation would result in hot spots, effecting conversion, yield, catalyst life, materials of construction and overall operability and control.

From the data illustrated in Figures 4-8, when operating at 25% conversion with a 15% Mn catalyst, approximately 600 cal/gm reactor feed must be removed.

The key criteria in evaluating a viable reactor design are

- the maximum and minimum temperature encountered during each phase of the redox reactor
- pressure drop
- temperature during the methane conversion step

In terms of demonstration work, a 76mm reactor system with nine side feed ports and a numerical methods computer model was used to examine the temperature profiles, reaction, and catalyst oxidation level within the context of a switching system.

EXPERIMENTAL

The fixed bed, adiabatic reactor option was studied by ACC in experiments using a three inch and a one inch reactor. The processes consist of alternate cycles of methane conversion and catalyst re-oxidation interspersed with purging and cooling steps. This development work is in addition to the smaller bench scale reactors used for much of the initial patent work.

In the three inch reactor system, which is described below, feeds are available in an upflow or downflow arrangement and also may be fed through nine side injection points. The purpose of the three inch reactor study was to determine the impact of catalyst hot spots on conversion, selectivity, cycle time and configuration, as well as system stability and overall operability. It is anticipated that large volume commercial fixed bed reactor systems would be adiabatic in nature and hence subject to degradation or other problems associated with hot spots.

THREE INCH UNIT

In Figure 15, a process flow diagram for the three inch reactor is presented. Air, nitrogen or methane feed lines are indicated. These materials may be introduced through the overhead, bottom and/or the nine side ports. The flow valve for the feeds is operated by a Brooks flow controller with set points provided by the computer. Uniform side feed flow rates are promoted by the restriction orifices.

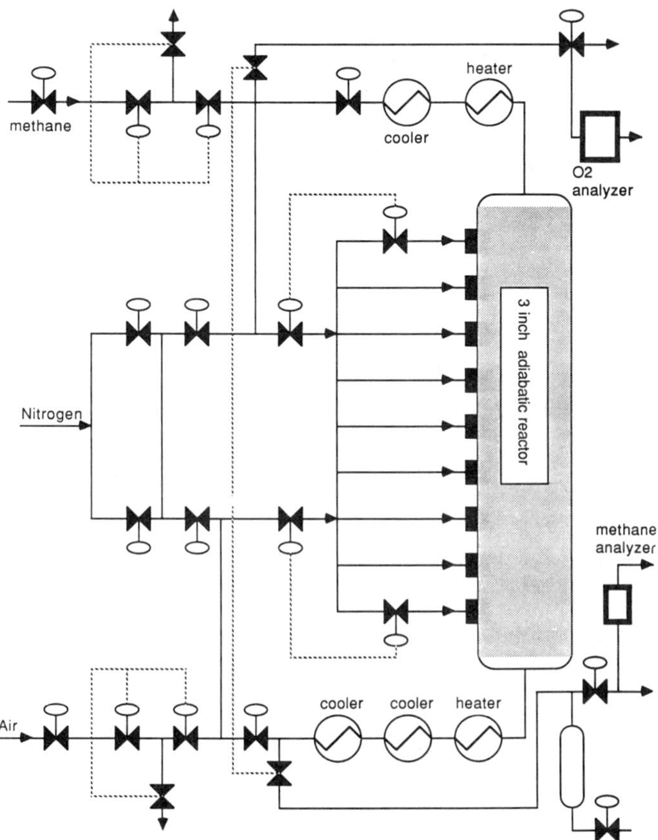

Figure 15. 3 inch adiabatic reactor PFD.

The reactor is a three inch diameter, schedule 40, 2.9 m long pipe. The top 30 cm and the bottom 15 cm are packed with inert catalyst support (SiO_2). The catalyst bed itself is eight feet long with a side feed port at the top and bottom and others spaced at one foot intervals down the bed. The reactor is mounted inside a cylindrical jacket constructed of 13 cm diameter, schedule 10 pipe; the annulus is filled with insulation. The entire assembly is positioned inside a furnace, containing five independent heater zones, each two feet long.

Temperatures are monitored with ten"fast response" thermocouples mounted in an axially oriented thermowell in the center of the bed. Additional temperature monitors are located in three of the side feed lines. Control of the bed temperatures is accomplished by the adjusting feed temperature and flow rate, and the initial bed temperature profile and oxidation state. The bed is maintained adiabatic by reactor surface heaters which balance heat losses by keeping identical temperatures on the outside metal and the outside of the reactor shell. There is thus a zero temperature gradient, which most nearly mimics the full scale unit.
Back pressure on the unit is controlled with a PCV located on the exist line from the overhead and bottom of the reactor.

The reactor cycle of bed reduction (methane oxidation), cold methane purging, nitrogen purging, bed re-oxidation (air reduction), and nitrogen smoothing is described below. As mentioned, one of the independent variables is the cycle pattern, flow directions and locations of the feeds.

The methane feed, after passing the flow control valve, flows through a series of three valves arranged in a standard *double block and bleed* manner. This prevents methane leakage into the reactor during purge, smoothing or reoxidation steps. Methane can be heated to 875°C (temperature is a study variable) in a home-made electric heater.

DESCRIPTION OF EXPERIMENTAL FLOW PATTERN: ONE OPTION:

Preheated methane flows downward through the reactor. Reactor products are cooled to ~200°F in a double pipe cooling water heat exchanger. Another cooling water exchanger in series cools the reactor effluent to room temperature. Condensate from the second cooling are collected and sampled in a knockout pot.

After the available oxygen in the bed has been depleted, cold methane is fed through the reactor from the overhead and sides and exits the bottom. A Brooks flow controller provides the proper methane flow; a by-pass is included to allow for a quick nitrogen purge in the event of any emergency. The cold methane purge is on a timed cycle; when complete, the reactor is purged of methane with nitrogen.

Nitrogen flow continues until all of the methane is purged from the reactor. An analyzer on the exit monitors the methane concentration. In addition to being an integral part of the smoothing of the temperature profiles, the flow of nitrogen is important in that methane is displaced and the bed prepared for reoxidation.

The reoxidation of the bed is accomplished by the introduction of air in an upflow manner through the bottom and the side ports. The introduction of cool air reduced the bed temperature near the inlet port and then rises throughout the bed section. The temperature continues to increase until the next feed port where cold air is encountered. Consequently, at the end of the air step when the bed has been re-oxidized, the bed will consist of nine cold zones near the side feeds and nine hot zones between the feeds. The goal of the experimental program and the modelling work was in part to determine the temperature profile in the reactor at the end of this step and whether this condition is stable.

After the air step, determined by the lack of change in the oxygen analyzer, another nitrogen purge step is performed to remove oxygen from the bed and to balance/smooth the temperatures. once the oxygen content in the exit gas has fallen below a safe level, the next cycle is initiated.

TABLE 1

(Hydrocarbon conversion over 15% Mn, 5% Na4P2O7/SiO2; Refs (1-43))

Redox Yields/Energy Consumption (Mol Balance Basis)

Feed -->	Methane	Ethane	Propane	n-Butane
Oxygen	-0.833	-0.441	-0.246	-0.214
Nitrogen	---------	---------	---------	---------
Hydrogen	---------	---------	---------	---------
Carbon Monoxide	---------	---------	---------	---------
Methane	-1.0	0.20	0.762	0.872
Carbon Dioxide	0.25	0.02	0.06	0.02
Ethylene	0.245	0.826	0.640	0.857
Ethane	0.085	-1.0	0.064	0.077
Propane	---------	---------	-1.0	0.003
Propylene	0.017	0.013	0.109	0.191

C4H6	--------	--------	0.035	0.032
Butenes	0.003	0.015	0.007	0.013
Butanes	--------	--------	--------	-1.0
C5's	--------	0.001	0.012	0.014
Heavy	0.003	0.003	0.032	0.049
Water	1.170	0.842	0.372	0.388
lb O2/lb Feed	0.418	0.343	0.17	0.116
lb O2/lb Reacted	1.670	0.470	0.179	0.118
Δ BTU/lb O2	-4833	-3023	-1862	130
conversion, %	25	73	95	98
Δ BTU/lb Feed	-2018	-1038	-316	15
Δ BTU/lb Ethylene	-4706	-1346	-776	36

MODEL

BACKGROUND:

A computer model has been developed to predict transient axial temperature profiles and the effective use of catalyst in a fixed bed reactor. The model as designed is specific to the conversion of methane to ethylene and other products operating in a reduction/oxidation mode with a reducible metal oxide catalyst. Such catalysts have been extensively described in the literature by ACC researchers and others (1-44).

As stated, the catalyst bed reduction (methane oxidation step) is approximately thermoneutral and the catalyst bed reoxidation (air reduction) is exothermic. The heat of reaction, which depends on the carbon dioxide selectivity, is released during the bed reoxidation. A typical product slate, heat release and oxygen requirements are listed in Table 1. This may be used for reference during the model description.

It is important to maintain a sufficient oxygen content of the catalyst in order to maintain high rates and catalyst productivity. Oxidation is first order in oxygen concentration, and reduction is zero order in oxygen concentration. The design to be modelled includes a reactor system operating in a three to five cycle arrangement as indicated in Figures 16, 17, and 18. Some of the possible flow configurations/patterns are also indicated for each reactor arrangement. The number of steps, the order, feed locations and directions are among the independent variables to be studied with the computer model.

COMPUTER MODEL

To describe the details and illustrate the workings of the model, a three-step cycle of oxidation (using air), reduction (using pure

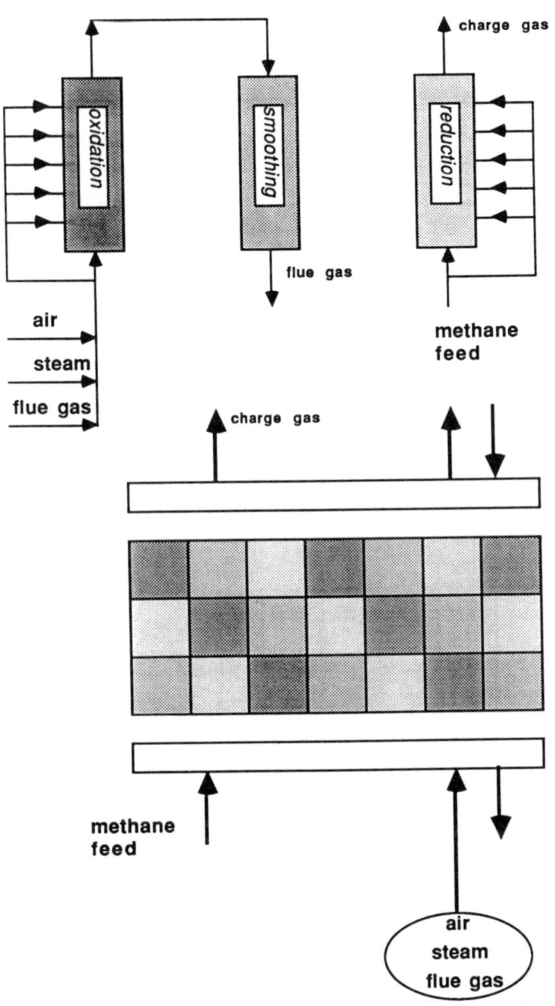

Figure 16. Generic Switching Bed Mode.

methane), and cooling will be used. This simple version of the system is shown in Figure 19. More complicated arrangements are similarly modelled by initializing the bed temperatures.

In the scheme of Figure 19, methane is preheated by cooling a bed which has been depleted of oxygen; that is, undergone reduction

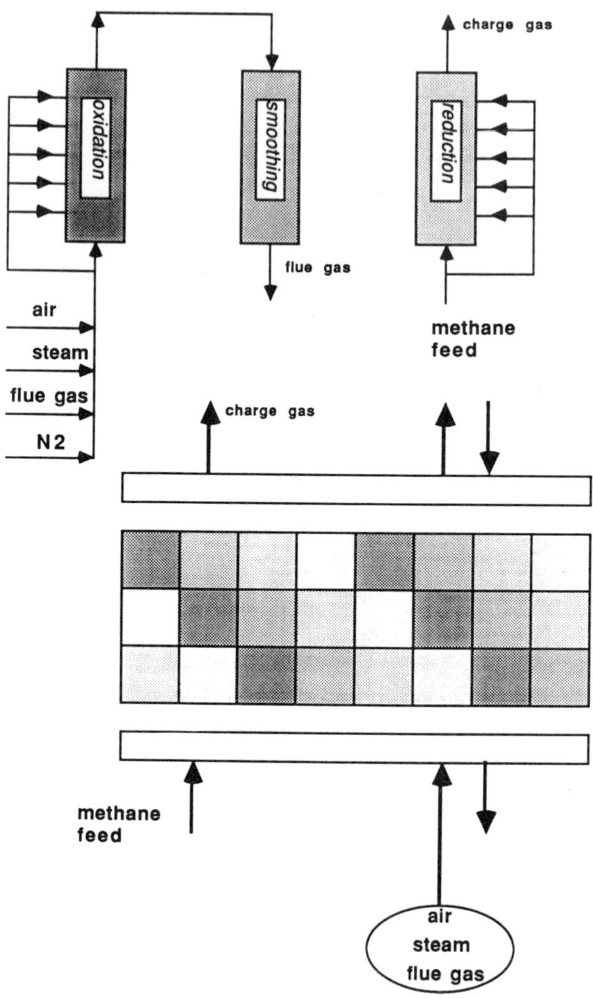

Figure 17. Expanded Four Cycle System.

in a previous cycle. As hot methane leaves this bed, it feeds the bed of the reduction step, which, as stated, is nearly thermoneutral. The third bed which has undergone cooling is being reoxidized with air. The model provides the flexibility of multiple, equally spaced feed injections and forward or reverse

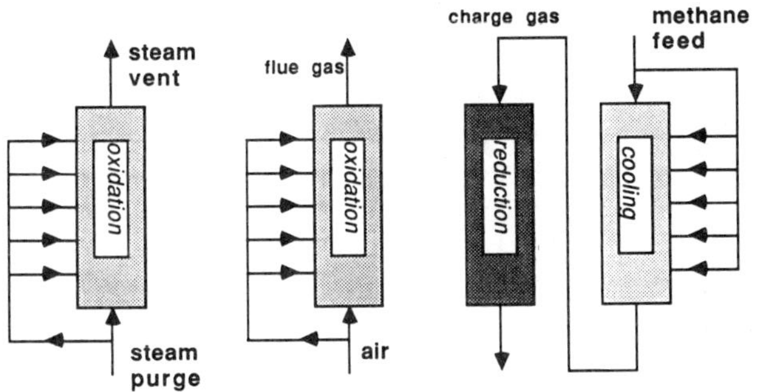

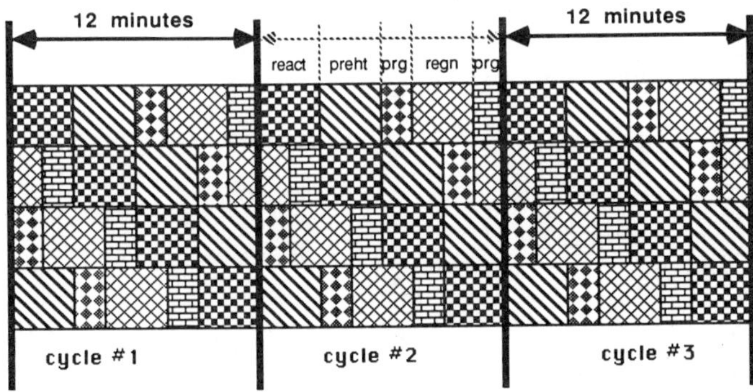

Figure 18. Five Stage Fixed Bed Reactor System (USP #4751055).

flow direction. The feeds are assumed to come to equilibrium with the bulk system flow within the cell they enter.

Originally, two programs were written by Caryl A. Fick (ACC Computer Services) to describe the three step system. Either can be used for any of the non-reactive steps; such as purge, smoothing, cooling, etc. However, one program is used for the oxidation step and the other for the reduction step. The program uses finite difference equations to approximate the differential equations. The rate equation for the oxidation step is as follows:

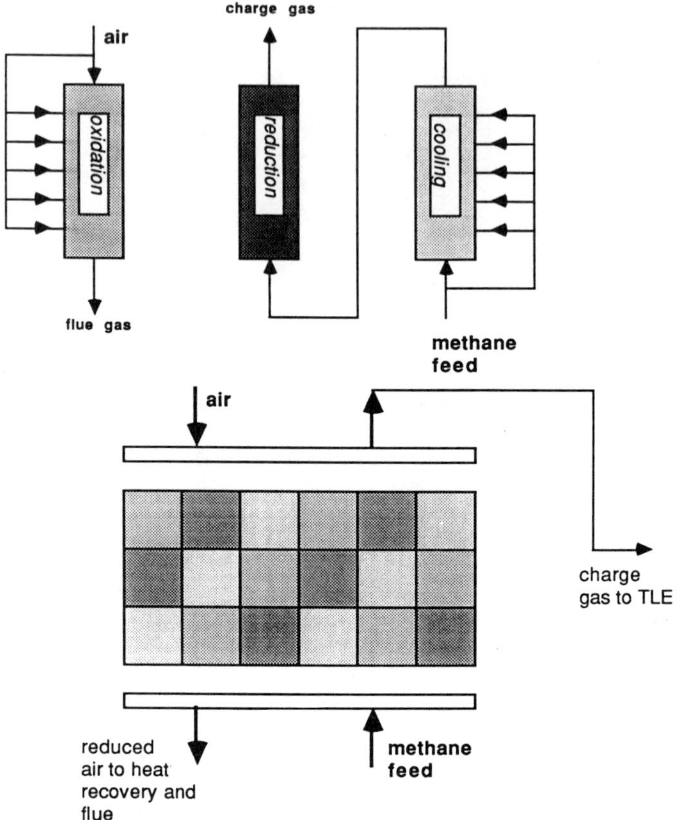

Figure 19. 3 Cycle System Used to Describe the Model.

$$\frac{df_{ox}}{dt} = k_{ox}\,(1 - f_{ox})\,PO_2$$

and for the reduction step:

$$\frac{df_{ox}}{dt} = -k_{reduction}\,f_{ox}$$

The equation of continuity is expressed as the total gas flowing out of each cell at any time. It is calculated as the total gas flow in, minus the oxygen that has just reacted with the bed.

The energy balance accounts for the heat of reaction, the heat transferred between the gas and the bed and the conductive heat flow. The programs assumes all the material expected to react during the time increment, reacts at the beginning of the increment. The heat generated by the reaction is absorbed immediately by the bed. It is assumed that there are no mass transfer or heat transfer limitations (diffusive resistances). Upon absorption of the heat, the bed then exchanges heat with the gas. Gas and bed temperatures are adjusted for this heat transfer. If the calculated heat transfer is large enough to cause a cross between the bed and gas temperatures, the final cell temperature is taken as the equilibrium temperature expected in an isolated cell containing the gas and the bed solids.

At the end of each time step, the temperature profile is adjusted for convective heat transfer according to the energy balance equation:

$$cL(T) = \frac{\partial T}{\partial t}$$, where L() is the Laplacian operator

A Crank-Nicholson technique is used to solve this equation and give a new bed temperature profile.

This technique is repeated for as many cycles as required in order to test the stability of the system. Computer time saving measures can be taken if the focus is merely on effective catalyst utilization and the conductive heat transfer correction can be dropped from the end of the algorithim during each step of distance and time.

RESULTS

MODEL RESULTS

The model was tested for many different arrangements of flow, variable feed temperatures and different protocols of cooling and bed temperature smoothing. As an example of the performance of the model, the results are plotted in Figures 20, 21, and 22 for the model test case of Figure 19.

Initially a bed at 760°C is oxidized with air at 82°C producing the temperature profile in Figure 20 at the indicated times of 1, 2 and 3 minutes; the total oxidation occurs in 3

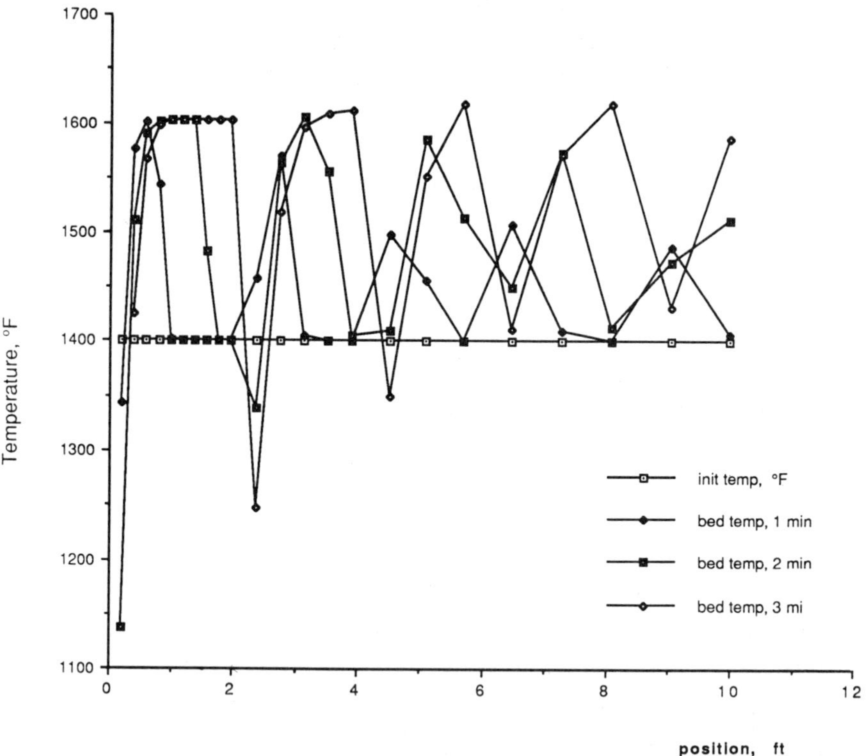

Figure 20. Oxidation Profiles.

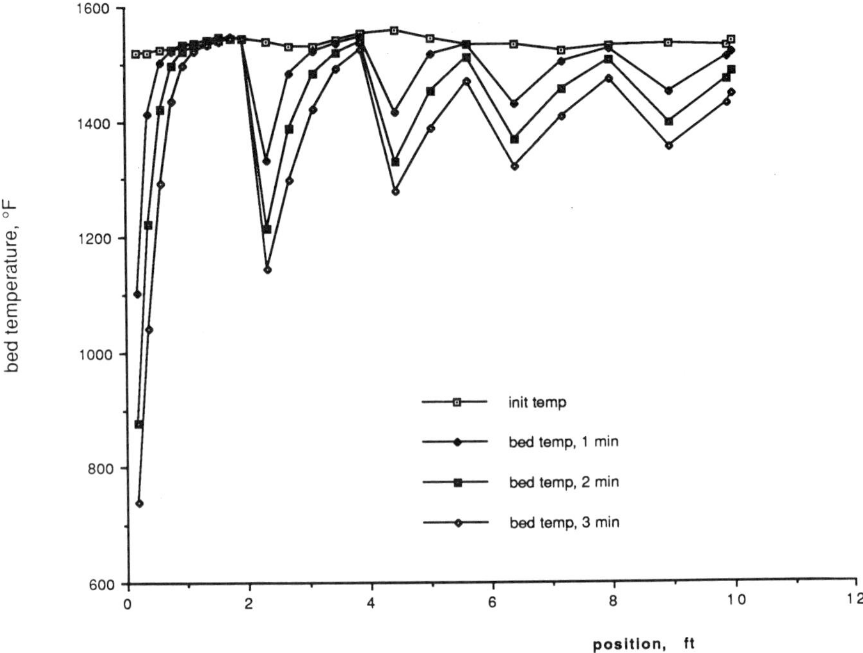

Figure 21. Cooling Profile.

minutes. The combination of the cold air input, at compressor discharge, and the large exotherm produces the large temperature differences across the bed. This bed then undergoes reduction and cooling, both of these operations tending to smooth the temperature profile as shown in Figures 21 and 22. In the cooling step methane is fed at 760°C.

It is important to understand the reactor profile within the context of the overall energy balance. The heat of reaction is indicated in Figure 8. Given a conversion, and the required heat recovery, the peripheral stream temperatures are set for a given flow pattern. That is, for example, if the air is preheated above the compressor discharge, then the methane feed temperature must be reduced. The result of this is to create wider swings in the bed

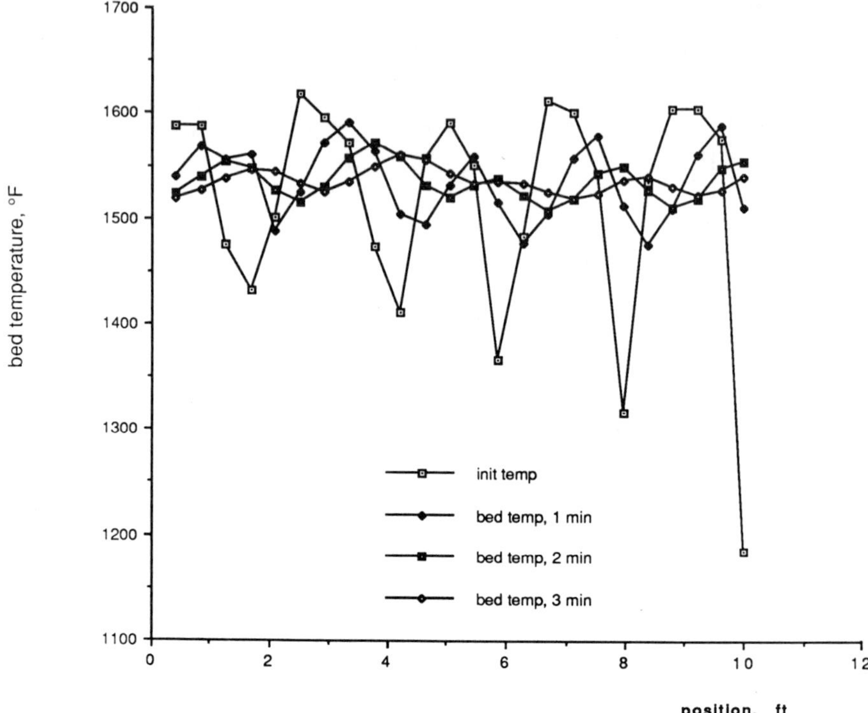

Figure 22. Reduction Profiles.

profile since the thermoneutral steps are running with feeds that are on the average lower than the desired reaction temperatures. If the feeds are not at temperatures that would produce an overall heat balance, the catalyst bed will be dependent variable and temperatures will rise or fall as the catalyst absorbs the additional heat.

The overall energy balance is illustrated in Figure 23 for the model sample case. Methane is preheated to 290°C in a steam heat exchanger; this material is then fed (via five separate feeds) to the cooling step and produces the profiles in Figure 21. Methane in the exit of the cooling step has been preheated to about

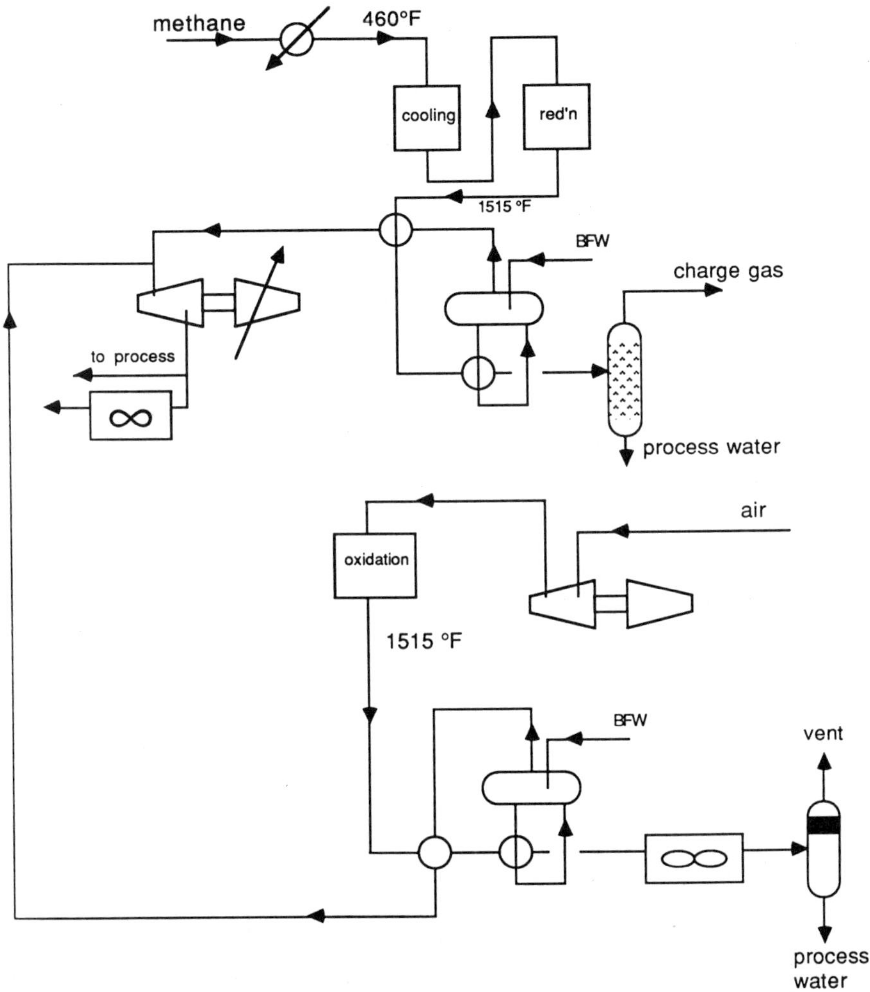

Figure 23. Overall Heat Balance on the Reactor Section.

825°C which is a desirable temperature for the coupling reaction. This stream is fed (single feed) in an upflow manner to the reduction reactor (methane coupling reactor) producing the profiles in Figure 22. Similarly, air is fed to the bed in the third cycle and produces exotherms similar to Figure 20.

Given the exit temperatures of the reduction step and the oxidation step both being about 825°C, and the flows set by the overall material balance of the plant, the feed temperatures of the incoming methane must be 290°C and air should be 32°C. This will guarantee an overall heat balance. The next level of design is then to see how the feeds can best be organized to avoid hot spots and large temperature swings.

One such flow pattern is that indicated in Figure 18 and was the subject of the US patent (#4751055) associated with that design. Flow patterns and reactor cycles are indicated in the figure and the associated temperature profiles are illustrated in the patent drawings. In this five mode operation, the bed stabilizes to steady state after about five complete cycles. This is typical of the amount of time to reach a steady state, albeit sometimes unacceptable operation.

PILOT PLANT RESULTS

For a stabilized catalyst (broken in), the model and bench scale data were nearly identical when both run in an "isothermal" mode. Pilot unit data and the model were consistent, selectivity to C2+ being the key variable and being nearly identical.

An automatic cycling mechanism was installed for the bed operation with the following protocol as an example: reaction, bed cooling with methane then oxidation with air. This is similar to the test case described for the modelling work. With this protocol the temperatures were controlled as also illustrated in the model example. However, due to the temperature swings, the C2+ selectivity was up to 10% lower than an isothermal operation and coke yields were higher.

The results of a forth cycle automatic cycling run are illustrated in Figure 24 and the bed temperatures during the

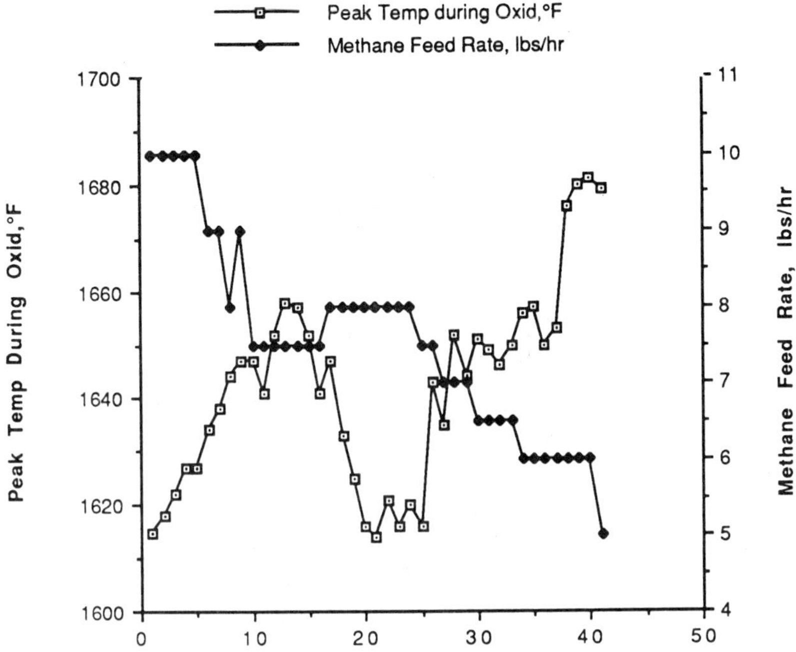

Figure 24. Results of Automatic Cycling Run; Peak Temperatures.

methane reaction mode during the thirty fifth cycle are illustrated in Figure 25.

The computer model accurately predicted the temperature profiles with respect to time and position. It was a useful tool in predicting the best way to operate the reactor.

ECONOMICS

There have been a number of recent studies that address the issue of the cost effectiveness of methane conversion to light olefins; for example see references 86 and 87. In terms of ARCO studies,

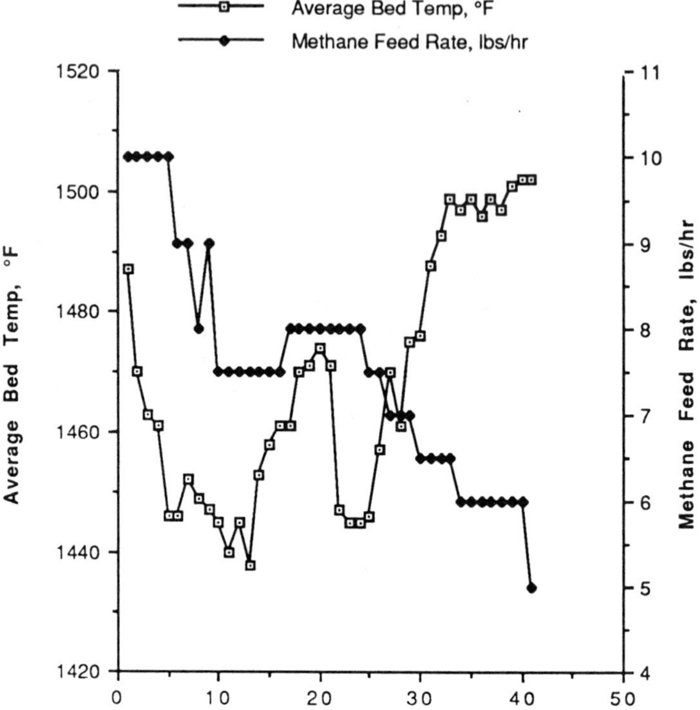

Figure 25. Results of Automatic Cycling Run; Average Bed Temperatures.

methane to gasoline has been studied as a function of feed gas price. This plot is shown in Figure 26 for the fluidized bed design case operating in a redox mode. The gas flow is 6 MMCFH and produces 12000 BPD of gasoline. Capital costs sensitivities are shown from the base of $300 MM to a plant costing twice as much due to severe location or other exogeneous parameters. Other economic assumptions are listed in the legend of the graph.

The conclusion from this graph is that with current oil prices, natural gas must be available in the $1/MSCF range to produce attractive rates of return.

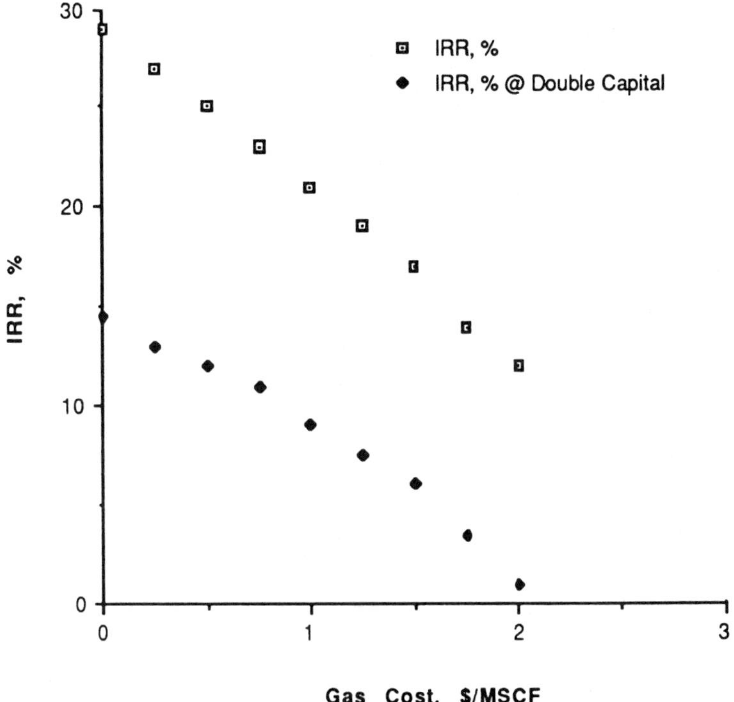

Figure 26. Gas Conversion Economics for 5.5 MMCFH of Natural Gas to 12000 BPD of Gasoline.

Relative to ethylene production, current margins for the traditional feed stocks are shown in Figures 27 and 28. Relative capital costs are shown in Figure 29. The implications of these charts are well known and have been studied intensely over the rocky past of ethylene pricing, shown in Figure 30.

Consequently, about 75% of the domestic olefins producers have plants with feed flexibility.

Gas (ethane/propane) feeds are the preferred feedstock for ethylene production producing the greatest margin per pound of feed. The downside of this is these plants have production that

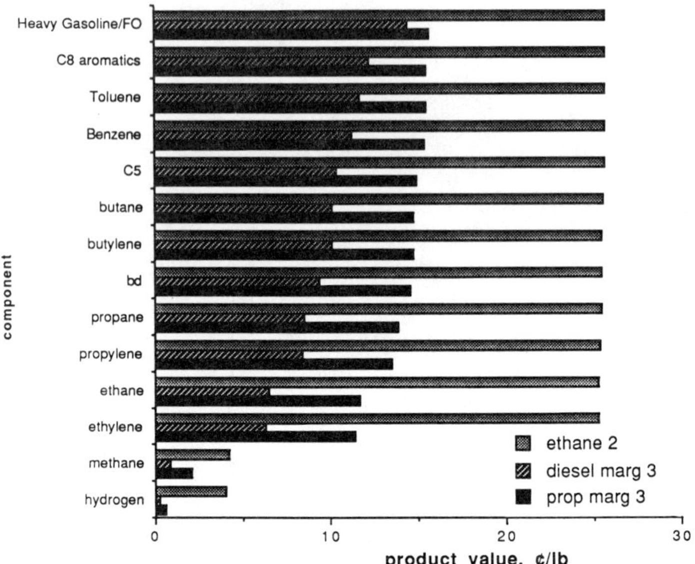

Figure 27. Product Value from Pyrolysis.

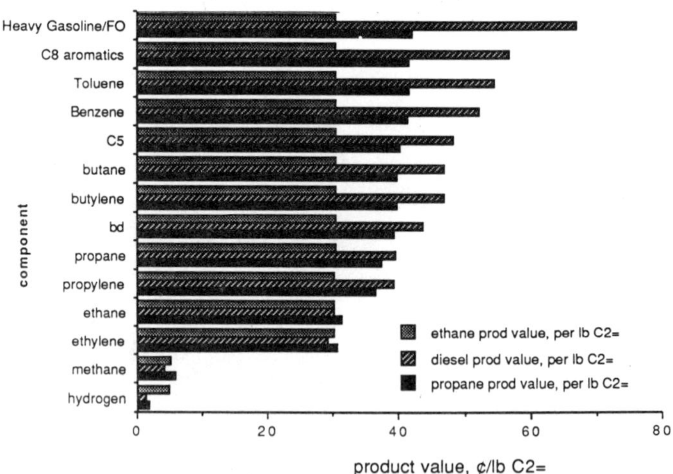

Figure 28. Product Value from Pyrolysis per pound of Ethylene.

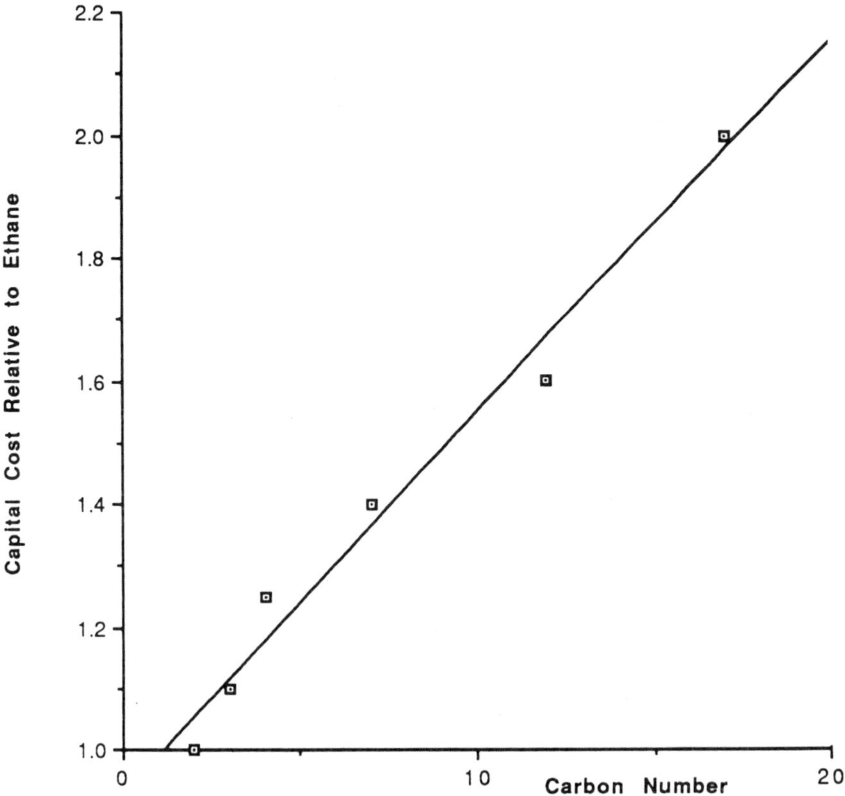

Figure 29. Pyrolysis Capital Costs on a Relative Basis.

are driven by one product and hence are subject to cycling with the supply/demand situation of ethylene. With increasing gas prices, these plants could quickly become high cost producers. This has not been the situation historically.

Relative to methane conversion to ethylene via a redox plant operating in a cycling fixed bed mode or fluid bed transport mode, raw materials margins are presented in Figure 31 along with those for ethane pyrolysis. It is important to realize that ethane crackers must import fuel gas in order to be in fuel balance.

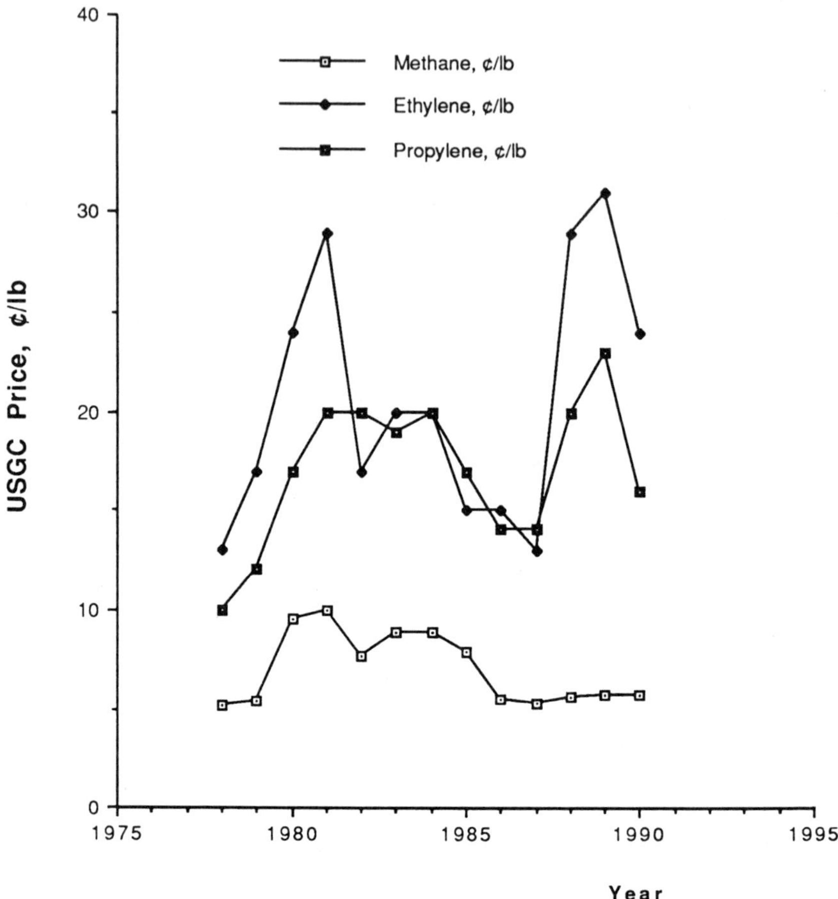

Figure 30. Historic Pricing.

These plants require 2800-6000 cal/gm ethylene for processing; the indicated range is due to the age of the plants. In methane conversion, nearly 2800 cal/gm ethylene is generated in the conversion step, thus making these plants self sufficient in fuel. Capital costs are not presented for the methane to ethylene case since these numbers will be site specific and several key processing calls such as the degree of methane recycle depend on local economics. Processing charge gas should be no more difficult

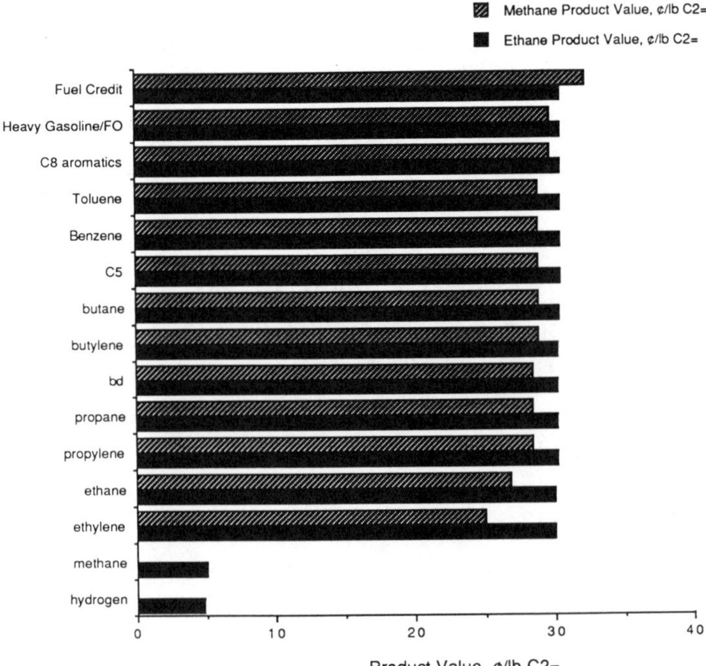

Figure 31. Methane Redox/Ethane Pyrolysis Comparison.

than ethane pyrolysis with the addition of CO2 removal via distillation rather than caustic scrubbing.

CONCLUSIONS

Methane conversion has been successfully demonstrated and modelled in a switching fixed bed mode. The key item is the temperature swing between peaks and valleys that occur during the various cycles.

The model has proved to be an effective tool in deciding which arrangements make for a practical design. This was then used to test the economic sensitivity of temperature swing reduction.

Methane conversion in a switching fixed bed mode is operable although probably not as economic as a circulating fluidized bed which allows near isothermal conversion.

REFERENCES

1. John A. Sofranko, U.S. Patent 4,879,427A (1989)

2. Anne M. Gaffney, John J. Leonard, John A. Sofranko, and Hsiang Ning Sun, Chem. Ind. 33: 87-100 (1988)

3. Anne M. Gaffney, C. Andrew Jones, and John A. Sofranko, U.S. Patent 4,795,849A (1989)

4. Anne M. Gaffney, C. Andrew Jones, and John A. Sofranko, U.S. Patent 4,795,842A (1989)

5. Harry Mazurek and John A. Sofranko, U.S. Patent 4,788,376A (1988)

6. C. Andrew Jones, John J. Leonard, and John A. Sofranko, U.S. Patent 4,737595A (1988)

7. J.A. Sofranko, J.J. Leonard, C.A. Jones, A.M. Gaffney, and H.P. Withers, Catal. Today 3: 127-35 (1988)

8. Robert G. Gastinger, Howard P. Withers, Jr., and John A. Sofranko, U.S. Patent 4,769508A (1988)

9. Anne M. Gaffney, C. Andrew Jones, John J. Leonard, and John A. Sofranko, J. Catal 114: 422-32 (1988)

10. A.M. Gaffney, C.A. Jones, J.J. Leonard, and J.A. Sofranko, Prepr.-Am. Chem. Soc., Div. Pet. Chem. 33: (1988)

11. A. Gaffney, C.A. Jones, J.J. Leonard, J.A. Sofranko, and H.P. Withers, Stud. Surf. Sci. Catal. 38: (1988)

12. Robert G. Gastinger, Andrew C. Jones, and John A. Sofranko, European Patent EP 253,522 A2 (1988)

13. C. Andrew Jones, John J. Leonard, and John A. Sofranko, U.S. Patent 4,665,260A (1987)

14. J.A. Sofranko, J.J. Leonard, C.A. Jones, A.M. Gaffney, and H.P. Withers, Prepr.-Am. Chem. Soc., Div. Pet. Chem. 32: (1987)

15. Howard P. Withers, Jr., C. Andrew Jones, John J. Leonard, and John A. Sofranko, U.S. Patent 4,634,800A (1987)

16. C. Andrew Jones and John A. Sofranko, U.S. Patent 4,650781A (1987)

17. John A. Sofranko, U.S. Patent 4,654,459A (1987)

18. A.M. Gaffney, J.J. Leonard, J.A. Sofranko and H.N. Sun, Prepr.-Am. Chem. Soc., Div. Pet. Chem. 32: (1987)

19. C. Andrew Jones, John J. Leonard, and John A. Sofranko, J. Catal. 103: 311-19 (1987)

20. John A. Sofranko, John J. Leonard, and C. Andrew Jones, J. Catal. 103: 302-10 (1987)

21. A. Andrew Jones, John J. Leonard and John A. Sofranko, Energy Fuels 1: 12-16 (1987)

22. C. Andrew Jones, John J. Leonard and John A. Sofranko, U.S. Patent 4,567,307A (1986)

23. John J. Leonard, John A. Sofranko, and C. Andrew Jones, U.K. Patent GS 2,156,842 A1 (1985)

24. C. Andrew Jones, John J. Leonard and John A. Sofranko, U.S. Patent 4,547,607A (1985)

25. C. Andrew Jones, John J. Leonard, and John A. Sofranko, U.S. Patent 4,554,395A (1985)

26. C. Andrew Jones, John J. Leonard, and John A. Sofranko, U.S. Patent 4,560,821A (1985)

27. John A. Jaecker, Andrew C. Jones, John A. Sofranko, Marvin F.L. Johnson, Howard P. Withers, Jr., John J. Leonard, and William E. Breder, Jr., PCT International Patent WO 8,504,821 A1 (1985)

28. Gennaro J. Maffia, Andrew C. Jones, John J. Leonard, and John A. Sofranko, PCT International Patent WO 8,504,867 A1 (1985)

29. John A. Sofranko and Howard P. Withers, Jr., U.S. patent 4,547,610A (1985)

30. Howard P. Withers and John A. Sofranko, U.S. Patent 4,544,785A (1985)

31. C. Andrew Jones, John J. Leonard, John A. Sofranko, and Howard P. Withers, U.S. Patent 4,523,049A (1985)

32. C. Andrew Jones and John A. Sofranko, U.S. Patent 4,523,050A (1985)

33. John A. Sofranko, U.S. Patent 4,517,398A (1985)

34. C. Andrew Jones and John A. Sofranko, PCT International Patent WO 8,500,804 A1 (1985)

35. C. Andrew Jones and John A. Sofranko, U.S. Patent 4,499,322A (1985)

36. C.A. Jones and John A. Sofranko, U.S. Patent 4,495,374A (1985)

37. C. Andrew Jones, John J. Leonard, and John A. Sofranko, U.S. Patent 4,443,646A (1984)

38. C. Andrew Jones, John J. Leonard, and John A. Sofranko, U.S. Patent 4,444,984A (1984)

39. C. Andrew Jones, John J. Leonard, and John A. Sofranko, U.S. Patent 4,443,644A (1984)

40. C. Andrew Jones, John J. Leonard, and John A. Sofranko, U.S. Patent 4,443,645A (1984)

41. C. Andrew Jones, John J. Leonard, and John A. Sofranko, U.S. Patent 4,443,647A (1984)

42. C. Andrew Jones, John J. Leonard, and John A. Sofranko, U.S. Patent 4,443,648A (1984)

43. C. Andrew Jones, John J. Leonard, and John A. Sofranko, U.S. Patent 4,443,649A (1984)

Texas A&M Work

44. Youdong Tong, Michael P. Rosynek, and Jack H. Lunsford, J. Phsy. Chem., 93: 2896-8 (1989)

45. J.H. Lunsford, Oxygen Complexes Oxygen Act. Transition Met., 265-72 (1988)

46. Karl C.C. Kharas, and Jack H. Lunsford, J. Am. Chem. Soc., 111: 2336-7 (1989)

47. Hong Sheng Shang, Ji Xiang Wang, Daniel J. Driscoll, and Jack H. Lunsford, J. Catal., 112: 366-74 (1988)

48. J.H. Lunsford, C.H. Lin, J.X. Wang, and K.D. Campbell, J. Catal., 111: 305-14 (1988)

49. Chie Hsun Lin, Ji Xiang Wang, and Jack H. Lunsford, J. Catal., 111: 302-16 (1988)

50. Jack H. Lunsford, Langmuir, 5: 12-16 (1989)

51. Jack H. Lunsford, Prepr. Pap.- Am. Chem. Soc., Div. Fuel Chem., 33: 357-62 (1988)

52. Kenneth D. Campbell and Jack H. Lunsford, J. Phys. Chem., 92: 5792-6 (1988)

53. Jack H. Lunsford, Stud. Surf. Sci. Catal,, 36: 359-71 (1988)

54. K.D. Campbell, H. Zhang, and J.H. Lunsford, J. Phys. Chem., 92: 750-3 (1988)

55. H.S. Zhang, J.X. Wang, D.J. Driscoll, and J.H. Lunsford, Prepr. Pap.- Am. Chem. Soc., Div. Fuel Chem., 32: 242-8 (1987)

56. K.D. Campbell, E. Morales, and J.H. Lunsford, Report,: 10 pp (1987)

57. Kenneth D. Campbell, Edrick Morales, and Jack H. Lunsford, J. Am. Chem. Soc., 109: 7900-1 (1987)

58. Chiu Hsun Lin, Tomoyasu Ito, Jixiang Wang, and Jack H. Lunsford, J. Am. Chem. Soc., 109: 4808-10 (1987)

59. John Henry Kolts and Jack H. Lunsford, EP 196,541 A1 (1986)

60. Chie Hsun Lin, Kenneth D. Campbell, Ji Xiang Wang, and Jack H. Lunsford, J. Phys. Chem., 90: 534-7 (1986)

61. Daniel J. Driscoll, Wilson Martir, Ji Xiang Wang, and Jack H. Lunsford, Stud. Surf. Sci. Catal., 21: 403-8 (1985)

62. Daniel J. Driscoll and Jack H. Lunsford, J. Phys. Chem., 89: 4415-18 (1985)

63. Tomoyasu Ito, Jixiang Wang, Chiu Hsun Lin, and Jack H. Lunsford, J. Am. Chem. Soc., 107: 5062-8 (1985)

64. Tomoyasu Ito and Jack H. Lunsford, Nature (London), 314: 721-2 (1985)

65. Daniel J. Driscoll, Wilson Martir, Ji Xiang Wang, and Jack H. Lunsford, J. Am. Chem. Soc., 107: 58-63 (1985)

66. P. Berkowitz, D.J. Driscoll, J.H. Lunsford, J.B. Butt, and H.H. Kung, Combust. Sci. Technol., 40: 317-21 (1984)

67. J.H. Lunsford, W. Martir, and D.J. Driscoll, Prepr.- Am. Chem. Soc., Div. Pet. Chem., 29: 920-6 (1984)

68. H.F. Liu, R.S. Liu, K.Y Liew, R.E. Johnson, and J.H. Lunsford, J. Am. Chem. Soc., 106: 4117-21 (1984)

69. J.H. Lunsford and M. Niwa, Prepr.- Am. Chem. Soc., Div. pet. Chem., 27: 429-39 (1982)

70. Stanley E. Moore and Jack H. Lunsford, J. Catal., 77: 297-300 (1982)

71. Miki Niwa and Jack H. Lunsford, J. Catal., 75: 302-13 (1982)

72. Bruce L. Gustafson and Jack H. Lunsford, J. Catal., 74: 393-404 (1982)

73. Francois Fajula, Ratford G. Anthony, and Jack H. Lunsford, J. Catal., 73: 237-56 (1982)

74. R.S. Liu, M. Iwamoto and Jack H. Lunsford, J. Chem. Soc., Chem. Commun.: 78-9 (1982)

75. Masakazu Iwamoto and Jack H. Lunsford, J. Phys. Chem., 84: 3079-84 (1980)

76. Miki Niwa, Tokio Iizuka, J. Chem. Soc., Chem. Commun.: 684-5 (1979)

77. C. Wilson, Report of the Workshop on Alternate Energy Strategies (1977)

78. United Nations Economic Commission for Western Asia, Arab Energy Prospects to 2000 (1982)

79. Independent Assessment by the authors based on the data in Ref (78)

80. Compiled by Dr. Hilmi Samara, Imperial College, UK in consultation with the ECWA (Ref 78)

81. J.D. Moody and R.W. Esser, An Estimate of the World's Recoverable Crude Oil Resources, 3: (1975)

82. G.J. Maffia, How to Guide to DeMethanization (1985)

83. G.J. Maffia, NGL Recovery (1985)

84. D.J. Victory et al, Hydrocarbon Processing: 44 (1987)

85. C.E. Orozco and D.L. Tiffin, Hydrocarbon Processing: 325 (1977)

86. Joseph Fox, Direct Methane Conversion Process Evaluations (1989)

87. Satish Nirula, Economic Prospects for Methane Conversion Technologies (1989)

11

ETHYLENE FROM NATURAL GAS: PROVEN AND NEW TECHNOLOGIES

J.W.M.H. Geerts, J.H.B.J. Hoebink and K. van der Wiele

Laboratory of Chemical Process Technology
Eindhoven University of Technology
P.O. Box 513, 5600 MB Eindhoven, The Netherlands

I. INTRODUCTION

Natural gas and mineral oil are very important raw materials in today's chemical industry. Figure 1 shows proven reserves over annual consumption (R/P ratios) calculated for the time span 1970-1987 [1]. Evidently, the R/P value of natural gas is increasing, while the ratio of mineral oil remains almost constant. So a relative scarcity of mineral oil may be expected in the future, resulting in a rise of oil prices. In vast deserted areas of Canada, the Middle East and the Soviet Union natural gas resources exceed oil reserves considerably. Low transport efficiency for uncompressed gas and the high costs of gas compression result in increased flaring of this valuable natural gas. Conversion of natural gas into transportable or more valuable products is therefore worthwhile to be considered. In this respect methanol, ammonia or fuel distillates can be mentioned, but, of course, ethylene is most interesting. The process under consideration has an enormous potential when an oligomerisation process is applied

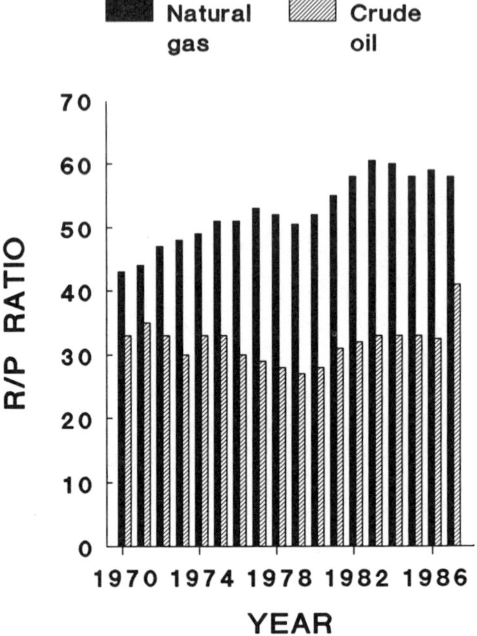

Fig.1 *Proven reserves over annual consumption for crude oil and natural gas.*

as consecutive step to produce gasoline. Moreover ethylene is the most important feedstock in the petrochemical industry. Nearly all polymers originate in part from ethylene. In 1983 the total world ethylene production, exceeded $50*10^6$ tons p.a.

Conventionally C_2H_4 is produced either by cracking of higher hydrocarbons in natural gas or by cracking of naphtha/gasoil. In the United States large amounts of natural gas are found. They may contain considerable fractions of higher alkanes, which are separated from the methane and cracked to obtain C_2H_4. In Western Europe most of the C_2H_4 is produced from naphtha or gasoil cracking.

Cracking is performed in tubular heaters, because the process is highly endothermic. Primary reactions produce olefins essentially via a free radical mechanism creating hydrogen abstraction and chain fracture. Secondary reactions involving primarily produced olefins become important at high levels of feedstock conversion. Ethane cracking has the highest C_2H_4 yield. Less C_2H_4 is obtained from naphtha cracking, while the C_2H_4 yield from gasoil is lowest. Cracking of hydrocarbons is today's production process with the lowest C_2H_4 production cost. This process involves high reaction rates without catalysts or solvents and it yields besides C_2H_4 a large variety of interesting by-products.

II. AIMS OF THIS STUDY

This work aims to compare ethylene production via methane oxidative coupling and via conventional naphtha cracking both from a technical and from an economical point of view. Therefore, a preliminary design of the new process was made, starting from the performance of methane coupling catalysts, as expected to be attainable within a few years. The economic feasibility of the process was studied by applying the same cost estimation techniques to the new process and to an existing naphtha cracker. This study also aimed to set targets for future research efforts in the methane oxidative coupling field.

III. KINETICS OF METHANE OXIDATIVE COUPLING

A literature study about catalysts for the oxidative coupling of methane was carried out in order to get an impression of the conversions and the ethylene selectivities that can be reached. The oxidative coupling of methane can occur without a catalyst too, see Fig. 2. However, the yields are very poor.

Performances of the best catalysts described in the literature are shown in Figure 2. Two groups of catalysts can be distinguished:
Type I: Catalysts that contain metal ions with a changing valency state. During the reaction they are reduced by methane and reoxidised by oxygen. These catalysts can function as oxygen carriers. Lead oxide is an example.
Type II: Catalysts that cannot really be reduced or oxidised. Those catalysts require a co-feed of methane and oxygen, because they are

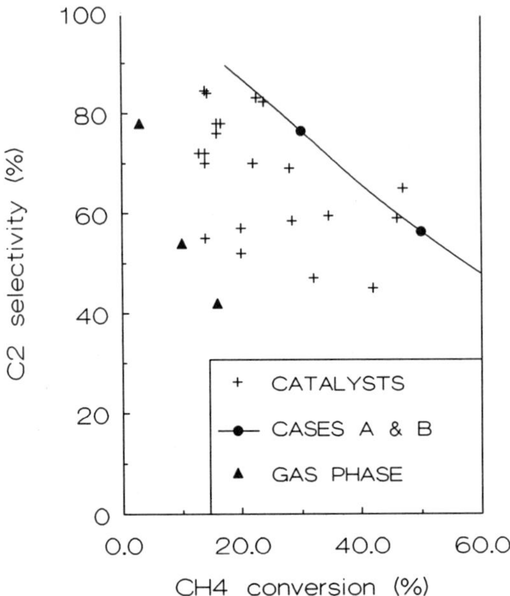

Fig.2 *Selectivity versus conversion as reported in the literature for homogeneous and catalytic oxidative methane coupling. Also shown are the starting points of the present study.*

only active when oxygen is present. Lithium doped magnesia belongs to this type.

In general the latter group of catalysts has a better performance than the former. Especially combinations of alkali and alkaline earth oxides make good catalysts.

In Fig. 2 two catalysts with an ethylene yield of ± 30% are indicated. They both contain chlorine, and it should be mentioned that the reactor feedstock contained a lot of inert gas (He= ± 80%). For catalysts containing a halide ion, the ability to get a good C_{2+} selectivity seems to be related to the ability of hydrogen abstraction by halogen [2]. As a consequence, chlorine was found to be the best halide ion for achieving a high yield of higher hydrocarbons. Disadvantages include the incorporation of halogen into the produced hydrocarbons and a very rapid deactivation of the catalyst because it loses the halogen.

The MgO supported catalysts have an especially high activity and selectivity. The reaction is favoured by higher temperatures, lower concentrations of oxygen or higher concentrations of the feed gas.

Finally it should be noted that the catalysts reported in literature usually had a very short runtime. Experimental evidence exists [3] that the activity of a Li/MgO catalyst becomes more or less constant after two days. However, the selectivity level then has dropped to half of its starting value. The longest runs that have been carried out with a Li/MgO catalyst lasted two weeks, during which time the activity of the catalyst stayed rather constant after the initial drop.

From Fig. 2 it can be concluded that it is a realistic assumption that a process can be developed with a 30% methane conversion and 80% selectivity with respect to the C_2-products (case A), or 50% conversion and 50% selectivity (case B), once the catalyst stability has been improved. These conversion and selectivity levels, both resulting in approximately 25 % yield, form the starting points for the present feasibility study.

The oxidative coupling of methane to ethylene over Li/MgO catalysts proceeds at circa 1073 K and involves both gas phase and catalytic reactions. The reaction mechanism consists of a complex scheme of parallel and consecutive reactions [4]. The initial step, the conversion of methane to a methyl radical, is very difficult, because the methane molecule is very stable. A mechanism has been proposed for the hydrogen abstraction [5]. The catalyst consists of a mixture of two metal oxides. It is characteristic that the radii of the metal oxide ions are almost identical. This is important, as the Li^+ can easily be built into the MgO lattice. By substituting a Mg^{2+} ion, Li^+ creates an O^- species. It has been demonstrated [6,7] that these O^- ions were formed on the surface of the catalyst. These [Li^+O^-] centres are capable of abstracting hydrogen from methane:

$$Li^+O^- + CH_4 \dashrightarrow Li^+OH^- + CH_3\cdot$$

The regeneration of [Li^+O^-] centres after this reaction is believed to occur with the following reactions:

$$2\ Li^+OH^- \dashrightarrow Li^+O^{2-} + Li^+\blacksquare + H_2O$$

$$Li^{+}O^{2-} + Li^{+}\blacksquare + \frac{1}{2} O_2 \quad \text{---->} \quad 2Li^{+}O^{-}$$

(the ■ denotes an oxygen vacancy)

A coupling of two CH_3 radicals, probably in the gas phase near the surface [5], produces C_2H_6:

$$2\ CH_3\cdot \quad \text{---->} \quad C_2H_6$$

A reaction mechanism has been proposed [4], see Fig. 3. In this scheme the vertical arrows denote abstraction of a hydrogen atom and the horizontal arrows denote the addition of a methyl radical. The formation of a C-O bond, which is irreversible, is shown by an oblique arrow. The thick lines represent the most important reactions. C_3 species are hardly formed and therefore less important.

Precautions should be taken that the reactor material has no influence on the conversion or the selectivity. A suitable inert material for the oxidative coupling of methane is quartz. It has been shown [8] that also other reactor materials, such as alumina and SiC, take no part in the methane coupling reactions. Metal reaction

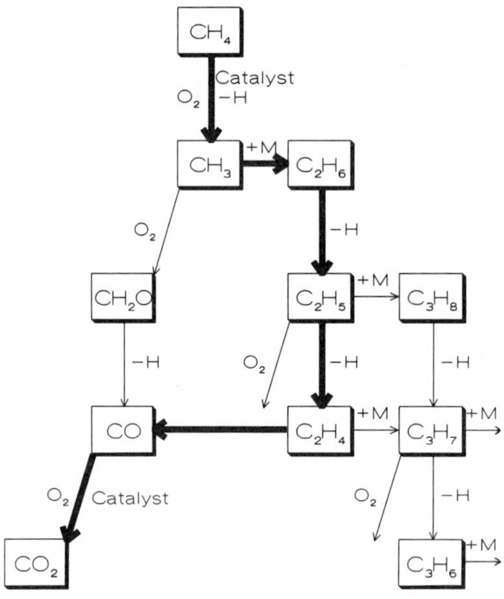

Fig.3 *Proposed reaction scheme [4].*

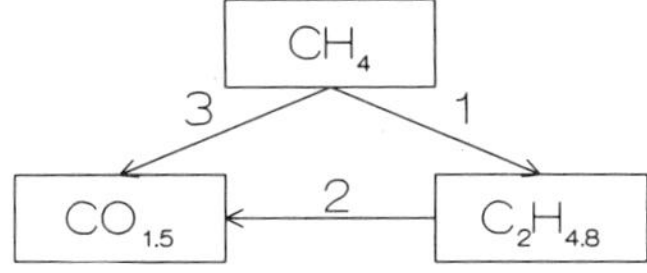

Fig.4 *Simplified reaction scheme, as applied in the present study.*

vessels (Fe, Ni, Ti) give rise to a higher CO and CO_2 selectivity and are therefore less suitable for this process, unless high wall temperatures can be avoided.

For the purpose of the present study the reaction scheme (see Fig. 3) has been simplified. The C_3 species, which are hardly formed, were neglected. In Fig. 4 CO_x means a mixture of CO and CO_2 in the ratio of one to one, and C_2H_x means a mixture of C_2H_4 and C_2H_6 in the ratio of $C_2H_4:C_2H_6 = 3:2$. Both ratios are realistic with respect to experimental results [3].

The kinetic data for these reactions, as they were used in this study for reactor design, are shown in Table 1. Most of the kinetic measurements were carried out with an aged lithium doped magnesium oxide catalyst. The yields from this catalyst are not as good as those of cases A and B. So, one of the measured frequency factors, k_3^0 had to be adapted to meet the desired conversion and selectivity levels, when an ideal plug flow reactor was applied. The reaction orders m and n were left unchanged. The activation energies E_A were obtained from literature and from our own experimental results.

TABLE 1 *The kinetic data used as a basis for the oxidative coupling reactor designs.*

	m	n	k_i^0 *	E_A *
$r_1 = k_1 PCH_4^m \cdot PO_2^n$	*0.9*	*0.3*	*0.016*	*100*
$r_2 = k_2 PC_2H_4^m \cdot PO_2^n$	*1.4*	*0.7*	*0.2*	*150*
$r_3 = k_3 PCH_4^m \cdot PO_2^n$	*0.5*	*1.5*	*0.02*	*70*

* The dimensions are for k: $mol/(g_{kat} \cdot s) * bar^{-(m+n)}$ and for E_A: kJ/mol.

IV. REACTOR DESIGN

A. *INTRODUCTION*

Three different types of reactors were designed to meet the requirements of cases A and B: a multi-tubular fixed bed reactor, a multi-stage reactor and a fluidised bed reactor. For each reactor design, the following assumptions were made:

inlet temperature : 973 K
outlet temperature : 1073K
inlet pressure : 1 bar
catalyst effectiveness : 100%

The reactors are fed with mixtures of pure oxygen and methane. Preliminary calculations carried out for reactor systems in which the catalyst is transported from a methane to an oxygen atmosphere showed that such incredibly huge quantities of catalysts had to be transported that those reactor systems were excluded in the rest of this study. Pure oxygen was chosen as the oxidator. The feed of oxygen was adjusted to achieve the desired methane conversion levels. The reactors were designed in such a way that the oxygen was completely consumed within the catalyst bed. The total ethylene production was taken as 350,000 tons C_2H_4 per year (=8000 operating hours). The enthalpies of the reactions, the specific heat, the densities, the viscosities and the heat conductivities of the gases were derived from literature data and reported elsewhere [9].

B. *THE MULTI-TUBULAR FIXED BED REACTOR*

For calculating the dimensions of the multi-tubular fixed bed reactor, a computer program was developed. The program uses Euler's method to solve simultaneously the differential equations for the mass balances of each component in the gas stream, for the heat balance and for the pressure drop.

The reactor was designed as a heat exchanger with the catalyst inside the tubes and a cooling agent flowing around the tubes. Ideal plug flow was assumed inside the tubes and ideal mixing at the outside. The radial temperature gradients were restricted to a maximum of 10 K difference between tube center and wall by adjusting the tube diameter. The heat transfer resistance was assumed to be concentrated in the catalyst bed and the tube wall. The external contribution to the total heat transfer coefficient was neglected. The heat transfer coefficient for transport from catalyst bed to the reactor wall was determined from packed bed relations [10]. The pressure drop was calculated from the well-known equation of Ergun.

Table 2 presents data concerning the reactor. Figure 5 shows the calculated temperature profiles for cases A and B. Figure 6 shows the conversion and selectivity profiles for the high conversion case: case B. In both cases A and B, a tube diameter of 4 cm is adequate to obtain uniform radial temperature profiles. The first part of the reactor is not cooled in order to allow the reaction to start. Cooling is performed approximately from the point where the temperature of the reaction gas reaches 1073 K onwards. For case A this results in an adiabatic zone of 0.30 m, for case B this zone has a length of 0.15 m. In case A, the temperature of the reaction gas immediately starts decreasing after the adiabatic zone. This results

TABLE 2 *The designed multi-tubular reactors for cases A and B.*

	Case A	Case B
Number of reactors	*17*	*20*
Number of tubes per reactor	*5000*	*5000*
Tube diameter (m)	*0.04*	*0.04*
Tube length (m)	*2.8*	*1.0*
Tube wall thickness (m)	*0.01*	*0.01*
Tube material	*Alumina*	*Alumina*
Cooling temperature (°C)	*690*	*636*
Outlet temperature (°C)	*693*	*707*
CH_4 conversion (%)	*29.5*	*47.0*
O_2 conversion (%)	*100*	*100*
C_2 selectivity (%)	*80.6*	*53.0*
CO_x selectivity (%)	*19.4*	*47.0*
Pressure drop (bar)	*0.27*	*0.11*

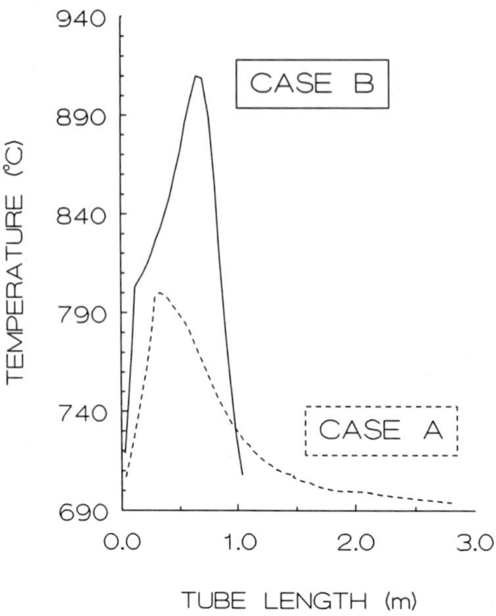

Fig.5 *Calculated temperature profiles for the multi-tubular reactor.*

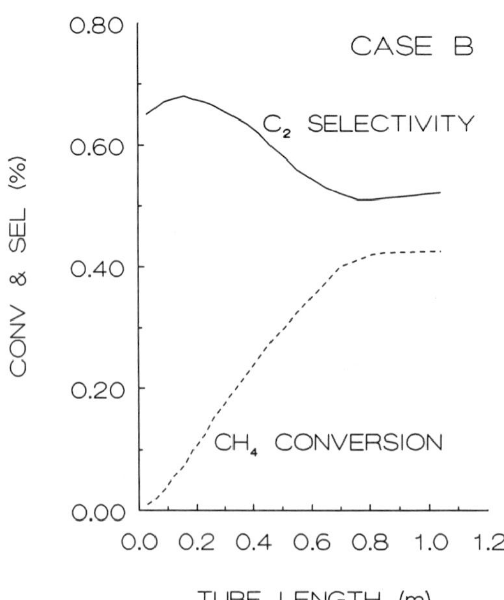

Fig.6 *Conversion and selectivity, calculated for the multi-tubular reactor (case B).*

in low reaction rates and in a reactor tube length of 2.8 m. The reactor pressure drop appears to be 0.27 bar.

In case B, the temperature of the reaction gas rises even after the adiabatic zone, to about 1183 K, before a decrease sets in, see Fig. 5. At this maximum temperature, almost 90 % of the oxygen has been converted already. This results in a smaller tube length and a smaller pressure drop (0.11 bar).

The temperature of the cooling agent has a value between 900 K and 1000 K; therefore, the cooling agent has to be a salt, or mixture of salts, with a low melting point and a low vapor pressure in this temperature range. Mixtures of salts containing lithium carbonate or eutectic mixtures of alkali fluorides like NaF, KF and LiF are suitable. One single salt mixture most probably cannot cover the temperature gap between the reactor and the water/steam side at 623 K, where the heat from the reactor is absorbed. Therefore, one or more additional heat exchangers in series have to be included when a multi-tubular fixed bed is chosen. Application of a multi-tubular fixed bed for this reaction has several disadvantages:

* The system is very sensitive to changes in the coolant temperature. Calculations show that a change in this temperature of only 5 K causes either runaway or extinction of the reactor.
* An expensive ceramic material has to be used for the tube material, because of the high temperatures of the tube wall. The catalytic activity of steel for the oxidation of hydrocarbons to carbon monoxide and carbon dioxide at the applied temperatures makes steel unsuitable as a reactor wall material. (The reactor wall of a fluidised bed reactor can have a much lower temperature and therefore can be made of steel).
* One or more salt baths have to be used in order to convert the reaction heat into steam.
* It is not possible to design a reactor for high conversions without considerable loss of selectivity.

An advantage of the multi-tubular fixed bed reactor is that the mechanical strength of the catalyst does not have to meet the severe requirements of a fluidised bed catalyst.

C. MULTI-STAGE REACTOR

The multi-stage reactor can be considered as a series of packed beds; therefore it allows inter-stage oxygen feeding and cooling. The feature of spreading the oxygen feed and cooling between the beds allows a better control of the conversion of methane into ethylene, when compared with a multi-tubular reactor. From the kinetic equations, it can be seen that the oxygen to methane ratio should be low in order to obtain the highest selectivity.

Multi-stage reactors have been designed for both cases A and B. The calculations for each stage were essentially the same as for the tubular reactor. Additional assumptions were made as there are adiabatic operation of each catalyst bed, negligible pressure drop in the packed beds and complete mixing of oxygen and reactants in between the beds. The packed bed length is set for each stage, whenever the oxygen conversion is complete or the adiabatic temperature rise is 200 K. The gas leaving the last catalyst bed is cooled down to the initial temperature 973 K.

The heat exchangers are horizontal pipes in the cross-section of the reactor in which saturated steam of 623 K is produced. The

TABLE 3 *Designed multi-stage reactor.*

	Case A	Case B	
maximum allowed temperature	*1173*	*1173*	*K*
cooling temperature	*623*	*623*	*K*
oxygen feed before bed	*1 and 3*	*1*	
PO_2 *gas mixture before bed 1*	*0.080*	*0.35*	*bar*
PO_2 *gas mixture before bed 3*	*0.094*	-	*bar*
ethylene selectivity	*81.0*	*53.3*	*%*
number of beds per reactor	*5*	*11*	
number of reactors	*11*	*11*	
diameter reactor	*6*	*6*	*m*
total bed length	*0.65*	*0.91*	*m*
total heat exchanger length	*0.48*	*1.32*	*m*
total reactor length *	*5.75*	*9.75*	*m*

* *includes manholes and feeds between the stages*

TABLE 4 *Specification per bed of the designed multi-stage reactors for both cases.*

CASE A:

Bed no.	bed length (m)	CH_4 conversion at end of bed (%)	C_2 selectivity at end of bed (%)
1	*0.12*	*8.7*	*91.3*
2	*0.17*	*16.4*	*90.7*
3	*0.08*	*21.6*	*84.6*
4	*0.11*	*26.5*	*81.1*
5	*0.17*	*30.0*	*81.0*

CASE B:

Bed no.	bed length (m)	CH_4 conversion at end of bed (%)	C_2 selectivity at end of bed (%)
1	*0.05*	*5.8*	*69.5*
2	*0.04*	*11.2*	*67.4*
3	*0.05*	*16.3*	*65.2*
4	*0.06*	*20.9*	*63.0*
5	*0.06*	*25.5*	*60.8*
6	*0.06*	*29.7*	*58.8*
7	*0.08*	*33.8*	*57.0*
8	*0.08*	*37.9*	*55.4*
9	*0.10*	*41.8*	*54.1*
10	*0.13*	*46.0*	*53.2*
11	*0.20*	*50.0*	*53.3*

designed heat exchanger satisfied both cases. Two layers of pipes with 0.08 m interspace between the layers were placed one above the other (pipe diameter 0.02 m, pipe spacing 0.006 m). The whole cross-section of the reactor is filled with those pipes, which results in a total heat exchange area of 1500 m^2.

Results of the calculations are shown in Tables 3 and 4. Here the advantages of a multi-stage reactor have not been fully utilised in order to allow a comparison of different reactor types on the same basis. One will obtain better selectivities if the oxygen feed is distributed over even more beds. This is illustrated in Fig. 7, where the selectivity at constant methane conversion is plotted versus the number of beds with an oxygen feed for both cases A and B.

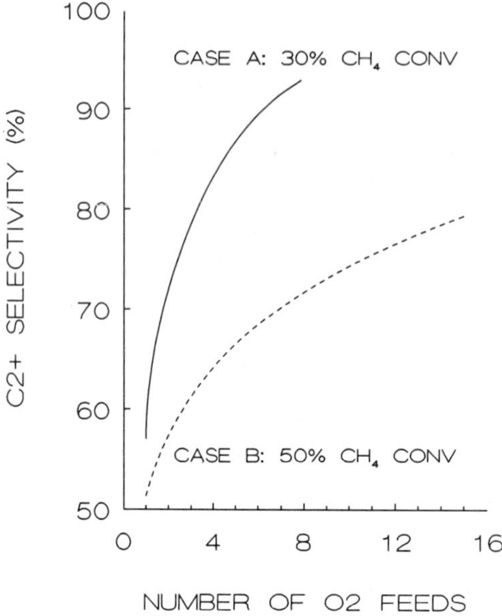

Fig.7 *Influence of oxygen distribution on selectivity in the multi-stage reactor.*

D. *FLUIDISED BED REACTOR*

For the design of a catalytic solid/gas fluidised bed reactor a simple two phase model was used [11]. The reactor content is then divided into a bubble phase and a dense phase. Basic assumptions were that the bubble phase, which contains no catalyst, passes the reactor in plug flow, and that the dense phase is ideally mixed. Mass transfer occurs from the bubble phase to the dense phase, and vice versa. The reactor is isothermal and reactions take place in the dense phase only. Analytical expressions for concentrations in the bubble phase as function of the height in the bed were derived, and inserted into the total mass-balances for each component. The resulting non-linear equations were solved by iteration for the unknown dense phase concentrations. Combination with bubble concentrations which now could be calculated led to conversions and

TABLE 5 *Parameter values used in the fluidized bed model.*

gas density	*0.25 kg/m3*	*bubble hold-up*	*0.1*
solids density	*1500 kg/m3*	*specific bubble surface*	*10 m^2/m^3 reactor*
particle size	*$2*10^{-4}$ m*	*dense phase porosity*	*0.45*
gas viscosity	*$3*10^{-5}$ kg/m*s*		

selectivities of the product gas. Table 5 shows the values of the used model parameters.

The superficial velocity in the bed was assumed to be ten times larger than the minimum fluidisation velocity of 0.06 m/s, as derived from well-known literature relations. The superficial velocity in the dense phase was assumed to be equal to the minimum fluidisation velocity. The superficial velocity in the bubble phase was the superficial velocity minus the superficial velocity in the dense phase. Finally, the mass transfer coefficient between bubble phase and dense phase, was determined as 0.067 m/s from a published relation [12] (assumed average bubble diameter 6 cm) and was assumed to be the same for each component.

Results of the calculations are shown in Table 6 and Fig. 8. The reactor can easily meet the required performance. As can be seen from the Table, the same reactor volume, i.e. the reactor height, is adequate for cases A and B by simply increasing the oxygen feed. In both cases the selectivity is higher than required. Increasing the

TABLE 6 *Designed fluidised bed reactors.*

	CASE A	CASE B	
reactor diameter	*4.60*	*4.57*	*m*
reactor height	*2.3*	*2.3*	*m*
conversion	*30.6*	*50.3*	*%*
selectivity	*89.7*	*59.4*	*%*
methane/oxygen ratio	*5*	*1.86*	
number of reactors	*36*	*43*	
reactor duty	*5.90*	*13.1*	*MW/reactor*
diameter HE tubes	*0.05*	*0.05*	*m*
length HE tubes	*2.3*	*2.3*	*m*
number of HE tubes	*83*	*147*	

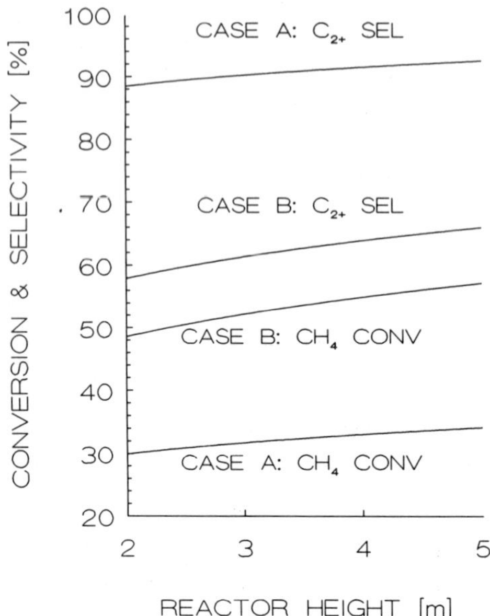

Fig.8 *Conversion and selectivity in the fluidised bed reactor versus total bed height.*

reactor volume improves the selectivity further. This is due to the assumption of ideal dense phase mixing. Although such behaviour is doubtful in very large reactors, it can be stimulated by applying a distributed oxygen feed.

If the reaction is carried out at a different temperature, the conversion and selectivity will change due to a disproportional change in the three reaction rate constants, see Table 7. The

TABLE 7 *Ratios of reaction rate constants at various temperatures.*

T (°C)	k_1 / k_2 / k_3
600	*1 / 2.5 / 3.3*
700	*1 / 1.7 / 6.9*
800	*1 / 1.3 / 12.5*
900	*1 / 0.9 / 20*

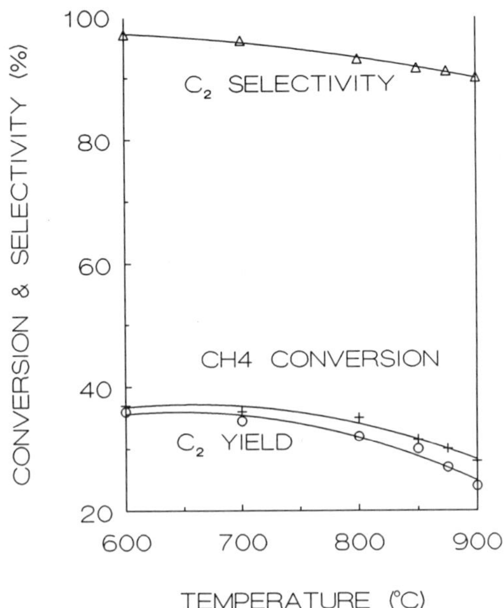

Fig.9 *Influence of bed temperature on conversion and selectivity in the fluidised bed reactor.*

temperature effect is shown in Fig. 9 for case A. The decreased selectivity was not as strong as might have been expected because of mass transfer limitation between both phases.

E. DISCUSSION ON REACTORS

A multi-tubular reactor is not attractive despite of the advantage that its scaling-up from laboratory to commercial size is relatively simple. A serious problem is runaway or extinction of the reaction, which may occur easily when the temperature of the cooling medium changes slightly. Also the selectivity that can be achieved is lower than with a multi-stage or fluidised bed reactor.

A possible problem with the multi-stage reactor could be the mixing of oxygen and methane in between two beds; unless this mixing is rapid, non-selective gas phase reactions will occur before the mixture reaches the catalyst bed.

For a fluidised bed reactor, scaling-up is known to be an important problem. However, this type of reactor has several advantages when compared to a multi-tubular or multi-stage reactor. In the first place it is possible to get a selectivity for the partial oxidation of methane to ethylene which is much higher than with a multi-tubular reactor. Secondly, it is possible to recycle the catalyst continuously, and to replace it if deactivation occurs. The temperature inside the reactor can be easily controlled too. This is beneficial for a good selectivity, and it allows safer plant operation. Finally, the construction of a fluidised bed reactor is relatively simple.

In conclusion a fluidised-bed reactor is preferable over a multi-tubular or multi-stage reactor for a process of methane oxidative coupling.

V. PROCESS DESIGN

A general outline of the designed process can be seen on the flowsheet (Fig. 10). All calculations were made with the flowsheet computer program 'CHEMCAD'. In the flowsheet the temperature of a stream (in degrees centigrade) is plotted in the rectangular box next to the flow, and its pressure (in bars) is shown in the rounded boxes below the temperature record.

The process downstream of the reactor is designed as a conventional separation section. This means that no further uncertainties have been introduced with respect to the technical feasibility of the process. Alternative separation techniques have not been considered.

In the reactor the catalytic reactions of methane with oxygen to ethylene, ethane, carbon monoxide, carbon dioxide and water take place according to the kinetic model at the conversion and selectivity levels mentioned before (Cases A and B). The post-catalytic part of the reactor serves as an adiabatic ethane cracker with 60 % C_2H_6 conversion and 82 % C_2H_4 selectivity (by-product being methane). In the flowsheet it is indicated as a separate reactor and

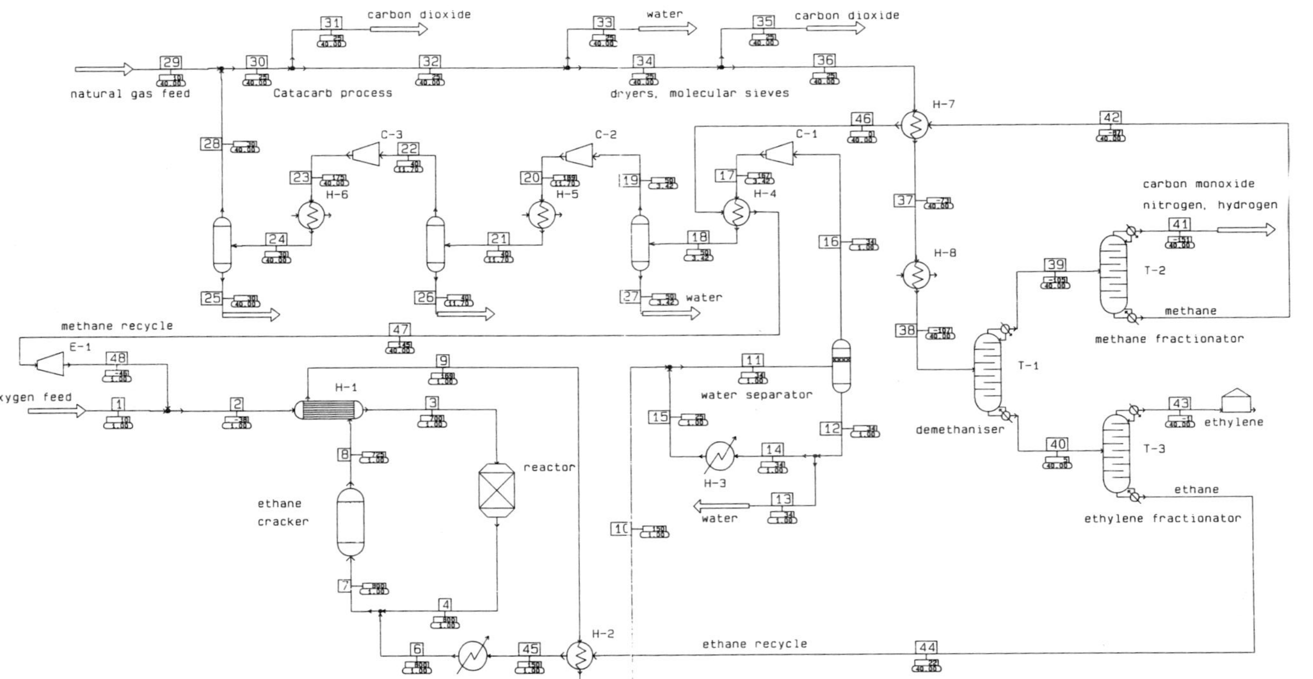

Fig.10 *Flow diagram for the methane coupling process.*

condensers and application of refrigeration cycles. For the remaining net energy required, an efficiency of 50% was assumed.

The ethylene product (purity 99.9%) contains 0.02 mol% methane, 0.07 mol% ethane and 5 ppm carbon dioxide and is delivered to an ethylene pipeline network. Details about the process flow diagram are reported elsewhere [9].

VI. COMPARATIVE INVESTMENT ESTIMATES

To determine the economic feasibility of the ethylene production through partial methane oxidation, total capital investment estimates using Miller's method [13] were made for the process and they were compared with similar calculations for naphtha cracking. For both processes the same plant capacity of 350,000 tons ethylene/year and plant location in the Netherlands were chosen.

The Miller method is based on the principle that links exist between the costs of main process items (= MPI's) and all the other costs of planning and building a chemical plant. These links are expressed as cost percentages which depend on the scale and complexity of the plant to be built. Miller's method is claimed to result in a 25% accurate estimate of the total capital investment (= TCI) of a plant with known capacity and complexity on the basis of only a simplified flowsheet containing the main process items.

The costs of the MPI's were obtained from two sources: the computer program ChemCost and a method which uses the equipment cost data published in Chemical Engineering [14]. The average of these methods was taken for estimating the costs of construction, planning, engineering etc. For estimation of equipment cost operating data were taken from the process flow diagrams (see Fig. 10 for methane coupling and Fig. 11 for naphtha cracking). The rest of the data required was provided by means of 'rules of thumb' [15].

Storage tanks were included in the investments. Ethylene storage tanks with the same capacity as used in the cracking process were

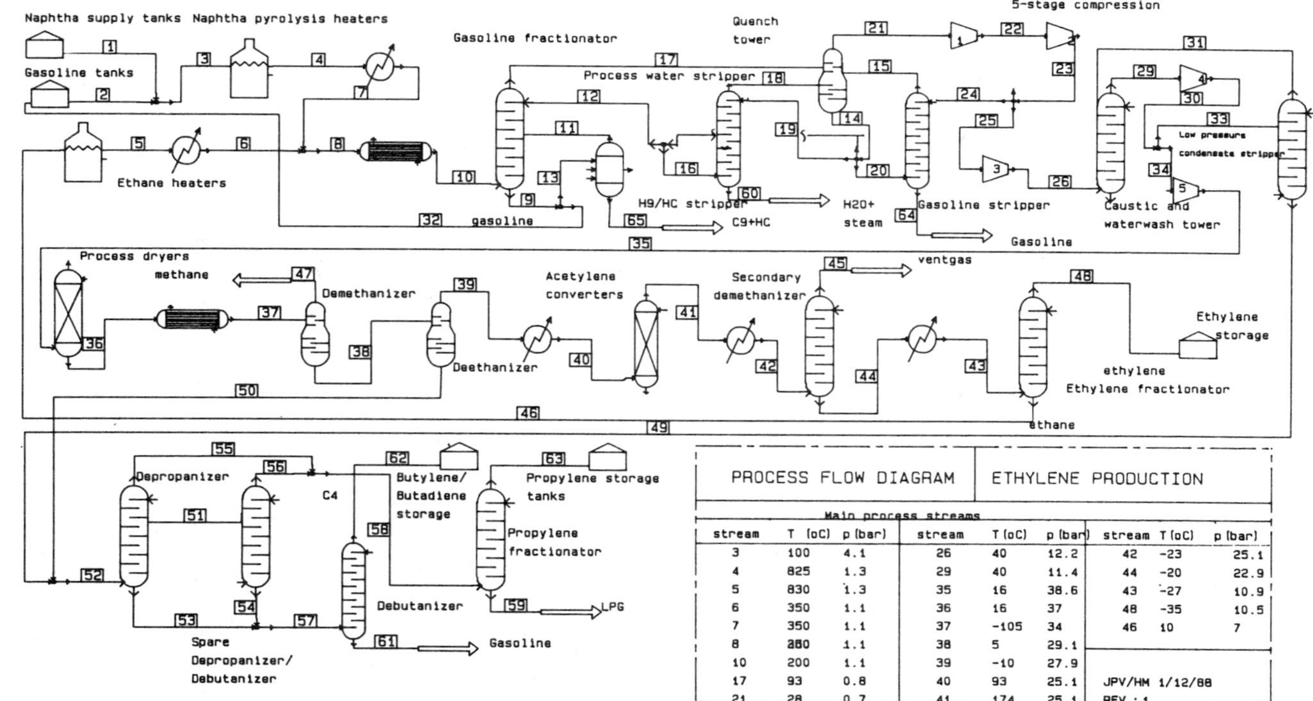

stream	T (oC)	p (bar)	stream	T (oC)	p (bar)	stream	T (oC)	p (bar)
3	100	4.1	26	40	12.2	42	-23	25.1
4	825	1.3	29	40	11.4	44	-20	22.9
5	830	1.3	35	16	38.6	43	-27	10.9
6	350	1.1	36	16	37	48	-35	10.5
7	350	1.1	37	-105	34	46	10	7
8	280	1.1	38	5	29.1			
10	200	1.1	39	-10	27.9			
17	93	0.8	40	93	25.1			
21	28	0.7	41	174	25.1			

Fig.11 *Flow diagram for naphta cracking, as used for estimating equipment costs.*

TABLE 9 *Composition of investment costs for the cracking process and the new process in millions of US Dollars in January 1989.*

Type of costs	*Cracking*	*Direct oxidation*	
		Case A	*Case B*
basic equipment	*85.7*	*41.1*	*53.4*
Direct costs	*324.3*	*98.9*	*128.3*
Indirect costs	*85.9*	*26.2*	*34.0*
Contingencies	*20.5*	*6.3*	*8.1*
Land	*4.8*	*3.0*	*3.0*
Working Capital	*34.0*	*34.0*	*34.0*
Total capital investment	*469.6*	*168.4*	*207.5*

chosen for the methane coupling process. No long-term stock facilities were taken into account.

Some of the design and cost estimation procedures as used for methane coupling were also used for the cracking process, when information on that process flow diagram was lacking. In the end, all the purchased equipment costs were updated by the average Chemical Engineering Plant Cost Index and the Marshall and Swift Equipment Cost Index for 1988 [16].

For the cost of all miscellaneous unlisted equipment (= MUE) a percentage of the MPI costs was taken, depending on the scale and complexity of the plant. The percentages were 30 for the naphtha cracking process and only 15 % for the direct oxidation process, due to its lower complexity. To obtain reasonable capital investment estimates, -10%, 0%, and +10% respectively were added to the costs of basic equipment (MPI plus MUE) for improbable, average and expensive circumstances. For estimation of the total fixed capital investment (= TFCI) the procedure as described by Ulrich [17] was followed. Factors and percentages chosen are reported elsewhere [9]. The costs of land and working capital have to be added in order to obtain the total capital investment. The working capital was taken as the revenues expected from two months ethylene sales at a revenue of US $ 584 per ton ethylene. The cost of land at the site location in The Netherlands is about US $ 27.5 per square meter, for areas of 4 - 8 hectares, including wharf-side facilities.

TABLE 10 *Prices in US $/ton, as used for various chemicals* [18].

Natural gas	*119*	*Oxygen*	*35*
Ethylene	*649*	*Propane*	*94*
Propylene	*452*	*Butane*	*85*
Butylene	*595*	*Butadiene*	*435*
Acetylene	*740*	*Gasoil*	*111*
Naphtha	*126*	*Hydrogen*	*110*
Ethane	*109*		

For the new process no distinction was made for the different reactors, the difference in prices being only marginal with regard to the total capital investment. The capital investment estimates for the conventional cracking process and the oxidation process, cases A and B are shown in Table 9. It can be readily seen that the total capital investment of the new process is substantially lower than for the existing one. This difference is due to both lower equipment costs as well as lower construction costs for the new, less complex plant.

TABLE 11 *Results of profitability analysis.*

estimate	net annual profit	POP[1]	ROR[2]	DCF[3]	Process
low	*72,200,000*	*5.12*	*17.9*	*12.2*	
average	*9,600,000*	*8.88*	*10.3*	*1.9*	*Naphtha-*
high	*-64,900,000*	*--*	*--*	*--*	*cracking*
low	*92,400,000*	*2.00*	*39.1*	*32.9*	
average	*70,500,000*	*2.47*	*31.7*	*26.7*	*Methane,*
high	*46,400,000*	*3.32*	*23.5*	*19.1*	*case A*
low	*59,000,000*	*3.36*	*24.4*	*19.8*	
average	*30,300,000*	*4.97*	*16.5*	*11.3*	*Methane,*
high	*-3,300,000*	*--*	*--*	*--*	*case B*

1 pay-out period, 2 rate of return, 3 discounted cash flow-rate of return.

VII. PROFITABILITY ANALYSIS

The manufacturing costs were calculated according to Holland's method [17], using low, average and high estimates for the percentages to be applied. Table 10 presents the prices of the chemicals as used in the analysis.

The cracking of naphtha is strongly endothermic, which requires an energy consumption of about $7.5*10^6$ kJ/ton ethylene, which introduces extra costs into the total manufacturing costs. The credits for by-products, which are of major importance for the cracking process, were taken into account.

The methane oxidation process produces energy. The net energy production is the sum of the duties of the reactor and the turbine minus the duties of the compressors, the distillation towers and the energy needed for heat exchangers.

The fuel gas stream is for a minor part used to heat up the ethane recycle. The remainder is to be sold as fuel gas (3.16 US $/GJ). The excess energy is converted into electricity with an

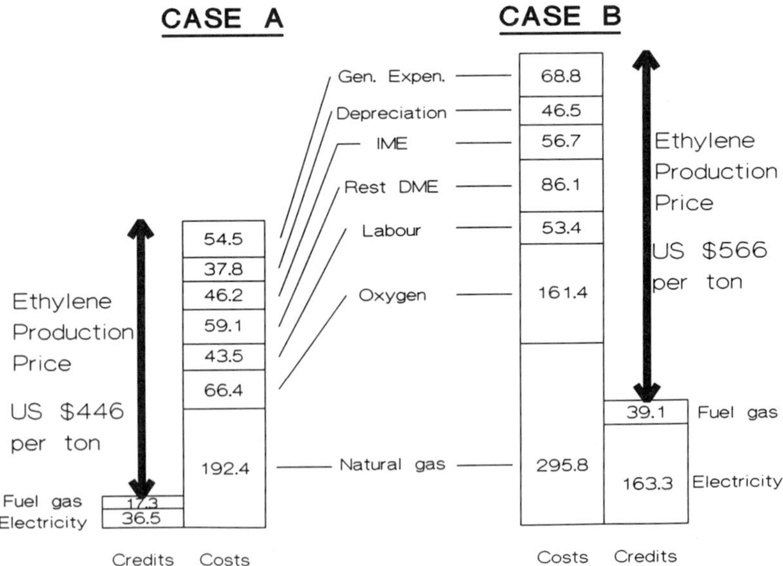

Fig.12 *Composition of the ethylene cost price (All figures in US $ per ton ethylene produced).*

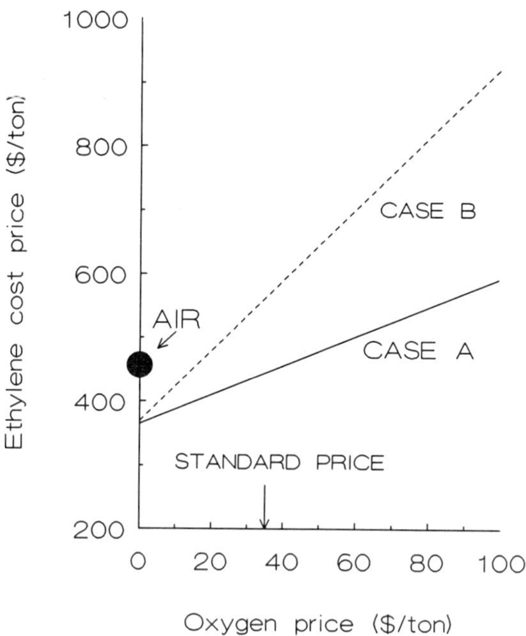

Fig.13 *Influence of the oxygen cost on the ethylene cost price.*

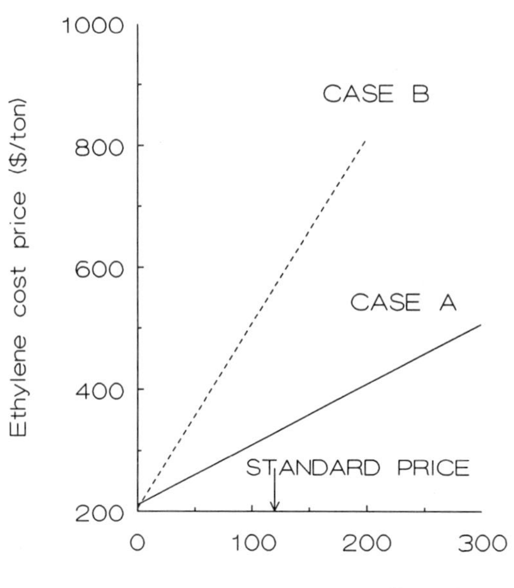

Fig.14 *Influence of the natural gas price on the ethylene cost price.*

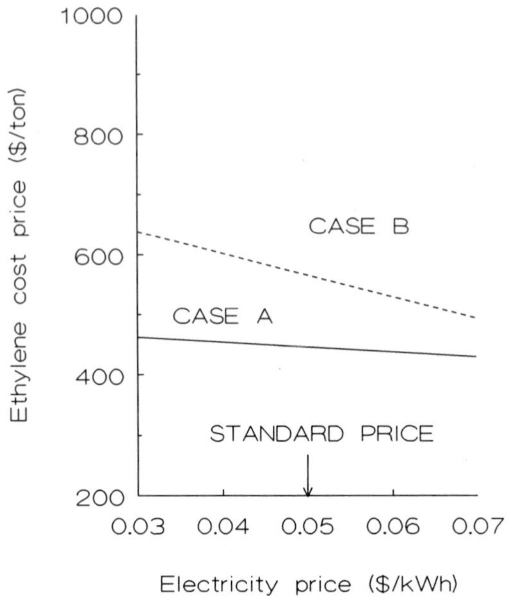

Fig.15 ***Ethylene cost price dependence on the price of sold electricity.***

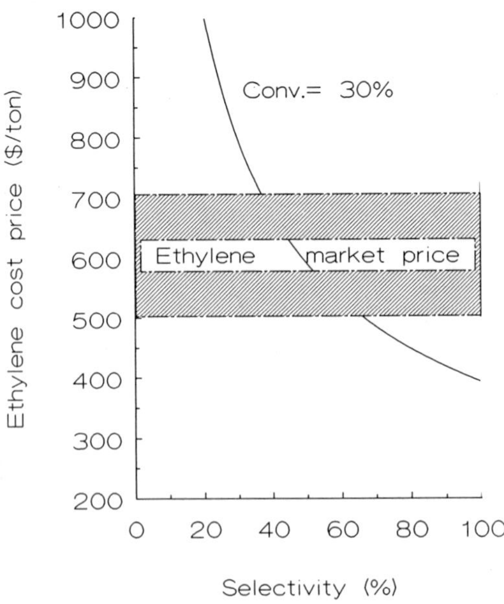

Fig.16 Relation between ethylene cost price and C_{2+}-selectivity in the coupling reactor. Comparison to market prices.

efficiency of 30% and sold at a price of 0.05 US $/kWh. The catalyst costs were assumed to be 500,000 US $/year, but they are of minor importance.

The estimations of the manufacturing costs for all studied situations are listed in Table 11, from which it follows that both case A and case B are more profitable than cracking, case A being the most advantageous.

Figure 12 shows how the price of the ethylene produced is composed. It is clear that the price mainly depends on the cost of the raw materials and to a smaller extent on the revenues from selling electricity and fuel gas.

The influences of some factors on the manufacturing expenses were investigated. In Fig. 13, the price of oxygen varies, while other costs remain constant. The open circle at a zero price for oxygen represents air as the oxidator instead of pure oxygen (for case A only). The use of air requires larger equipment; the corresponding higher capital investment was estimated. It follows that the profitability of a plant using air instead of oxygen is higher only when the oxygen price exceeds US $50 /ton.

Figure 14 shows the influence of the natural gas price on the manufacturing expenses. From Fig. 13 and Fig. 14 it can be seen that case A is not as much influenced by the prices of feedstock as case B. This is due to the fact that case B consumes much more oxygen and natural gas in order to produce the same tonnage of ethylene. As shown in Fig. 15 the influence of the electricity price on ethylene manufacturing costs is negligible in case A. In case B, where considerably more electricity is produced, the effect is more pronounced.

Finally, the influence of selectivity on the production price of ethylene has been examined. In Fig. 16 the ethylene manufacturing price is plotted versus selectivity at a constant methane conversion of 30%. It appears that a high selectivity is to be preferred over a high conversion per pass. The effect of a lower conversion per pass on the ethylene price is not very large, because all the unconverted natural gas and ethane are recycled. Increasing costs due to recycling are balanced partly by lower raw material consumption. Figure 16 also shows a band which represents ethylene market price fluctuations during 1988. It can be seen that 30% conversion and 80% selectivity (approximate yield 25%) should result in a profitable process.

VIII. CONCLUSIONS

Ethylene production via methane oxidative coupling appears to be a feasible process.
A fluidised bed seems the most suitable reactor for this process. The process itself is much simpler than naphtha cracking because of a limited number of product components, which allows a rather simple separation section.

As a consequence, the investment cost for a methane coupling plant is significantly lower and the profitability higher than those of naphtha cracking, if catalysts are available which achieve approximately 25 % yield at a C_{2+} selectivity of at least 65%. Presently known catalysts show such performance only initially, so their stability should be improved.

IX. ACKNOWLEDGMENTS

The financial support for this work was provided , in part, by the European Communities under the Non-Nuclear Energy R&D programme in the sub-programme "Hydrocarbons", contract no. EN3C-0038-NL (GDF), and partly by the Netherlands Organization of Scientific Research (NWO). We wish to thank for their invaluable contributions to this work: M. Campman, W. Hoeben, L. van Gruijthuijsen, C. Kooijman, E. Mallens, J. Mornout, M. Sagis, R. Snoeren, D. Swinckels, J.P. Vissers and I. de Witte.

X. REFERENCES

1 B.P., Statistical Review of World Energy, British Petroleum Company PLC, London, June 1988.

2 Ueda W., Thomas J.M., Proceedings 9th ICC, Calgary, 1988, 960.

3 Kasteren J.M.N. van, Geerts J.W.M.H., Wiele K. van der, Proceedings 9th ICC, Calgary, 1988, 830.

4 Geerts J.W.M.H., Chen Q., Kasteren J.M.N. van, Wiele K. van der, Catal. Today, 6, 1990, 519.

5 Lunsford J.H., Ito T.,J. Am. Chem. Soc., 107, 1985, 5062.

6 Lunsford J.H., Driscoll D.J., J.Phys.Chem., 89,1985,4415.

7 Abraham M.M., Chen Y., Phys. Rev. Lett.,37,1976,89.

8 Geerts J.W.M.H., Ph.D-thesis, Eindhoven University of Technology, 1990.

9 Geerts J.W.M.H., Hoebink J.H.B.J., Ethylene from natural gas, internal report, Eindhoven University of Technology, 1989

10 Leva M., Chem.Eng. 56,1949,115.

11 May W.G., Chem. Eng. Progr., 55 (12), 1959, 19.

12 Davidson J.F., Clift R., Harrison D., Fluidization, Academic Press, London, 1985, 614.

13 Miller C.A., Chem. Eng., Sept 1965, 226.

14 Hall R.S., Matley J., McNaughton J., Chem. Eng., April 1982, 82.

15 Walas S., Chem. Eng., March 1987, 75.

16 Chemical Engineering, Indicators, Jan.1989.

17 Ulrich G.D., A guide to chemical engineering process design and economics, 1984, J. Wiley and Sons.

18 Chemical Market Reporter, 1988, May - October.

12

THERMAL COUPLING OF METHANE

Paul BROUTIN, Christian BUSSON, Jérome WEILL

IFP CEDI,
BP. N° 3, F-69390 VERNAISON FRANCE

François BILLAUD

ENSIC, DCPR URA 328 (CNRS)
BP. N° 451, F-54001 NANCY Cedex FRANCE

I. INTRODUCTION

Since the 1978 energy crisis, which brought about an increase in crude-oil prices as well as a temporary shortage of fossil energy, a huge effort has been devoted to opening up other channels for the use of natural gas as a substitute for transportation fuels and as a source of petrochemical building blocks. This effort has been enhanced by recent estimates showing that world reserves of natural gas are now comparable to those of crude oil.

On this topic the Institut Français du Pétrole (IFP) has developed, either alone or in collaboration, many original approaches to both direct and indirect conversions or else is still working on them. For example, mention can be made of:

- CO, H_2 production by partial combustion (1) or partial oxidation (2)
- methanol synthesis and C_2 + alcohol synthesis (3)
- methanol conversion to propene (4)
- oxidative coupling (5)
- thermal cracking.

This last point is the subject of this paper.

The new approach we have developed is the combination of two reflections: (a) the result of research by IFP on steam cracking, and (b) the possibility of reconsidering methane pyrolysis to produce higher hydrocarbons as a function of the appearance of new technologies.

II. MAIN RESULTS ON STEAM CRACKING WITH IFP NEW TECHNOLOGY

First of all, it is well known that hydrocarbon cracking reactions are highly endothermic reactions for which the limiting factor is generally heat transfer, *i.e.* how to bring in the energy required for this cracking which, moreover, may have to be done at a very high temperature. For steam cracking, which is done industrially and in which naphtha or ethane is cracked at a temperature of about 900° C, IFP has recently made important improvements by using new reactor designs inspired by heat exchangers. These new reactors are built with new materials such as ceramic materials, so that original reactor designs can be used, with which higher cracking temperatures can be reached so that the olefin yield can thus be increased.

Indeed, for testing these new technological approaches in the case of steam cracking, a pilot plant with a production

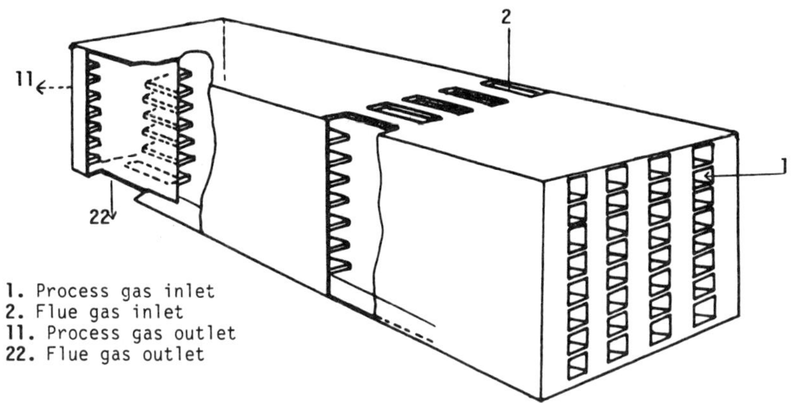

Figure 1

Drawing of the IFP reactor for thermal cracking heated by combustion gas

capacity of 5 kg/h of naphtha was built at the Centre d'Etude et de Développement Industriel (CEDI) at Solaize, near Lyon. The idea was to use a heat-exchanger technology of the compact gas-gas type by using combustion fumes as a way of providing heat (6).

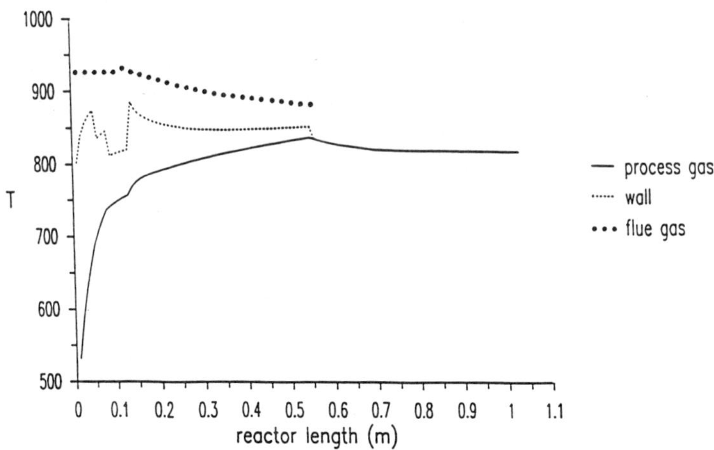

Figure 2

Temperature (°C) profiles

A diagram of the reactor is given in Figure 1. The general principle of this reactor is based on a honeycomb type of ceramic heat exchanger (7). Silicon carbide was chosen because of its high conductivity and its resistance to thermal shocks.

With such a technology, as indicated in Figure 2, a suitable temperature profile was obtained along with a perfectly controlled residence time.

As shown in Table A, excellent results were obtained in the steam cracking of naphtha and ethane with a sharp increase in

Table A ***IFP results in the steam cracking of naphtha with the new ceramic reactor (8)***

Comparison of steam cracking yields obtained with the same naphtha in CERAMIC HONEYCOMB REACTOR (exp. results) and in industrial coils (simulation results)

		HONEYCOMB REACTOR	*Industrial Coils*		
			single	*millisecond*	*split*
Residence time	*ms*	*175*	*480*	*90*	*350*
Outlet temperature	*°C*	*890*	*855*	*935*	*845*
Yields	*%*				
Fuel gas		*16.1*	*16.10*	*13.0*	*15.0*
C_2-cut		*39.3*	*33.1*	*32.6*	*31.1*
C_3-cut		*15.6*	*14.0*	*16.6*	*15.5*
C_4-cut		*9*	*7.2*	*13.6*	*8.8*
Gasoline		*17.4*	*27.1*	*21.8*	*27.2*
Fuel oil		*2.6*	*2.5*	*2.4*	*2.4*

conversion, so that the economic cost effectiveness was very good. This research as a whole is described in Reference (8).

At the same time as obtaining important results in steam cracking, we felt it to be important to see whether methane pyrolysis might again be an interesting approach to the direct chemical conversion of methane if it ever becomes possible to find new reactor-design technologies, considering that the challenge would be more difficult because of the cracking temperature of methane.

Indeed, as was demonstrated in the early 1930s, the greater stability of methane requires cracking to be performed at a higher temperature than in steam cracking (for a review of methane pyrolysis, see (9)). To date, the available technology has not allowed the industrial application of this reaction as is done in steam cracking, and only other routes have been tried out such as going via combustion as is done in the BASF process, or via pyrolysis in an electric arc as is done in the Huls process. Each of these processes has disadvantages such as CO, H_2 production by the BASF process or energy consumption for the Huls processs (for a complete description of these processes, see reference (10)).

This is why IFP decided in 1987 to investigate thermal pyrolysis in a micropilot plant to see if this reaction could be suitable for the chemical conversion of methane.

III. PRELIMINARY RESEARCH ON A MICROPILOT

The aim of this project was to perform a quick screening test of some of the most important parameters such as the nature and concentration of the diluent gas, residence time and cracking temperature, and also to find good running conditions to obtain an accurate material balance and a good energy balance so as to be able to make a preliminary economic study.

OPERATING CONDITIONS

A photo of the micropilot plant we used for this research is shown in Figure 3 and a simplified flow-sheet in Figure 4. The flow rate was about 15 l/h. The reactor was an alumina tube (length = 610 mm, inside diameter = 12 mm, outside diameter = 18 mm), which was heated by an electric furnace. The

Figure 3

Photo of micropilot plan for methane pyrolysis

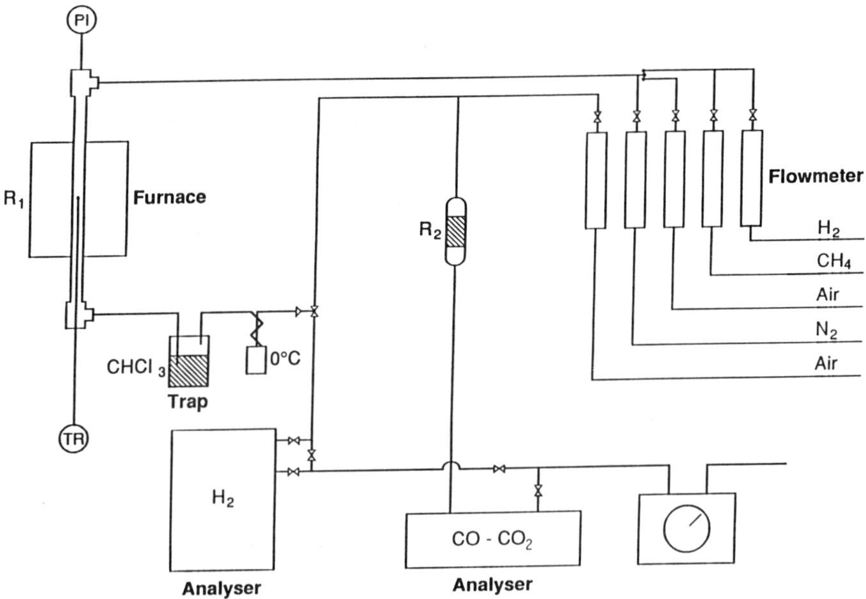

Figure 4

Micropilot: simplified flowsheet

temperature was controlled by a thermocouple (0-1600°C) protected by an alumina tube (6/8) inside the main tube.

For reasons independent of our control, we had to change the furnace during experimenting. These results vary slightly with the nature of the furnace, particularly concerning the pyrolysis temperature. According to the nature of the furnace used, the results will be indicated as "furnace A" or "furnace B." The differences may be explained either by calibration differences of the thermocouples at this temperature level or, as seems more probable, by the differences in heat transfer according to one or the other furnace. It is thus essential to compare the results in a relative way with the same installation.

During the pyrolysis step, an analysis was performed by gas chromatography. Hydrogen was determined online by GC, with an external calibration by a thermal conductivity detector (TCD).

By frequently measuring the amount of hydrogen present, we had a good indication of the stability of the plant. Hydrocarbons were analyzed offline by a GC equipped with a capillary column (HP special PONA; 50 meters). The duration of a run was 2 to 4 hours.

The reactor was decoked by passage of a mixture of air and nitrogen at 1000°C. The carbon monoxide produced during decoking was catalytically converted online to carbon dioxide, which was measured online by an infrared spectrometer.

The material balance was determined by integrating total flow rates and analyses. The difference, taken as the methane equivalent between the initial methane before cracking and the cracked gas and coke, was added as "heavy ends" to the coke ratio.

PRINCIPAL RESULTS

The principal results are shown in Table B. So as to be able to compare the effects of the different operating parameters, we decided to try to work at the same methane conversion of about 30%.

Figure 5 shows the influence of the hydrogen content as a diluent gas causing a sharp decrease in coke formation. In Figure 6, if we compare the effect of hydrogen and helium as dilution gas on coke formation, we can see that the effect of helium is much weaker and is quickly limited. The chemical influence of hydrogen is confirmed in Figure 7 where we can see that, to remain at isoconversion (of course with the same residence time), the pyrolysis temperature has to be increased, whereas this is not necessary with helium.

From this preliminary study, we chose average conditions that seemed to be a good compromise between the chemical

Table B ***MAIN RESULTS: EFFECT OF HYDROGEN ON YIELDS AND SELECTIVITIES - furnace A***

TEMPERATURE °C	*CONV. % DILUANT*	*CONVERSION %*	*SELECTIVITY IN % CH_4 CRACKING*				
			Coke	*Ethylene*	*Acetylene*	*Benzene*	*Naphthalene*
1150	20	28.0	23.1	20.2	24.5	19.2	2.2
1170	0	34.8	43.6	13.1	15.9	15.7	5.3
1160	0	29.6	34.9	15.2	16.8	17.3	7.6
1160	10	27.9	30.6	17.6	20.7	17.7	3.8
1160	20	27.4	23.0	19.3	24.6	18.7	5.0
1175	30	28.8	25.7	22.0	28.7	19.7	2.8
1190	40	29.3	21.0	22.6	29.0	18.1	1.9
1210	50	31.0	18.4	23.4	31.5	14.3	3.7
1160	30	29.7	25.9	20.4	26.2	18.9	1.0
1175	60	24.6	7.0	31.0	41.3	9.3	2.7
1160	20	27.7	25.0	19.4	24.1	20.7	2.0
1250	65	32.4	9.8	28.9	39.1	9.8	1.9
1250	65	33.0	9.5	29.8	39.9	9.7	1.7
1210	60	23.6	10.2	30.7	38.8	9.9	2.4
1190	40	24.9	16.2	22.6	32.3	15.6	5.1

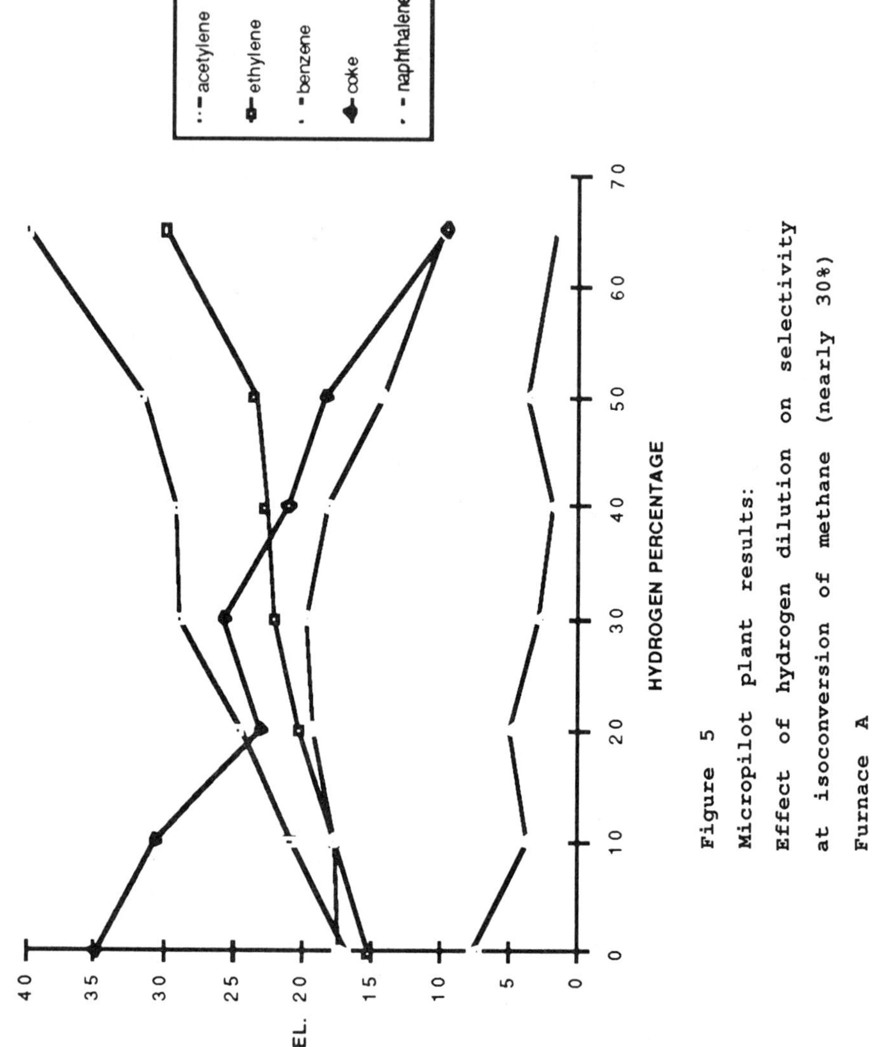

Figure 5
Micropilot plant results:
Effect of hydrogen dilution on selectivity at isoconversion of methane (nearly 30%)
Furnace A

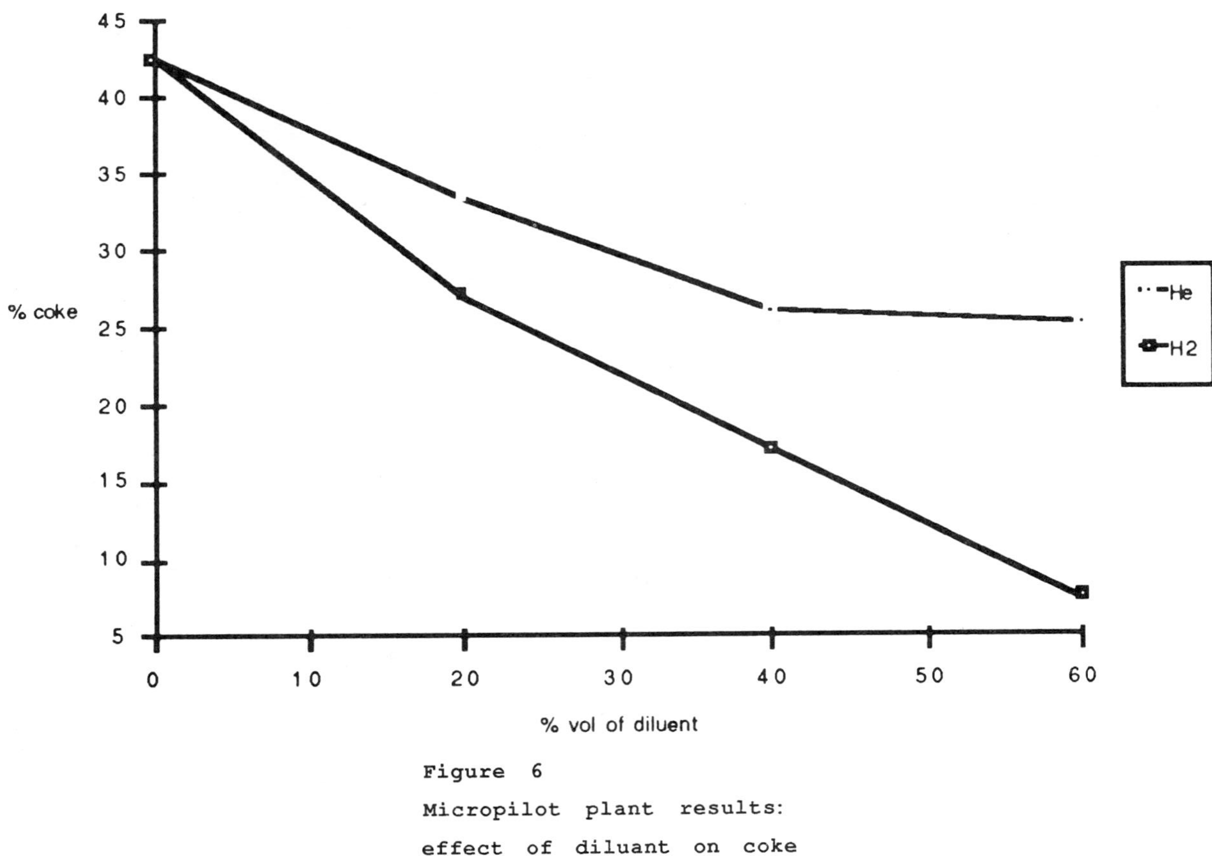

Figure 6
Micropilot plant results:
effect of diluant on coke

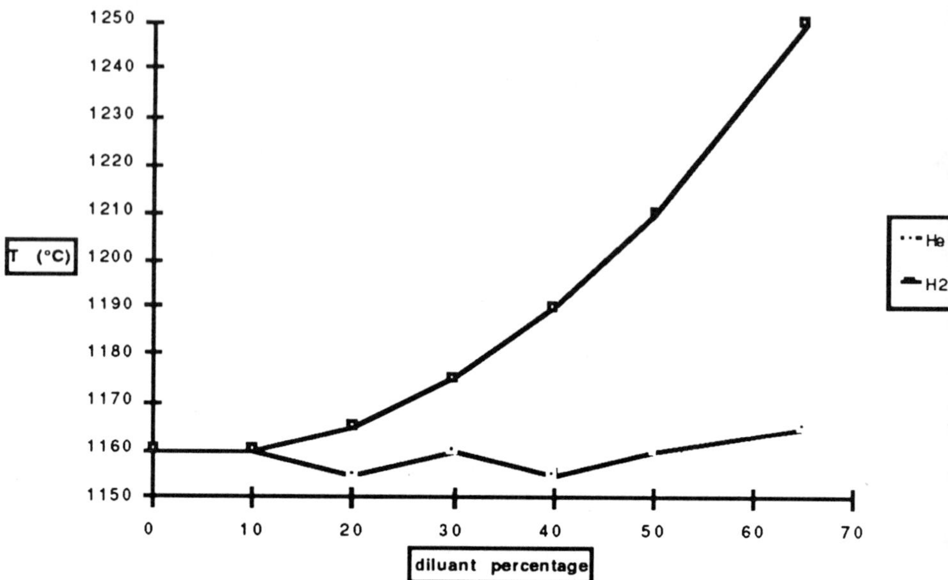

Figure 7 Micropilot plant results:
Effect of hydrogen and helium dilution on the pyrolysis temperature at isoconversion of methane Furnace B

requirements (large amount of hydrogen) and the economic requirements (price of separation, etc.) for the preliminary economic study and as a goal to be reached during process development. These conditions and the accompanying material balance are as follows:

OPERATING CONDITIONS			SELECTIVITY (in % of CH_4 converted)	
Pyrolysis temperature	^oC	*1200*	C_2H_2	*32*
Residence time	*ms*	*200*	C_2H_4	*23*
Diluent gas		*H2*	C_6H_6	*15*
Percent of diluent	*vol %*	*50*	C_3-C_5	*4*
Conversion	*%*	*31*	C_7-C_{12}	*8*
			Heavy ends + coke	*18*

With this material balance, the chemical equation can be summarized as:

$$CH_4 \text{ -------------> } CH_{1.1} + 1.45\ H_2$$

PROPOSAL OF A MECHANISM

The primary formation of ethane and hydrogen in the decomposition of methane at high temperature proceeds through a radical mechanism by stages (or open sequences):

$$CH_4 \longrightarrow CH_3\cdot + H\cdot$$
$$H\cdot + CH_4 \longrightarrow H_2 + CH_3\cdot$$
$$2\ CH_3\cdot \longrightarrow C_2H_6$$

This mechanism accounting for the primary formation of ethane and hydrogen is compatible with the following stoichiometric equation:

$$2\ CH_4 = H_2 + C_2H_6$$

Secondary reactions of ethane by dehydrogenation via a radical chain mechanism accounts for the formation of secondary hydrogen and ethylene. Such a mechanism involves propagation steps which can be divided into:

H· transfer reactions (or metathesis) that are bimolecular reactions in which H atoms are abstracted from a molecule (ethane or methane) by a radical yielding another one and transforming the abstracting radical into the corresponding molecule)

$$CH_3\cdot + C_2H_6 \longrightarrow C_2H_5\cdot + CH_4$$
$$H\cdot + CH_4 \longrightarrow CH_3\cdot + H_2$$

Unimolecular radical b elimination reactions by carbone-hydrogen bond breaking to yield an H atom and an olefin:

$$C_2H_5\cdot \longrightarrow H\cdot + C_2H_4$$

This mechanism is compatible with the following stoichiometric equation:

$$C_2H_6 = H_2 + C_2H_4$$

The chain radical secondary reactions of ethylene by dehydrogenation mechanism ($C_2H_4 = H_2 + C_2H_2$) or by methylation of ethylene ($CH_4 + C_2H_4 = H_2 + C_3H_6$) explain the formation of acetylene and propene.

The methylation of acetylene yields propyne:

$$CH_4 + C_2H_2 = H_2 + C_3H_4$$

Propadiene (C_3H_4, $CH_2=C=CH_2$) and 1-butene (C_4H_8, $CH_3-CH_2-CH=CH_2$) can be interpreted by the dehydrogenation or methylation of propene:

$$C_3H_6 = C_3H_4 + H_2$$
$$CH_4 + C_3H_6 = C_4H_8 + H_2$$

It seems that there is a general agreement to explain the formation of C_1-C_4 hydrocarbons by the above mentioned mechanisms. On the contrary, the formation of liquid hydrocarbons and coke is not completely understood and still open to controversy.

This fundamental research is continuing at present, in particular to finish making a complete parametric study of the same installation and to be able to determine kinetic data.

Table C COMPARISON OF IFP RESULTS IN METHANE PYROLYSIS WITH RESULTS FROM THE LITERATURE (11)

		IFP	TOKYO UNIVERSITY	TRONDHEIM UNIVERSITY
Reactor type		Alumine tube heated by radiation	Quartz tube heated by radiation	Graphite tube heated by Joule effect
Residence time	ms	200	41	15
Temperature	°C	1 200	1 330	1 500
H_2 / CH_4	vol.	1	2.8	1
Conversion	%	30	63	40
Coke selectivity	% mol	18	20	35
Acetylene/ethylene		1.3	8	12.5

However, the results described here confirm the importance of hydrogen as a moderating agent of coke, so that average operating conditions could be determined for the development of the process, and this is coherent with results obtained previously (11) as shown in Table C.

IV. PROCESS DEVELOPMENT

As we have already said, hydrocarbon cracking is a very endothermic reaction. The enthalpy of this reaction can be estimated to be 30 kcal per mole of cracked methane.

The development problem raised was (a) to find a design that could be scaled up while respecting short residence times, and (b) to chose a way of providing energy.

For technological reasons, we felt in the beginning that going via the technology used during steam cracking at a temperature of 950 to 1200° C was an important jump, and we decided to focus our efforts in the beginning on achieving a reactor design using electrical energy, and we chose a configuration of the shell-and-tube heat exchanger type (12).

PYROLYSIS BY ELECTRIC FURNACE

The possibility of using electric power was considered for several reasons:

heating by the Joule effect is quantitatively efficient;

the technological advance to be made seems to be less difficult since high-temperature electric furnaces exist; the problem is to adapt them to the petrochemical industry;

depending on where installations are located, they can either use electrity from the grid or produce it on the spot from the hydrogen produced during methane pyrolysis.

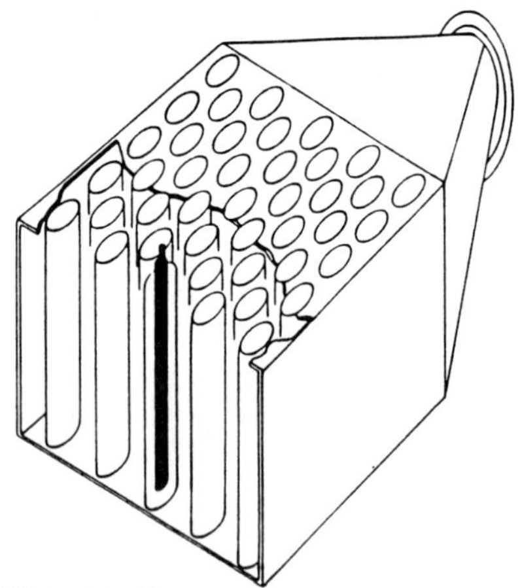

Figure 8

General principle of IFP electric furnace for methane pyrolysis

The general principle of the technology chosen is shown in Figure 8. It is a shell-and-tube type heat exchanger in which the hydrogen/methane mixture circulates between the tubes. High-temperature electric heating elements made of silicium carbide are protected by silicium-carbide sheaths. An inert gas circulates inside the sheaths.

On this principle, we built a pilot furnace with a production capacity of 10 m^3/h. This consists of an alignment of 21 heating units (tube + heating element) so that the gas could be heated from 500 to 1200° C, and then pyrolysis could be done at 1200° C. A photo of the installation is given in Figure 9.

During operations since the middle of 1989, the following main technical points were resolved:

checking and installation of high-temperature seals;

in the presence of inert gas, checking the thermal performances, checking heat transfers, and determining several thermal heating profiles;

Figure 9
Photo of IFP pilot plant with the electric furnace

for the natural gas, resolving several technological problems linked to the nature of the environment at high temperature (*e.g.* the effect of hydrogen on the control thermocouples).

Since January 1990, a great many cracking tests of natural gas in the presence of hydrogen have been performed, and the results obtained in the micropilot plant were reduplicated.

With these results and by applying an optimized process flowsheet, particularly concerning the separations (13), economic estimates show that ethylene (or acetylene) can be produced at costs that are comparable to those of naptha steam cracking, with the production cost varying with the installation and the use of the hydrogen produced (for a complete study, see Reference (14)).

V. CONCLUSION

At the same time as fundamental research on understanding the mechanism of methane pyrolysis, IFP has developed a technology of an electric furnace that can be scaled up, and it has already been possible to obtain interesting conversions and yields. With these results, if we can resolve all the technological problems inherent in a totally inovative approach, it will be able to start with natural gas and produce higher hydrocarbons such as ethylene or acetylene at competitive costs.

REFERENCES

1. Gateau P., Maute M., Feugier A. and Perthuis E., Fr. Patent 2 608 581 (1986)
   Quanq D.V., Gateau P., Feugier A. and Poussin B. , Fr. Patent 2 617 577 (1987)

2. Busson. C, and Alagy J., Fr. Patent Appl. 2 588 773 (1985)
Alagy J., Broutin P., Busson C., Gougne Y., and Weill J., Fr. Patent Appl. 89/12016 (1989)
3. Courty P., Durand D., Freund E. and Sugier A., J. of molecular catalysis, 17: 241(1982)
Courty P., Chaumette P., Raimbault C. and Travers P., European Applied Research Conference on Natural Gas, Trondheim, 28 - 30 Mai (1990)
4. Martino G. and Juguin B. , Fr. Patent 2 577 549, (1985)
Juguin B., Hugues F. and Hamon C., 199 th A.C.S national meeting BOSTON (USA) 22 - 27 April (1990)
5. Cameron C.J., Mimoun H., Robine A., Bonnaudet S., Chaumette P. and Quanq D.V., Fr. Patent Appl. 88/04588, (1988)
Cameron C.J., Quanq D.V., Lepage J.F., and Mimoun H., Fr. Patent Appl. 88/11312, (1988)
Cameron C.J., Mimoun H., Robine A., Bonnaudet S., Chaumette P. and Quanq D.V., Fr. Patent Appl. 89/00 188,(1989)
Mimoun H., Robine A., Bonnaudet S. and Cameron C.J., Appl. Catal. 58:269 (1990)
6. Alagy J., Busson C., and Chaverot P., Fr. Patent 2 584 733, (1985)
7. Minjolle L. (1978) Fr. Patent 2 436 958,(1978)
8. Broutin P., Busson C., Weill J. , Heynderickx G. and Froment G., Spring National AIChE Meeting, 19-23 March, Orlando, USA (1990)
9. Billaud F., Baronnet F., Freund E., Busson C. and Weill J., Rev. Inst. Franç. du Pétrole 44 :813 (1989)
10. Chauvel A. and Lefebvre G., Petrochemical processes. volume 1. TECHNIP ed Paris (1989)
11. Holmen A., Rokstad O.A. and Solbakken A. Ind. Eng. Chem, Process Des. Dev. 15 :439 (1976)
Kunugi T., and Tamura T., Chem. Eng. Progress 57 :43 (1961)

12. Alagy J. , Busson C., Broutin P., and Weill J., European Patent Appl. 323287, 5 July 1989
13. Juguin G., Collin J.C., Larue J. and Busson C., Fr. Patent Appl. 89/ 15767, (1989)
14. Busson C., Weill J. and Raimbault C., European Applied Research Conference on Natural Gas, Trondheim, 28 - 30 Mai (1990)

13

ETHYLENE, ACETYLENE, AND BENZENE FROM METHANE PYROLYSIS

O.A. Rokstad, O. Olsvik, B. Jenssen and A. Holmen

Dept. of Industrial Chemistry, Norwegian Institute of Technology and SINTEF, Applied Chemistry, N-7034 Trondheim, Norway.

I. SUMMARY

Equilibrium calculations indicate that pyrolysis of methane at high temperatures may give ethylene, acetylene, benzene and hydrogen as the main products if carbon formation is excluded. Kinetic investigations in a tubular flow reactor in the temperature range 1000–1200°C at 1 bar pressure show that these components are in fact the main gas products formed from methane. Ethane was observed as a product only at very short reaction times. To avoid excessive carbon formation the residence time was usually less than one second and the methane feed was diluted with helium or hydrogen. Hydrogen was found to be especially effective in suppressing carbon formation and a high selectivity for C_2 products was obtained. The ratio of acetylene to ethylene increased with increasing temperature. Formation of the products is explained by a free radical mechanism.

II. INTRODUCTION

Methane is the main component of natural gas and there is much interest in finding ways to convert methane directly to valuable chemicals. Thermal coupling of methane to ethylene and acetylene has been investigated in many laboratories and several review articles have been published (Refs. 1,2,3).

Thermal coupling of methane may be described as a stepwise dehydrogenation at high temperature:

$$2\ CH_4 \rightarrow \underset{+H_2}{C_2H_6} \rightarrow \underset{+H_2}{C_2H_4} \rightarrow \underset{+H_2}{C_2H_2} \rightarrow \underset{+H_2}{2C} \qquad (1)$$

The hydrocarbons are thermodynamically unstable at high temperature and the only products would be carbon and hydrogen if the reaction time is long enough. Time is therefore an important variable to obtain the products ethylene and acetylene from methane.

III. THERMODYNAMICS

Equilibrium calculations indicate that pyrolysis of methane at high temperatures may give ethylene, acetylene, benzene and hydrogen as the main products if we are able to stop the reaction before carbon is formed. The yield of these products depends on the temperature.

Simultaneous gas equilibria were calculated on a UNIVAC 1108 computer using thermodynamic data published by TRC at Texas A&M University (Ref. 4). Calculated gas equilibrium distribution of carbon atoms as a function of the temperature is shown in Fig. 1. The equilibrium yield of ethylene is seen to be less than 5% over the whole temperature range. The equilibrium yield of acetylene is also small below 1100°C, but increases strongly with increasing temperature. The curve for benzene has a broad maximum above 60% between 1100 and 1220°C.

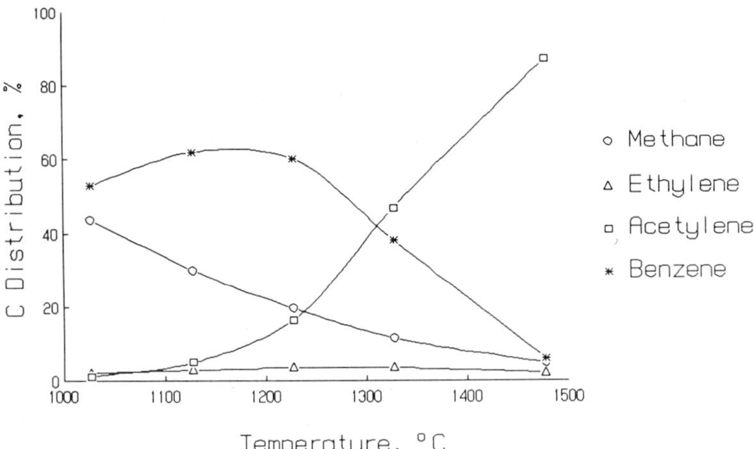

FIG. 1 *Calculated gas equilibrium distribution of carbon atoms from methane. Carbon was not allowed to form. Pressure 1 bar.*

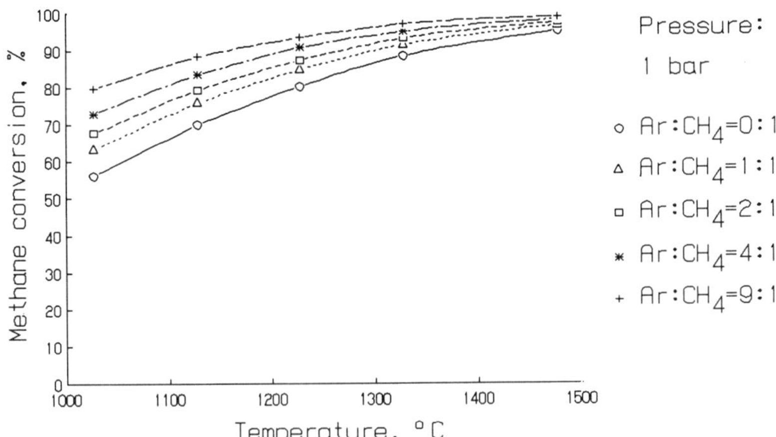

FIG. 2 *Calculated gas equilibrium conversion of methane as a function of temperature at different Ar : CH_4 ratios in the feed.*

Reducing hydrocarbon partial pressure by dilution with argon increases the equlibrium conversion of methane (Fig. 2) and increases the equilibrium yield of the gas products. If the methane feed is diluted with hydrogen the effect on the equilibrium conversion of methane is much stronger and opposite that of argon as shown in Fig. 3.

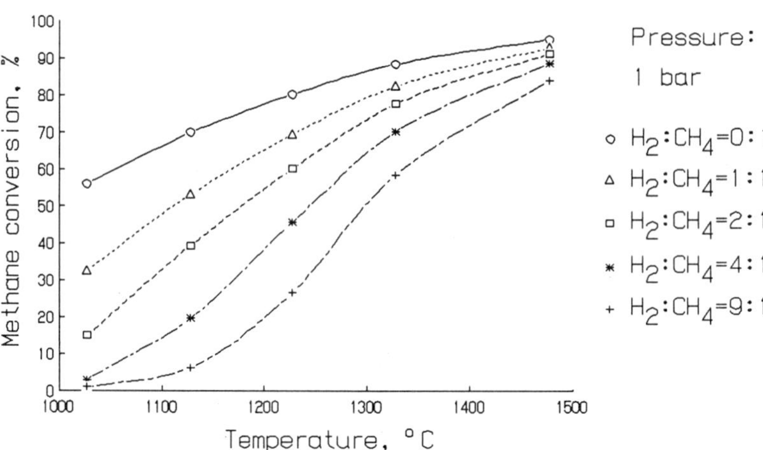

FIG. 3 *Calculated gas equilibrium conversion of methane as a function of temperature at different H_2 : CH_4 ratios in the feed.*

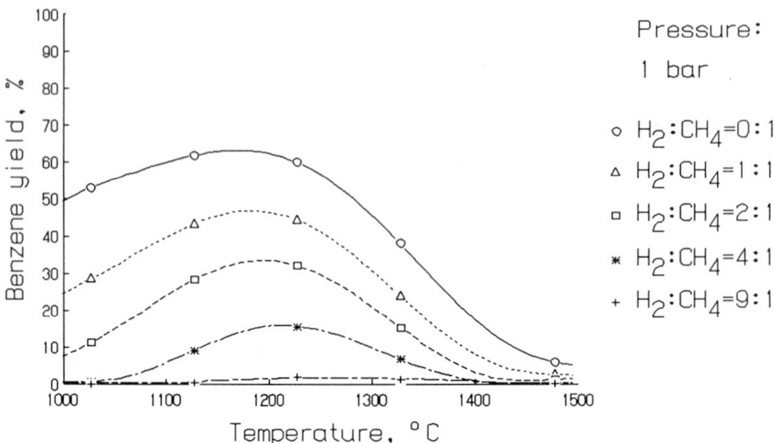

FIG. 4 *Calculated gas equilibrium yield of benzene from methane as a function of temperature at different H_2 : CH_4 ratios in the feed.*

The effect of hydrogen dilution on the equilibrium yield of benzene is very strong as shown in Fig. 4. The effect of hydrogen on ethylene and acetylene equilibrium yield is not so strong (Figs. 5 and 6).

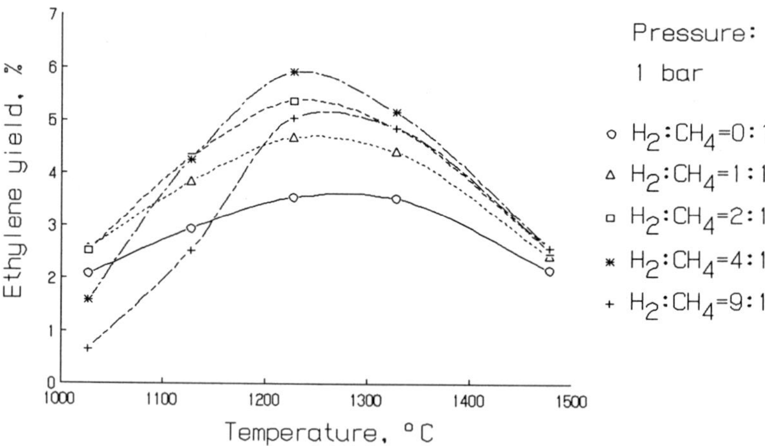

FIG. 5 *Calculated gas equilibrium yield of ethylene from methane as a function of temperature at different H_2 : CH_4 ratios in the feed.*

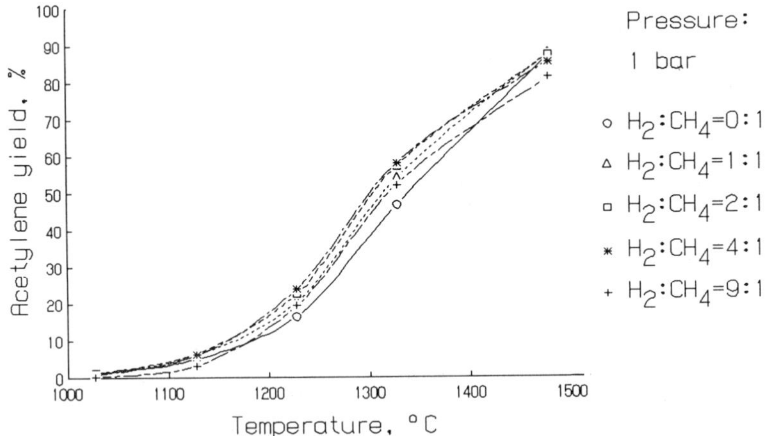

FIG. 6 *Calculated gas equilibrium yield of acetylene from methane as a function of temperature at different H_2 : CH_4 ratios in the feed.*

IV. EXPERIMENTAL

A schematic diagram of the experimental apparatus is shown in Fig. 7. A methane mixture (93% methane and 7% argon) and a diluent (helium or hydrogen) were metered from storage cylinders into a continuous flow reactor. The reactor tube was made of alumina (Alsint). The reactor tube was 1000 mm long with an inner diameter of 9 mm. Methane (99.9% minimum purity), hydrogen (99.7%), helium (99.99%) and argon (99.99%) were obtained from Norsk Hydro. The reactor was electrically heated by Kanthal elements. Power input to the furnace was controlled manually with a variable transformer connected to a 220 V supply. The furnace allowed experiments to be made at temperatures up to about 1250°C. The temperature inside and outside the reactor was measured with platinum–rhodium thermocouples.

The product gas was rapidly quenched at the outlet of the reactor. Indirect water cooling with a cold–finger was used. The cold–finger system consists of an additional cooler placed inside the other, thus forming a narrow annulus. The distance between the reactor outlet and the top of the cold finger (conical shape) was about 5 mm.

The quenched product gas passed through a filter where solid particles and liquids were removed. Manometers (Bourdon, XM–801) were installed at the inlet and at the outlet of the reactor. The reaction pressure was close to one atm. The gas flow through the reactor was regulated by mass flow meters/controllers (HI–TEC).

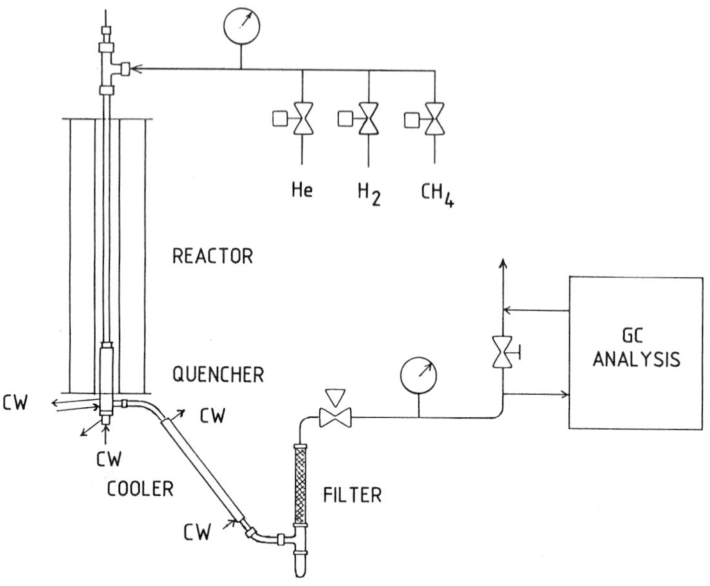

FIG. 7 *Experimental apparatus.*

On-line gas analysis were carried out using a Hewlett-Packard gas chromatograph, model 5890 equipped with two detectors and two columns in parallel. Hydrocarbons were separated on a 30 m megabore GS-Q capillary column connected to a flame ionization detector (FID). Argon, methane, ethylene and acetylene were especially separated on a 3 m column (1/8") packed with Carbosive S-2 (100-120 mesh) connected to a thermal conductivity detector (TCD). The gas chromatograph was connected to two Hewlett-Packard integrators, model 3396A. The gas chromotograph was temperature programmed at 40°C for 7 min., then increasing 32°C/min up to 225°C.

V. RESULTS AND DISCUSSION

Data for methane conversion and product yields were obtained from the gas chromotographic analysis. The effect of varying residence time, temperature and feed composition at atmospheric pressure was investigated.

The main products were ethylene, acetylene, benzene, hydrogen and carbon. Minor products observed in the product gas were ethane, propane, propylene, propyne, propadiene, butenes, butadiene and toluene. To avoid excessive carbon formation the residence time usually had to be less than one second and the methane feed was diluted with helium or hydrogen.

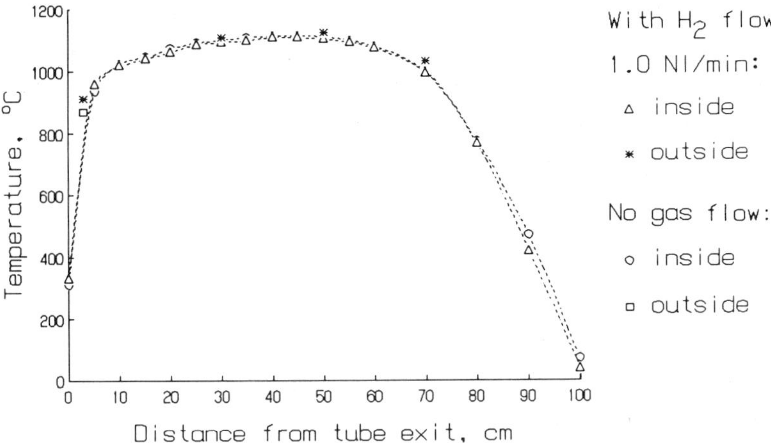

FIG. 8 *Temperature profiles of the reaction tube with and without gas flow.*

At low conversion of methane no carbon formation was observed and the conversion was calculated from the FID analysis data. At higher conversion of methane the conversion was calculated from the TCD analysis data using argon as an internal standard. The carbon formation was estimated from the carbon balance using the combined FID and TCD data. Product selectivities S_i (%) were calculated from product yields U_i (% of methane in the feed) and methane conversion η (%):

$$S_i = \frac{U_i}{\eta} \cdot 100 \tag{2}$$

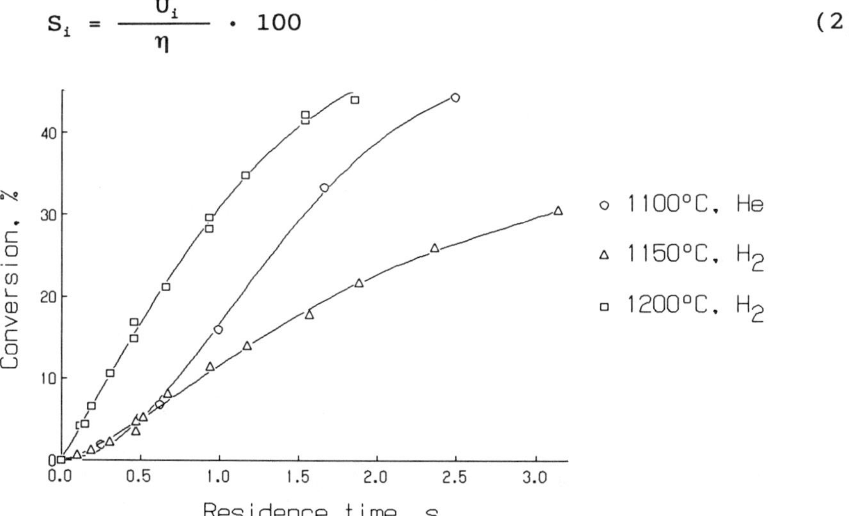

FIG. 9 *Conversion of methane as a function of residence time at different temperatures. Methane was diluted with He or H_2 in the ratio 1:2. Pressure 1 atm.*

Measured temperature profiles of the reaction tube with and without gas flow are shown in Fig. 8. From the temperature profile the reaction zone was defined to be 60 cm long. Outside this zone the temperature falls steeply from about 1000°C, and the conversion outside the reaction zone would be negligible. Within the reaction zone an effective isothermal temperature T_{eff} may be calculated from the temperature profile assuming first order kinetics:

$$- \ln (1-\eta) = \int k dt = A \cdot \exp \left(- \frac{E}{RT_{eff}}\right) \cdot t \qquad (3)$$

where η is the conversion, t is the residence time and k is the first order rate konstant. A and E are the Arrhenius parameters. By summing over a finit number of elements s the logarithmic form of Eq. (3) may be written:

$$\frac{E}{RT_{eff}} = - \ln \left[\frac{1}{s} \cdot \sum_{s=1}^{s} \exp \left(- \frac{E}{RT_s}\right)\right] \qquad (4)$$

The activation energy E was given a value of 85 kcal/mol typical for methane pyrolysis in flow reactors (Refs. 5,6,7,8).

T_{eff} calculated from Eq. (4) was found to be 35–40°C lower than the maximum measured temperature. The true temperature in the flowing gas will be lower than the measured tube wall temperature, but this difference is probably small in this case since the measured temperature profiles with and without gas flow are close thogether (Fig. 8).

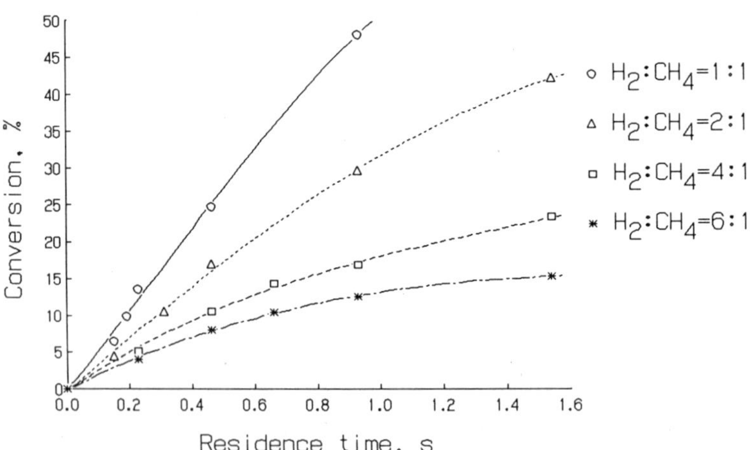

FIG. 10 *Conversion of methane as a function of time at different $H_2 : CH_4$ ratios in the feed. Temperature 1200°C. Pressure 1 atm.*

The conversion, yield and selectivity data (in Figs. 9–15) are correlated with maximum reactor temperature. The residence time τ (s) was calculated from the volume of the reaction zone V (dm^3) and the inlet gas flow F (dm^3s^{-1}) at reaction temperature and pressure:

$$\tau = \frac{V}{F} \tag{5}$$

Thus, the change in gas volume during the reaction was neglected since the feed is diluted and the conversion is in most cases low.

To avoid excessive carbon formation the methane feed was diluted with helium or hydrogen. With helium dilution the conversion curve has a distinct S shape as seen in Fig. 9. Similar behavior has been reported by others (Refs. 7,9,10). This is typical for free radical reactions. With hydrogen dilution the S form is less distinct at 1150°C (Fig. 9), and is not observed at 1200°C, although Holmen et al. (Ref.8) reported S-formed curves with hydrogen dilution at higher temperatures.

From Fig. 9 it is obvious that hydrogen inhibits the methane conversion after an induction period. Figure 10 shows that the methane conversion decreases with increasing hydrogen dilution of the methane feed. A similar effect of hydrogen has been reported by Eisenberg and Bliss (Ref.9). When methane is diluted with hydrogen the kinetics is far from first order, more like second order. With helium dilution the kinetics is closer to first order, but in this case too we observed an increase in conversion when the initial concentration of methane was increased.

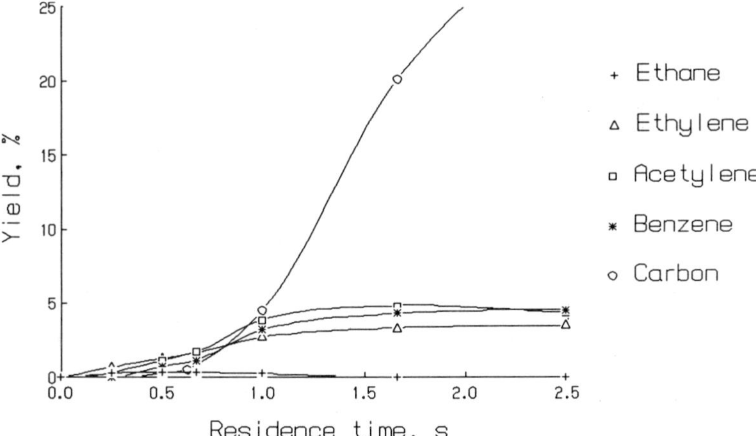

FIG. 11 *Yield of products from methane pyrolysis as a function of time. Temperature 1100°C. Pressure 1 atm. Feed composition He : CH_4 = 2:1.*

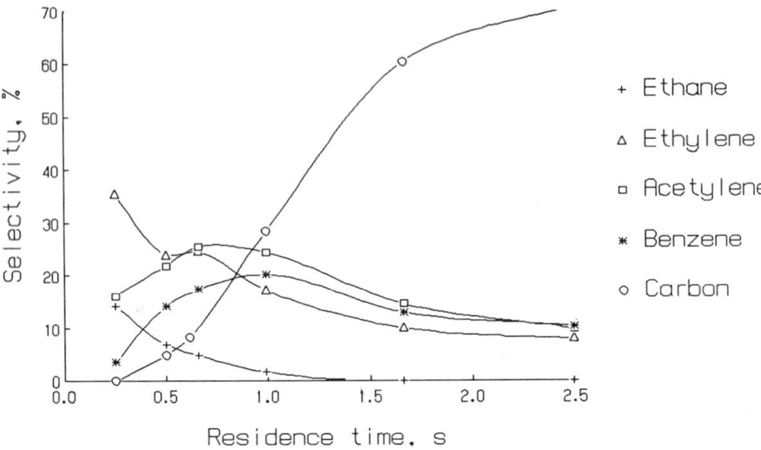

FIG. 12 *Selectivity of products from methane pyrolysis as a function of time. Temperature 1100°C. Pressure 1 atm. Feed composition He : CH_4 = 2:1.*

The time-dependent yields of the principal hydrocarbon products at 1100°C with helium dilution are shown in Fig. 11. Carbon becomes the dominating product after about one second. Shortly after one second the yields of acetylene, ethylene and benzene are leveling out. For acetylene and ethylene the plateaus closely correspond to the calculated gas equilibrium yields (5 and 2.5 % respectively). The plateau for benzene is much lower than the calculated gas equilibrium yield (70 %). We therefore believe that benzene is an important intermediate in the formation of carbon. Ethane production is very small in accordance with the calculated gas equilibrium yield (0.02%).

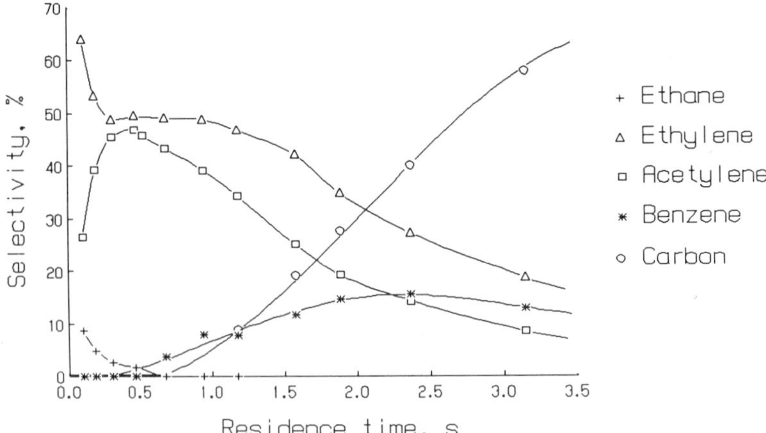

FIG. 13 *Selectivity of products from methane pyrolysis as a function of time. Temperature 1150°C. Pressure 1 atm. Feed composition H_2 : CH_4 = 2:1.*

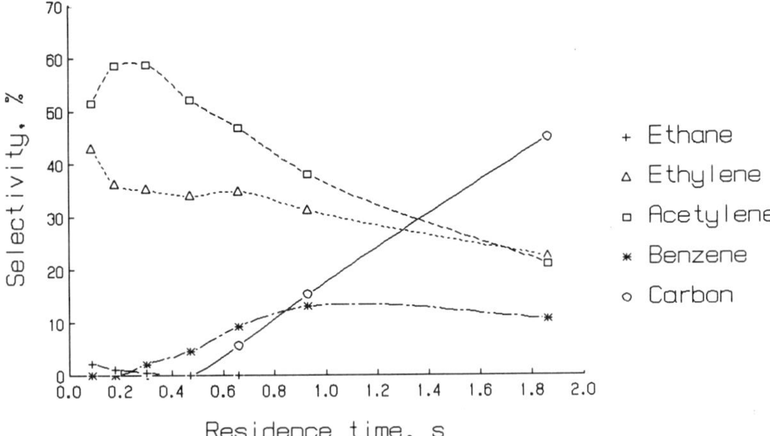

FIG. 14 Selectivity of products from methane pyrolysis as a function of time. Temperature 1200°C. Pressure 1 atm. Feed composition $H_2 : CH_4 = 2:1$.

The selectivity for ethane increases as the residence time becomes very small indicating that it is a primary product. Selectivities for ethylene and acetylene at short residence times are much higher when the methane feed is diluted with hydrogen instead of helium (comparing Figs. 12 and 13). Carbon and benzene formation are delayed by hydrogen. Higher temperatures increases the selectivity of acetylene relative to ethylene (comparing Figs. 13 and 14). Selectives of ethylene, acetylene and benzene as a function of the H_2:CH_4 ratio in the feed at 1200°C and 10% conversion of methane are shown in Fig. 15.

The formation of the gas products from methane pyrolysis is explained by a free radical mechanism (Refs. 3,11). Ethane is formed as a primary product from methane:

$$CH_4 \rightarrow CH_3 + H \qquad \Delta H^{o}_{25} = 104 \text{ kcal/mol} \qquad (6)$$

$$H + CH_4 \rightarrow CH_3 + H_2 \qquad (7)$$

$$2\ CH_3 \rightarrow C_2H_6 \qquad (8)$$

Net reaction:

$$2\ CH_4 \rightarrow C_2H_6 + H_2 \qquad \Delta H^{o}_{25} = 16 \text{ kcal/mol}$$

Reaction (6) has a high activation energy and is rate controlling. A high concentration of hydrogen may depress the methyl radical concentration (reverse of reaction 7), thus, slowing down the rate of ethane production. This may partly explain the inhibition of the methane pyrolysis by hydrogen.

Ethylene is a secondary product formed from ethane by a radical chain reaction:

$$CH_3 + C_2H_6 \rightarrow CH_4 + C_2H_5 \quad (9)$$

$$C_2H_5 \rightarrow C_2H_4 + H \quad (10)$$

$$H + CH_4 \rightarrow CH_3 + H_2 \quad (7)$$

Net reaction:

$$C_2H_6 \rightarrow C_2H_4 + H_2 \qquad \Delta H^0 = 33 \text{ kcal/mol}$$

Acetylene is a secondary product formed from ethylene by a radical chain reaction:

$$CH_3 + C_2H_4 \rightarrow CH_4 + C_2H_3 \quad (11)$$

$$C_2H_3 \rightarrow C_2H_2 + H \quad (12)$$

$$H + CH_4 \rightarrow CH_3 + H_2 \quad (7)$$

Net reaction:

$$C_2H_4 \rightarrow C_2H_2 + H_2 \qquad \Delta H^0 = 42 \text{ kcal/mol}$$

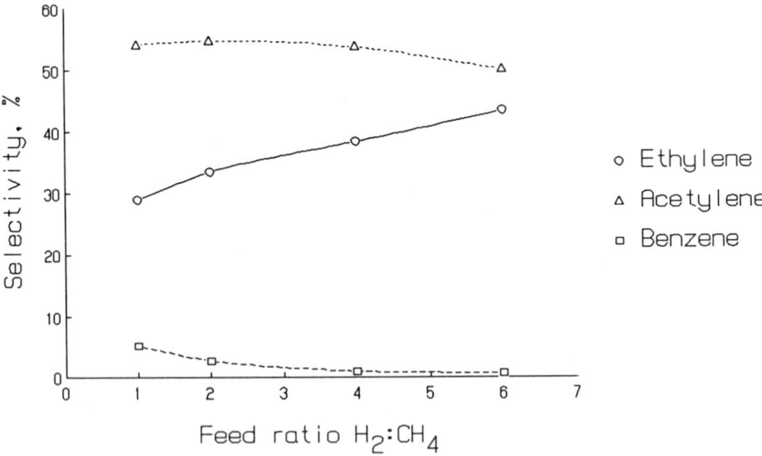

FIG. 15 *Product selectivities as a function of H_2 : CH_4 ratio in the feed. Methane conversion was 10%. Temperature 1200°C. Pressure 1 atm.*

Benzene is probably formed by a trimerization of acetylene:

$$M + C_2H_2 \rightarrow M + C_2H_2^* \quad (13)$$

$$C_2H_2^* + C_2H_2 \rightarrow C_4H_4^* \quad (14)$$

$$C_4H_4^* + C_2H_2 \rightarrow C_6H_6^* \quad (15)$$

$$C_6H_6^* + M \rightarrow C_6H_6 + M \quad (16)$$

Net reaction:

$$3\, C_2H_2 \rightarrow C_6H_6 \qquad \Delta H = -143 \text{ kcal/mol}$$

In reaction (13) acetylene is activated to a biradical by a collision and in reaction (16) the very hot benzene molecule is stabilized by giving away energy to a second molecule. We belive that hot benzene molecules also are intermediates for carbon formation.

VI. CONCLUSION

Ethylene, acetylene and benzene may be obtained as the main gas products from methane pyrolysis at short reaction times in the temperature range 1000–1200^{0}C. The yields of ethylene and acetylene are approaching the calculated gas equilibrium values during the reaction. Observed benzene yield is far less than the calculated gas equilibrium value. Hydrogen has a strong effect on the kinetics of the methane pyrolysis. It is especially effective in depressing carbon formation. With hydrogen dilution of the methane feed more than 90 % selectivity for C_2 hydrocarbons may be obtained at 1200^{0}C and 10 % conversion of methane.

ACKNOWLEDGMENT

The support of this work by the SPUNG program of the Royal Norwegian Council for Scientific and Industrial Research and by Statoil is gratefully acknowledged.

REFERENCES

1. M. S. Khan, B.L. Crynes, *Ind. Eng. Chem. 62(10):* 54 (1970).
2. K. I. Omar, *Chem. Eng. Research Bull. 6:* 29 (1982).
3. M. H. Back and R. A. Back, "Thermal Decomposition and Reactions of Methane". In *"Pyrolysis: Theory and Industrial Practice."* (L.F. Albright, B.L. Crynes, and W. H. Corcoran, eds.) *Academic Press:* 1 (1983).
4. *TRC Thermodynamic Tables. Hydrocarbons. Vol IX. Thermodynamic Research Center, Texas A & M University* (1981–89).

5. I. A. Schneider, *Z. Phys. Chem. 223:* 234 (1963).
6. P. S. Shantarovich and B.V. Pavlov, *Intern. Chem. Eng. 2:* 415 (1962).
7. H. B. Palmer, J. Lahaye, and K. C. Hou, *J. Phys. Chem. 72:* 348 (1968).
8. A. Holmen, O. A. Rokstad and Å. Solbakken, *Ind. Eng. Chem. Process Des. Dev. 15:* 439 (1976).
9. B. Eisenberg, H. Bliss, *Chem. Eng. Progr., Symp. Ser. 63:* 3 (1967).
10. Y.–T. Tu and C.–T. Yeh, *J. Chinese Chem. Soc. 30:* 179 (1980).
11. C.–J. Chen, M. H. Back, and R. A. Back, "Mechanism of the Thermal Decomposition of Methane". In *"Industrial and Laboratory Pyrolysis"* (L.F. Albright and B.L. Crynes, eds.) *ACS Symp. Ser. 32: Am. Chem. Soc., Washington, D.C.:* 1 (1976).

14

OLEFIN PRODUCTION BY CATALYTIC HYDROGENATION OF CARBON MONOXIDE OVER Co/Al_2O_3 AND HX MIXED CATALYST

Wei-Jye Wang and Yu-Wen Chen*

Department of Chemical Engineering
National Central University
Chung-Li, Taiwan, R.O.C., 32054

ABSTRACT

The synthesis of hydrocarbons from catalytic hydrogenation of carbon monoxide was investigated over a Co/Al_2O_3 and HX mixed catalyst at 1 atm, 493-553K, $H_2/CO=1$, and WHSV=1200. This combination resulted in the formation of lower olefins in amounts as high as 42 wt% under our reaction conditions. Two catalysts, Co/Al_2O_3 and Co/HX, were also included for comparison. The results indicated that the Co/Al_2O_3 is much more active than the Co/HX catalyst. This is attributed to the structure-sensitive nature of the reaction. Addition of HX in Co/Al_2O_3 did not influence its activity. However, it produced changes in secondary reactions of primary olefin products and resulted in a high olefin yield. In addition, 2-butene is the major product in C_4 component,

*To whom all correspondence should be addressed.

indicating that the predominate secondary reaction is isomerization rather than oligomerization and cracking reactions in this composite catalyst system.

INTRODUCTION

An important goal in recent research on hydrocarbon synthesis by the catalytic hydrogenation of carbon monoxide has been the selectivity control [1]. Unsaturated hydrocarbons, especially those with two to four carbon atoms, constitute a large proportion of the feedstocks for the petrochemical industry. To date, the main problem has been the Fischer-Tropsch(FT) catalysts which have high selectivity to lower olefins do not give high activity [2-5].

The development of highly selective cobalt/zeolite catalysts for the FT reaction has been studied extensively. Nazar et al.[6] reported that the main products are C_4 olefins on finely dispersed cobalt metal in NaY zeolite. Cobalt metal in A zeolite yielded propylene as the sole product under specific conditions [7]. Taktchenko et al.[8-9] showed that small metallic aggregates in NaY zeolite are prerequisite to inducing a hydrocarbon chain length limitation. Lee and Ihm [10-11] reported that the Co/NaY catalysts prepared by carbonyl complex impregnation has a higher selectivity to olefins than those prepared by ion-exchange method. Unfortunately all of these catalysts showed a low activity compared to Co/Al_2O_3 catalyst.

Modification of FT catalysts by mixing with zeolites is another means of enhancing the selectivity with respect to the desired products [2-15]. However, most of the studies were on ZSM-5 system to produce aromatic products. Nevertheless, this provided the motivation for an investigation of the composite catalyst systems to achieve the desired olefin yield.

For the preparation of Co/HX catalyst, $Co(NO_3)_2 \cdot 6H_2O$ was dissolved in a weakly acidic hydrochloride solution (pH=4.5) of 0.04 N. NH_4X zeolite was then mixed with the solution by stirring continuously for 48 h. After ion exchange, the catalyst was filtered and washed with deionized water and dried in air at 373 K for 50 h.

In the present study, an investigation was carried out on the FT synthesis over a mixed catalyst consisting of Co/Al_2O_3 and HX zeolite. HX zeolite has been known to be active for the isomerization of 1-olefins. Two catalysts, Co/Al_2O_3 and Co/HX have also been included for comparison. The main idea is that the HX zeolite would provide a modification of the normal product distribution of the Co/Al_2O_3 catalyst.

EXPERIMENTAL

I. *Catalyst Preparation*

The Co/Al_2O_3 catalyst was prepared by incipient-wetness impregnation of 3 wt% cobalt on a r-Al_2O_3 support (Strem Chemicals) with an aqueous solution of $Co(NO_3)_2$. After impregnation, the sample was dried in air at 373 K for 50 h.

The composite catalyst was prepared by grinding together Co/Al_2O_3 catalyst and HX zeolite in a volumetric ratio of 1:1 and screening to 80-100 mesh. The NH_4X zeolite was obtained from NH_4-ion exchanged with NaX (Strem Chemicals). The X-ray diffraction measurements showed that the Ru/HY sample was highly crystalline.

After preparation all catalysts were reduced with hydrogen by heating to 673 K at 5 K/min and maintaining the temperature for 5 h. The X-ray line broadening analysis indicated that the cobalt particle size in all samples is less than 4 nm.

II. *FT Reaction*

A study of the catalyst activity and selectivity was conducted under typical FT reaction conditions, viz., 1 atm, 493 - 533K, GHSV=1200, and $H_2/CO=1$. All the reactant gases were of high purity (99.9%, Matheson) and were further purified by passing them through Drierite and molecular seive traps. The product distribution was determined by using an on-line gas chromatograph (HP 5890A) with a Porapak QS column. However, the C_4 components were not well separated by this column. Therefore, a Carbopack C/0.19% Picric acid column was used. The CO conversion was kept below 10% to prevent heat and mass transfer effects.

RESULTS AND DISCUSSION

I. *Activity*

The activities and product distributions presented in this paper represent the steady-state performance of the catalysts system and refer to a time frame of about 1-3 h after the syn-gas was introduced into the reactor.

Table 1 lists the activities of all the catalysts at 533 K. Fu and Bartholomew [16] reported that the FT synthesis over Co/Al_2O_3 is a structure-sensitive reaction. The activity is decreased with the decrease in

Table 1. *Activity and Product Distribution of the Catalysts**.

Catalyst	Co/Al_2O_3	Co/HX	Co/Al_2O_3+HX
$-r_{CO}$(μmole/g.sec)	$79.6X10^{-3}$	$6.21X10^{-3}$	$81.3X10^{-3}$
selectivity(wt%)			
C_1	71.52	48.68	42.4
C_2	9.39	12.69	17.4
C_3	10.19	19.54	21.1
C_4	5.26	13.52	16.2
C_{5+}	2.20	5.41	2.9
$C_2^=$/total C_2 (%)	13.0	61.0	63.0
$C_3^=$/total C_3 (%)	65.0	92.0	95.0
$C_4^=$/total C_4 (%)	38.0	79.0	86.0
% olefin in C_2-C_4	39.7	79.7	82.1
r_{olefin} (μmol/g.sec)	$7.85X10^{-3}$	$2.27X10^{-3}$	$36.5X10^{-3}$

**reaction condition: 1 atm, H_2/CO = 1, 260^0C, GHSV= 1500.*

cobalt crystallite size. In this study, the activity of Co/Al_2O_3 is an order higher than that of Co/HY due apparently to the structure-sensitive nature. The size of cobalt metal on the Co/Al_2O_3 catalyst is believed somewhat larger than that of the Co/HX catalyst.

Several workers [17-20] have observed a trend of increasing acidity for CO hydrogenation with increasing acidity of the support. The result was attributed to an increase in the surface concentration of less strongly bound carbon monoxide species, resulting from an electron-deficient character of the supported metal which increased with the acidity of the support. A

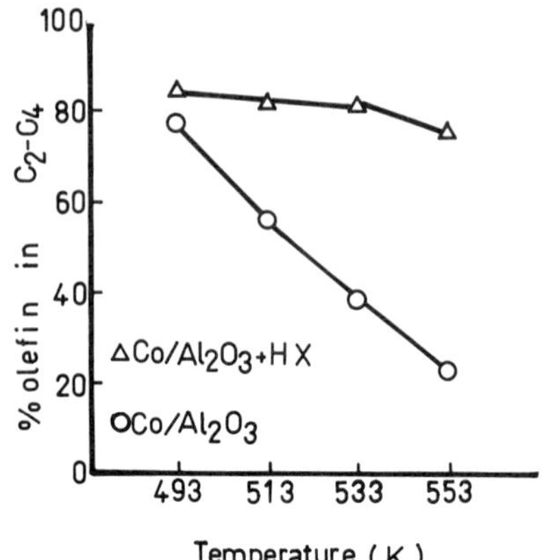

Figure 2. *Effect of Temperature on the Selectivity for Olefins*

significantly change activity. However, it produced changes in secondary reactions of primary olefinic products. The composite catalyst has a high olefin selectivity and high activity of CO conversion. Therefore, it has a high olefin yield (shown in Table 1).

Figure 2 shows the effect of temperature on the selectivity for olefins. The selectivities for C_2-C_4 olefins on Co/Al_2O_3 are decreased with increasing temperature. This behavior is normal as the olefinic hydrocarbons are less stable under the reaction conditions and can be hydrogenated successfully to alkanes. In contrast, the composite catalyst (Co/Al_2O_3+HX) did not show any dependence of olefin selectivity on temperature.

It has been reported [21-24] that olefins can adsorb on the acid site and undergo carbonium ion-type reactions. These includes isomerization, oligomerization, polymerization, and cracking. The double bond and cis/trans isomerizations can occur on sites of any strength while the oligomerization reactions accompanied by cracking and hydrogen transfer can take place only on

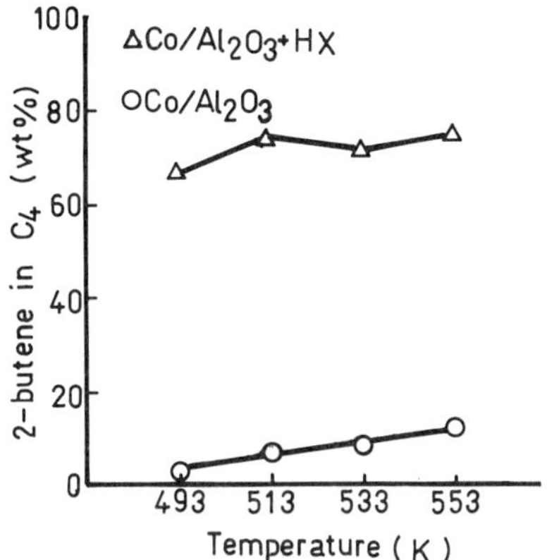

Figure 3. *Effect of Temperature on the Selectivity for 2-Butenes*

the strong acid sites. Figure 3 shows the addition of HX in Co/Al_2O_3 has a significant influence on the 2-butene selectivity. This can be attributed to the acidity of HX zeolite. In addition, 2-butene is the major fraction of C_4 products on the composite catalyst. This provides evidence that the primary olefinic products desorbing from F-T metal sites are being readsorbed on the acid sites where they can undergo a number of transformations. This also indicates that the isomerization rather than the oligomerization is the predominant secondary reaction in this HX-mixed catalyst system. Various researchers [22,25] have found that the Si/Al ratio of the zeolite appeared to be important in affecting the fraction of C_4-isomers. Oukaci et al. [22] concluded that the isomerization is predominant in HX system while oligomerization is predominant in HY, HL, and mordenite systems. Our results are in accord with this.

ACKNOWLEDGMENT

This study was supported by the National Science Council of the Republic of China.

REFERENCES

1. D. L. King, J. A. Cusumano, and R. L. Garten, *Catal. Rev.-Sci. Eng. 23:* 233 (1981).
2. R. L. Varma, N. N. Bakhshi, J. F. Mathews, and S. H. Ng, *Canad. J. Chem. Eng. 63:* 612 (1985).
3. J. M. Stencel, J. R. Diehl, L. J. Douglas, C. A. Spitler, J. E. Crawford, and G. A. Melson, *Colloids and Surfaces, 4:* 331 (1982).
4. T. J. Huang, and W. O. Haag, *ACS Symp. Ser. 152:* 301 (1981).
5. R. L. Varma, K. Jothimurugesan, N. N. Bakhshi, J. F. Mathews, and S. H. Ng, *Canad. J. Chem. Eng. 64:* 141 (1986).
6. L. F. Nazar, G. A. Ozin, F. Hugues, and J. Godber, *J. Mol. Catal. 21:* 313 (1983).
7. D. Fraenkel, and B. C. Gates, *J. Amer. Chem. Soc. 102:* 2478 (1983).
8. D. B. Taktchenko, N. D. Chan, H. Mozzanega, M. C. Roux, and I. Taktchenko, *ACS Symp. Ser. 152:* 187 (1981).
9. D. B. Taktchenko, and I. Taktchenko, *J. Mol. Catal. 13:* 1 (1981).
10. D. K. Lee, and S. K. Ihm, *Appl. Catal. 32:* 85 (1987).
11. D. K. Lee, and S. K. Ihm, *J. Catal. 106:* 386 (1987).
12. C. D. Chang, W. H. Lang, and A. J. Silvestri, *J. Catal. 56:* 268 (1979).
13. C. D. Chang, *Catal. Rev.-Sci. Eng. 25:* 1 (1983).
14. P. D. Caesar, J. A. Brennan, W. E. Garwood, and J. Ciric, *J. Catal. 56:* 274 (1979).
15. R. L. Varma, N. N. Bakhshi, J. F. Mathew, and S. H. Ng, *Appl. Catal. 32:* 197 (1987).

16. L. Fu, and C. H. Bartholomew, *J. Catal. 92:* 176 (1985).

17. I. R. Leith, *J. Catal. 91:* 283 (1985).

18. Y. W. Chen, W. C. Hsu, and C. S. Lin, *React. Kinet. Catal. Lett. 40:* 301 (1989).

19. F. Fajula, R. G. Anthong, and J. H. Lunsford, *J. Catal. 73:* 237 (1982).

20. M. A. Vannice, *J. Catal. 40:* 129 (1975).

21. R. Oukaci, A. Sugari, and J. G. Goodwin, Jr., *J. Catal. 102:* 126 (1986).

22. R. Oukaci, J. C. S. Wu, and J. G. Goodwin, Jr., *J. Catal. 107:* 471 (1987).

23. J. Datka, *Catalysis by Zeolites* (B. Imelik et al., eds.), Elsevier, Amsterdam, p. 121. (1980).

24. M. Guisnet, *Catalysis by Acids and Bases* (B. Imelik et al., eds.), Elsevier, Amsterdam, p. 283. (1985).

25. Y. W. Chen, H. T. Wang, and J. G. Goodwin, Jr., *J. catal. 83:* 415 (1983).

15

HETEROGENEOUS-HOMOGENEOUS PATHWAYS IN CATALYTIC OXIDATIVE DEHYDROGENATION OF PROPANE

K. T. Nguyen and H. H. Kung

Ipatieff Laboratory and Dept. of Chemical Engineering
Northwestern University
Evanston, IL 60208-3120

ABSTRACT

A mixed magnesium vanadium oxide has been found to be a quite selective catalyst for the oxidative dehydrogenation of light alkanes. It was established that during the oxidative dehydrogenation of propane above 556°C on this catalyst, the conversion of propane in a reactor with a large postcatalytic void volume was substantially higher than that in one without such a volume. This enhanced conversion was mainly due to desorption of reactive intermediates from the catalyst surface into the postcatalytic volume. Results of the modelling of the gas phase reactions in the void volume suggested that the desorbed species was most likely propyl or hydroxyl radicals, with the former being more plausible.

Supported by the Dept. of Energy, Chem. Sci. Div.

outer diameter of 15.5 and 17 mm, respectively. The catalyst, in the form of a wafer, was positioned upright at one end of a quartz holder, which was a short quartz cylinder with inner and outer diameters of 13 and 14 mm, respectively and a length of 25.4 mm. On one side of the catalyst wafer the reactor volume was packed with a 3 mm layer of quartz wool followed by quartz chips. Quartz chips were found to be very efficient in quenching homogeneous reactions [8]. About 2.5 cm length of the reactor on the other side of the wafer could be left empty as the void volume, or filled with quartz chips. The reactant feed could be controlled to enter the reactor from either end so that the void volume could act as either a precatalytic (before the catalyst wafer) or postcatalytic (after the wafer) volume. The temperature profiles in the void volume were measured with thermocouples. Since the temperature inside the reactor depended on the position in the void volume and the catalyst used, and was higher than the external temperature that was controlled with a temperature controller, the experiments were identified with the external temperature.

The reaction feed was 4 to 8 vol% propane, 8 to 16% O_2, and the balance He. The products were analyzed by gas chromatography [7,8].

RESULTS

Each set of experiments of a particular experimental condition such as temperture, feed composition and the catalyst used consisted of four experiments:

(i) Reactions in the void volume without catalyst. The corresponding conversions are labeled X_{vol}.

(ii) Reactions on the catalyst wafer but no void volume, and the conversions are X_{cat}.

(iii) Reactions in the precatalytic arrangement,

and the conversions are X_{pre}.

(iv) Reactions in the postcatalytic arrangement, and the conversions are X_{post}.

It should be noted that iii and iv were conducted consecutively by switching a valve to change the direction of the feed flowing into the reactor. Thus except for the change in the feed direction, all other experimental variables were kept absolutely identical. For i and ii, the reactor had to be dissembled and reassembled to remove the catalyst wafer or to fill up the void volume with quartz chips.

The extent of enhancement of propane conversion, E, is defined as follows:

$$E\ (\%) = 100 \times (X_{post} - X_{pre})/X_{vol} \qquad (1)$$

Ideally, if all other experimental variables are identical, the numerator of this equation measures the conversion due to homogeneous reactions in the postcatalytic volume induced by the desorption of radicals or other reactive intermediates. The experimental procedure mentioned above were designed to minimize any differences in the experimental conditions.

Table 1 shows a set of typical data obtained at 570°C and Figures 2 and 3 show the corresponding temperature profiles in the pre- and postcatalytic arrangement, respectively. If the catalyst wafer and the void volume operate independently, the conversions and the product yields would be the same in the postcatalytic (column 4, Table 1) and precatalytic (column 3) arrangements, and would be equal to the sum of the values due to the catalyst wafer only (column 1) and the void volume only (column 2). The data clearly showed that they were not. The conversion and product yields in the precatalytic arrangement were higher than

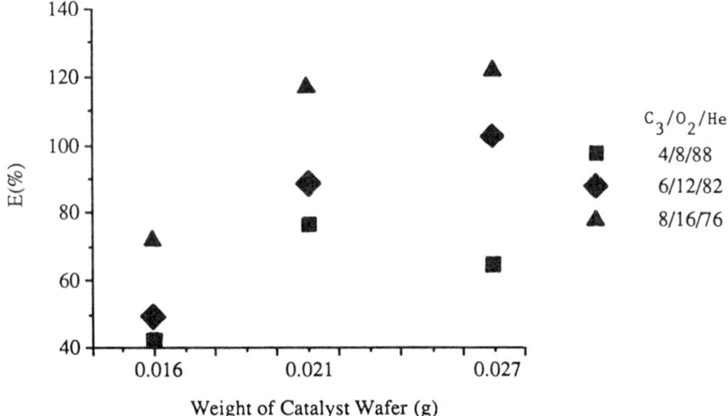

FIG 4. *Extent of enhancement (E) as a function of the weight of catalyst wafer at 570°C.*

is higher in the postcatalytic than the precatalytic arrangement. The other is that the products of the catalytic reaction undergo further reactions in the void volume that cause an enhanced conversion of propane.

The temperature profiles shown in Figures 2 and 3 indeed show that the temperature in the void volume in the postcatalytic arrangement was slightly higher than that in the precatalytic arrangement, and both were higher than when no catalyst was present. However, this higher temperature could not account for the high values of E observed due to the results of the following two experiments.

When a ZnO wafer was used as the catalyst instead of V-Mg-O, it was found that the increase in temperature in the void volume was larger (Figures 2 and 3) because ZnO was not a selective catalyst, and combustion was the only reaction detected. In spite of the higher temperature and the larger temperature difference between the pre- and postcatalytic arrangements, the conversions of propane in the two arrangements were the same within experimental uncertainties. That is, E was zero.

In the second experiment, the reaction was conducted with the catalyst wafer and the void volume separated into two different reactors in series. If the enhancement effect was entirely due to the increase in temperature in the void volume, the same enhancement should be obtained in this configuration as long as the temperature in the void volume was reproduced. In this experiment, the first reactor contained only the catalyst wafer and no void volume. The second reactor contained only the void volume and its temperature was independently adjusted. The products from the catalyst wafer in the first reactor was first quenched to room temperature before being sent to the second reactor. The reaction was then studied using the condition as that shown in Table 1. The catalyst wafer was heated to 570°C, while the second reactor was heated to 570 or 585°C. At 570°C, the total propane conversion was 27.9%, which was comparable to the sum of the conversions due to the catalyst wafer and the void volume measured independently. At 585°C, the total conversion was 33.9%. This 6% increase in conversion represented the results of heating of the void volume due to the heat of reaction released from the catalyst. It also represented the maximum increase in the conversion caused by the thermal effect, since the high temperature was maintained throughout the second reactor in this control experiment and did not fall off at one end as in Figure 3. This value was lower than the 16.4% increase shown in Table 1.

The results of these two experiments showed that thermal effect alone cannot account for the increase in enhancement observed when the catalyst and the void volume were in direct contact. The results of the second experiment further showed that the observed enhancements were not due to reactions of stable products, since the same reactions should occur even when the catalyst and the void volume were separated. Thus it can be concluded that the enhancement must be due to reactions in the void volume as a result of desorption of reactive intermediates from the catalyst. Furthermore, the product distributions of the enhanced portion of the reaction were very similar to the product distributions of the homogeneous oxidative pyrolysis reaction, which suggested that the enhancement was due to an increase in the extent of homogeneous oxidative pyrolysis. No new reactions were involved.

MODELLING

The experimenal results described above showed that reactive intermediates were desorbed from the catalyst wafer. These intermediates generated additional homogeneous oxidative pyrolysis in the postcatalytic void volume. However, there was no information on the nature of the desorbed intermdiates. In order to obtain this information, the surface-initiated gas phase reactions were modelled with elementary reactions of the oxidative pyrolysis of propane. The set of elementary reactions and their rate constants are known [9,10] and were used after suitable modifications. The reactions and the rate parameters used are listed in Table 2. There are a total of thirty seven elementary reactions and twenty chemical species. The rate parameters were the same as those in the references except for reactions 1, 5, 19, 24, and 25 which were adjusted so that the calculated conversions and product yields fit the experimental results better as is shown in Table 3.

The conversions and product distributions in the postcatalytic void volume were calculated assuming different desorbed reactive intermediates which included propyl, ethyl, methyl, and OH radicals. In each of these calculations, since the rate parameters of all elementary reactions were fixed, the only adjustable parameter was the rate of desorption of one of these radicals. The calculated results were then compared

TABLE 2 *Elementary reactions and rate parameters for the oxidative pyrolysis of propane*

	A (s^{-1} or cm/mol s)	E (kJ/mol)
1 - $C_3H_8 \rightarrow \cdot C_2H_5 + \cdot CH_3$	7.5×10^{15}	326
2 - $.C_3H_7 + O_2 \rightarrow C_3H_6 + \cdot HO_2$	1×10^{6}	0
3 - $C_3H_8 + \cdot CH_3 \rightarrow CH_4 + \cdot C_3H_7$	1.5×10^{12}	20.5
4 - $\cdot C_3H_7 \rightarrow C_2H_4 + \cdot CH_3$	4×10^{10}	133.7
5 - $\cdot CH_3 + O_2 \rightarrow CH_2O + \cdot OH$	6×10^{9}	0
6 - $\cdot HO_2 + \cdot HO_2 \rightarrow H_2O_2 + O_2$	1×10^{11}	0
7 - $\cdot HO_2 + C_3H_8 \rightarrow H_2O_2 + \cdot C_3H_7$	1×10^{7}	50
8 - $\cdot C_2H_5 + O_2 \rightarrow C_2H_4 + \cdot HO_2$	1×10^{6}	0
9 - $\cdot C_2H_5 + C_3H_8 \rightarrow C_2H_6 + \cdot C_3H_7$	4×10^{5}	25.0
10 - $\cdot C_2H_5 \rightarrow C_2H_4 + \cdot H$	3×10^{11}	146
11 - $\cdot H + C_2H_4 \rightarrow \cdot C_2H_5$	7.5×10^{11}	25.5
12 - $\cdot C_3H_7 \rightarrow C_3H_6 + \cdot H$	3.6×10^{10}	133.7
13 - $\cdot H + C_3H_6 \rightarrow \cdot C_3H_7$	1.8×10^{10}	4.8
14 - $\cdot CH_3 + \cdot C_2H_5 \rightarrow C_3H_8$	4.2×10^{12}	0
15 - $\cdot C_2H_5 + \cdot C_2H_5 \rightarrow C_2H_6 + C_2H_4$	1.0×10^{5}	0
16 - $\cdot H + C_3H_8 \rightarrow H_2 + \cdot C_3H_7$	1.8×10^{12}	19.2
17 - $\cdot CH_3 + \cdot C_3H_7 \rightarrow CH_4 + C_3H_6$	1.0×10^{14}	0
18 - $H_2 + \cdot C_3H_7 \rightarrow C_3H_8 + \cdot H$	5.6×10^{7}	83.6
19 - $C_2H_6 \rightarrow \cdot CH_3 + \cdot CH_3$	1.2×10^{12}	284.2
20 - $\cdot CH_3 + C_2H_6 \rightarrow CH_4 + \cdot C_2H_5$	8×10^{8}	50.0
21 - $\cdot H + C_2H_6 \rightarrow H_2 + \cdot C_2H_5$	1.3×10^{11}	37.6
22 - $\cdot OH + C_3H_8 \rightarrow H_2O + \cdot C_3H_7$	1×10^{6}	40.0
23 - $H_2O_2 + M \rightarrow \cdot OH + \cdot OH + M$	1×10^{6}	46.0
24 - $CH_2O + O_2 \rightarrow \cdot HCO + \cdot HO_2$	1×10^{12}	62.7
25 - $\cdot HCO + O_2 \rightarrow \cdot HO_2 + CO$	1×10^{10}	0
26 - $CO + \cdot OH \rightarrow CO_2 + \cdot H$	1.5×10^{3}	-3.21
27 - $CO_2 + \cdot H \rightarrow CO + \cdot OH$	1.7×10^{9}	90.2
28 - $CO + \cdot HO_2 \rightarrow CO_2 + \cdot OH$	1.5×10^{10}	98.86
29 - $CO_2 + \cdot OH \rightarrow \cdot HO_2 + CO$	1.7×10^{15}	357.4
30 - $CO + O + M \rightarrow CO_2 + M$	5.9×10^{11}	17.1
31 - $CO_2 + M \rightarrow CO + O + M$	5.5×10^{21}	550
32 - $CO_2 + O \rightarrow CO + O_2$	2.8×10^{8}	183.2
33 - $CO + O_2 \rightarrow CO_2 + O$	3.2×10^{11}	157.2
34 - $\cdot H + O_2 \rightarrow O + \cdot OH$	2.2×10^{14}	70.2
35 - $O + \cdot OH \rightarrow \cdot H + O_2$	1.7×10^{13}	2.84
36 - $C_3H_8 + O_2 \rightarrow \cdot C_3H_7 + \cdot HO_2$	4.0×10^{11}	198.6
37 - $C_3H_8 + O \rightarrow \cdot C_3H_7 + \cdot OH$	3.0×10^{13}	24.24

with the experimental data to determine which desorbed species and which rate of desorption of reactive intermediates could best fit the data.

Assuming that the postcatalytic void volume was a plug flow reactor, the steady state component balance equation over a differential reactor volume dV is given by the following:

$$\frac{dF(i,t)}{dV} = rate(i,t) \tag{2}$$

where F(i,t) is the molar flow rate and rate (i,t) is the rate of reaction of component i at a position that corresponds to a reaction time t. Since the volumetric flow rate of the gas mixture changes because of the change in the number of moles in the reactions, it can be shown that this equation when written in terms of

TABLE 3 *Comparison of experimental and calculated conversions and product distributions due to the oxidative pyrolysis reactions (at 570°C, τ = 1.18 seconds, and $C_3H_8/O_2/He$ = 4/8/88)*

	Experimental	Layokun's, Westbrook & Pitz's Parameters	Parameters in Table 4
C_3H_8 Conv. (%)	8.1	6.7	8.3
O_2 Conv. (%)	2.5	0.1	0.3
Carbon Product Distribution (% C_3H_8 Basis)			
CO	1.2	0	1.1
CO_2	0.8	0	0.3
CH_4	6.2	12.4	8.0
C_2H_4	19.8	23.5	19.5
C_2H_6	0	1.9	0
C_3H_6	71.9	62.2	71.1

concentration C and conversion X becomes as follows:

$$\frac{dC(i,t)}{d\tau} = rate(i,t) - \frac{C(i,t)\ \varepsilon}{(1 + \varepsilon X)} \frac{dX}{d\tau} \qquad (3)$$

where τ is the space time, and ε is the fractional change in volume at 100% conversion. Since:

$$\frac{dX}{d\tau} = \frac{-(1+\varepsilon)}{C(i,o)\ ((1 + \varepsilon(C(i,t)/C(i,o))^2} \frac{dC(i,t)}{d\tau} \qquad (4)$$

equations (3) can also be written as:

$$\frac{dC(i,t)}{d\tau} = \frac{rate(i,t)}{1 + C(i,t)\varepsilon K/C(i,o)(1+\varepsilon X)} \qquad (5)$$

where $K = \dfrac{-(1 + \varepsilon)/\ C(i,o)}{(1 + \varepsilon\ C(i,t)/C(i,o))^2}$

Since there were twenty different chemical species, there were twenty such equations. Equation (5) was used to calculate the concentrations of propane and oxygen, which was then used in Equation (4) to calculate $dX/d\tau$. These were then used in Equation (3) to calculate the concentrations of the remaining species.

The system of twenty coupled nonlinear differential equations was integrated numerically using the IVPAG subroutine [12], which solves initial value nonlinear ordinary differential equations using the Gear's method [13].

The Gear's method solves initial value first order differential equations using the predictor-corrector method. It uses the values of three previous steps to calculate the next step. The global error bound of this method was $O(h^6)$ where h is the step size. The Gear's method obtains its high order of error by using the recorrected values of the function and derivative values. In the first three steps, the concentrations were calculated using Euler's method which used the value of one previous step to calculate the next step. After the first three steps, the Gear's method was used to calculate the concentrations for the remaining steps.

The postcatalytic void volume was divided into 1000 steps. Since this set of equations was very "stiff", it was neccessary to reduce the step size to this magnitude to obtain results that were independent of the step size. In each step, the set of twenty equations was solved and the resulting concentrations of the various species were used for subsequent steps.

The computer program was run with a CYBER computer. In the program, the composition of the entering feed and the measured temperature profile were inputs. For each step, the rate constants, the rates of thirty seven elementary reactions, and the concentrations of the twenty species were calculated. These steps were then repeated for all of the remaining steps. Details of the computation was given in references 8 and 11.

The feed to the postcatalytic reactor consisted of unreacted propane, oxygen, helium, products from reactions on the catalyst wafer, and desorbed species. The concentrations of the stable products entering the postcatalytic reactor were calculated from the experimental product compositions obtained from reactions on the catalyst wafer. Corrections to these were made to the species that would have been derived from the desorbed reactive intermediate. For example, the assumed rate of desorption of propyl radicals was subtracted from the rate of flow of propene into the void volume, that of ethyl radicals from ethene and ethane, and methyl radicals from methane. For the OH radicals, it was assumed that its rate of formation was the same as the rate of the reaction of propane on the catalyst, since the breaking of the first C-H bond in propane was accompanied by the formation of a surface OH group.

Using these assumptions, the measured temperature profile and the equations, the concentrations of the radicals, stable products, and unreacted propane and oxygen were calculated for each experimental condition along the reactor.

RESULT OF MODELLING

The set of data shown in Table 4 was used to identify the likely candidate for the desorbed reactive intermediate. The desorbed species tested included methyl, ethyl, propyl, and hydroxyl radicals. In the calculation, the rate of desorption of each species was varied until the calculated propane conversion was comparable to the experimental value. The entering

TABLE 4 *Conversions and product distributions in the post-catalytic reactor obtained for different desorbed species at 570°C (0.016 g catalyst, τ = 1.18 s, $C_3H_8/O_2/He$ = 6/12/82)*

		Carbon Product Selectivity (% C_3H_8 Basis)				
	Conv.%	CO	CO_2	CH_4	C_2H_4	C_3H_6
Catalyst only	12.3	15.8	20.3	2.2	5.5	58.7
Postcatalytic reactor	22.6	7.6	0.9	15.3	27.1	49.1
No Radical	11.3	3.7	5.9	12.4	22.2	55.8
$f(C_2H_5)$=30%	21.[illegible]	3.1	4.7	11.5	47.3	33.4
$f(CH_3)$=5%	21.6	0.7	4.0	32.9	26.3	36.1
f(OH)=30%	21.3	2.0	3.3	13.9	28.7	52.1
$f(C_3H_7)$=23%	22.9	6.6	0	15.8	25.2	52.4

feed consisted of the propene, ethene, methane, carbon dioxide, carbon monoxide, water, and the unreacted propane and oxygen, and its composition was estimated using the data in the first row.

Row 2 shows the experimental conversions and product distributions in the postcatalytic volume. Row 3 shows the corresponding calculated data but with no radicals desorbed. The assumption of no radicals desorbing was equivalent to assuming that the catalyst wafer and the void volume acted independently, and the propane conversion in the postcatalytic reactor should be the same as that due to the reactor in the absence of the catalyst. Indeed, this was observed in the calculation. The results in row 3, when compared with the experimental results in row 2, however, showed that the calculated conversion was much lower than the experimental value. This indicated that there must be reactive intermediates that desorbed from the V-Mg-O catalyst surface and further reacted to enhance the propane conversion in the postcatalytic reactor.

The nature of the desorbed species that caused the postcatalytic reaction was investigated next. The calculated conversions and product distributions for different desorbed species are shown in Table 4. As shown in the table, desorbed methyl, ethyl, propyl and OH radicals were all effective radical propagators that substantially increased the conversion. From the results, ethyl and methyl radicals could be excluded as possible desorbed species since the calculated selectivities to ethene and methane, respectively, were much higher than those obtained experimentally. The calculated conversions as well as product distributions obtained with OH and propyl radicals as desorbed species were very similar to those obtained experimentally. From the data presented, it was impossible to distinguish them.

The concentrations of propyl and OH radicals and other major reaction intermediates in the postcatalytic volume when propyl radicals were the desorbed species are shown in Figure 5. The concentration of propyl radicals dropped rapidly from $5.7X10^{-8}$ mol/ml to a constant value of $6X10^{-10}$ mol/ml within 0.012 s of reaction time or 1% of the reactor volume. During the initial transient, the concentration of OH radicals increased by about 50%, and then increased further slowly throughout the reactor. This much smaller percentage change in the concentration of OH radicals than propyl radicals was expected since the OH concentration was three orders of magnitude higher than the concentration of propyl radicals. The propane conversion due to the gas phase reactions in the postcatalytic void volume increased rapidly from zero at the beginning of the volume to 11.2% at about 5% of the volume length, and then to 20% at the end of the volume.

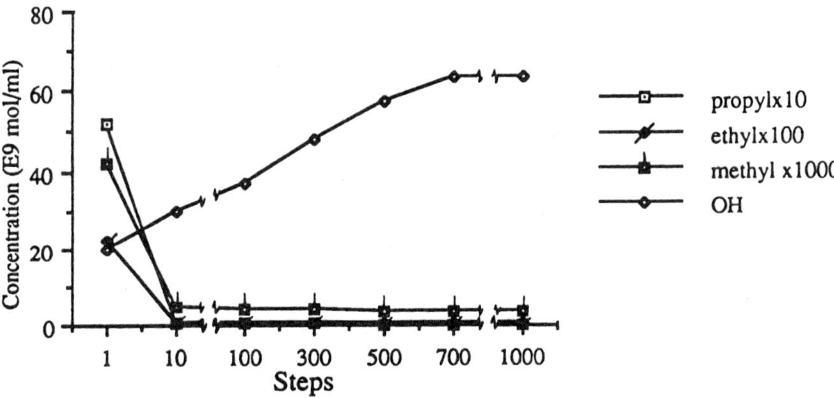

FIG 5. *Concentrations of reaction intermediates in the postcatalytic reactor with 23% of the propyl radicals desorbed, T=570°C, reactor length=1000 steps, 0.016 g catalyst,propane/oxygen/helium= 6/12/82*

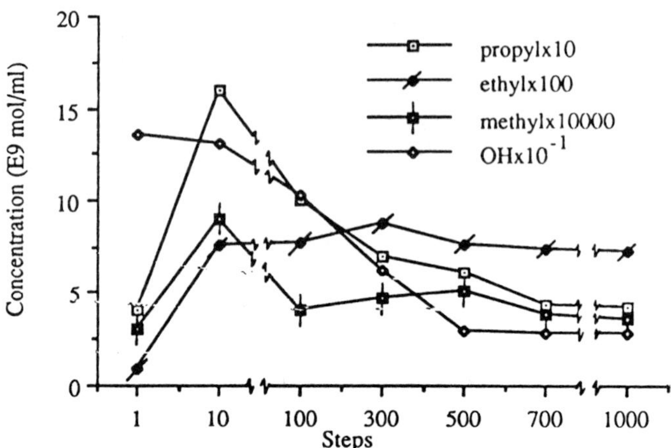

FIG 6. *Concentration of reaction intermediates in the postcatalytic reactor with 30% of the OH radicals desorbed at 570°C, 0.016 g catalyst, reactor length=1000 steps, propane/oxygen/helium=6/12/82*

However, this rapid transient in the calculated concentration profiles was not observed when OH was the desorbed species. This is shown in Figure 6. The concentration of propyl radicals increased from $6X10^{-10}$ mol/ml (at 0.1% of void volume) to $1.8X10^{-9}$ mol/ml (at 1% void volume), then decreased slowly to the end of the volume, while the OH concentration decreased slowly monotonically. The corresponding propane conversion increased from zero at the beginning to about 13.4% at half the length of the volume, and this accounted for most of the propane conversion in the reactor.

Since the conversion profiles and concentration profiles of OH and propyl radicals differed between the cases where the OH or the propyl radicals were the desorbed species, it might be possible to discriminate these two cases experimentally by measuring their concentration profiles, or profiles of other reaction intermediates, or propane conversion as a function of position in the reactor. Unfortunately, our apparatus did not have the capability to measure concentrations of radicals. The principal differences in the conversion profiles occurred in the first 0.1 of the reactor volume: the conversion reached 11.5% if propyl radicals were the desorbed species and 5.1% if OH radicals were the desorbed species. This difference could be measurable. However, this region of the reactor was most affected by the uncertainties in the entrance effect, the flow pattern and the heat transfer effect, which made quantitative comparisons in conversions in this region difficult.

Although the results of the calculation did not distinguish OH or propyl radicals as the desorbed species, we believe that propyl radicals are the more likely species based on chemical arguments. In order for OH radicals to desorb, the propane molecule has to first dissociatively adsorb on the V-Mg-O catalyst surface. This involves breaking a C-H bond in propane to form a propyl species and an OH on the surface followed by dissociation of the OH from the oxide surface. This latter step should be quite improbable energetically. On the other hand, a gaseous propyl radical could be formed by breaking a C-H bond in propane with simultaneous formation of a surface hydroxyl (a Rideal-Eley mechanism).

TABLE 5 *Conversions and product distributions for reaction in the postcatalytic reactor calculated for different fractions of C_3H_7 desorbed at 570°C (same conditions as Table 4)*

	Carbon Product Selectivity (% C_3H_8 Basis)					
	Conv.%	CO	CO_2	CH_4	C_2H_4	C_3H_6
Experimental	22.6	7.6	0.9	15.3	27.1	49.1
f = 30%	29.4	7.6	3.7	12.8	18.9	56.7
f = 23%	22.9	6.6	0	15.8	25.2	52.4
f = 10%	16.9	10.9	0	16.3	24.0	48.8

The rate of desorption of propyl radicals was examined. Table 5 shows the calculated conversions and product distributions for 10%, 23% and 30% of the propyl species generated desorbed into the gas phase. As expected, the conversions increased with the rate of desorption of propyl radicals, and the selectivities to CH_4 and C_2H_4 decreased, while C_3H_6 increased. As shown, the calculated conversion and product distributions were in good agreements with the experimental values when 23% of the reacted propane desorbed as propyl radical. In fact, this value was found to generate results obtained with different feed compositions at 570°C. For a reaction temperature of 556°C, a value of 5% desorption gave a satisfactory fit [8,11].

CONCLUSION

The results of this study showed that under the conditions used, reactive intermdiates desorbed from the surface of a V-Mg-O catalyst during oxidative dehydrogenation of propane. The desorbed intermediates greatly enhanced the homogeneous oxidative pyrolysis reaction in the postcatalytic void volume, which could be modelled with a mechanism that involves thirty seven elementary reactions. The results showed that propyl radicals were the likely species that desorbed from the catalyst surface, although the desorption of OH radicals could not be excluded. The desorption of CH_3 or C_2H_5 radicals were found to be unlikely. The rate of propyl radicals desorbing from a V-Mg-O catalyst was estimated to be 23% of the rate of reaction of propane on the catalyst surface at 570°C, and 5% at 556°C.

REFERENCES

1. M. Chaar, D. Patel, and H. Kung, J. Catal., 109, 463 (1988).
2. M. Chaar, D. Patel, M. Kung, and H. Kung, J. Catal., 105, 483 (1987).

3. D. Patel, M. Kung, and H. Kung, Proc. 9th Intern. Cong. Catal., 4, 1554 (1988).

4. D. Driscoll, W. Matir, J. Wang, and J. Lunsford, J. Amer. Chem. Soc., 107, 58 (1985).

5. W. Matir, and J. Lunsford, J. Amer. Chem. Soc., 103, 3728 (1981).

6. H. Swift, J. Bozik, and J. Ondrey, J. Catal., 21, 212 (1971).

7. K. Nguyen, and H. Kung, J. Catal., 122, 415 (1990).

8. K. Nguyen, PhD thesis, Northwestern University, 1989.

9. S. Layokun, Ind. Eng. Chem. Process Des. Dev., 18, 241 (1979).

10. C. Westbrook, and W. Pitz, Combust. Sci. Technol., 37, 117 (1984).

11. K. Nguyen, and H. Kung, Ind. Eng. Chem. Res., submitted.

12. IMSL INc., Edition 10, 1988, Houston, TX.

13. C. W. Gear, Numerical Initial Value Problems in Ordinary Differential Equations, Prentice Hall, Englewood Cliffs, 1971.

16

A NONTRADITIONAL METHOD OF PRODUCING LOWER OLEFINS BY PYROLYSIS OF HYDROCARBON FEED IN THE PRESENCE OF HETEROGENEOUS CATALYSTS

S.P.Chernykh, S.V.Adelson, T.N.Mukhina

All-Union Institute for Organic Synthesis,
Institute of Oil and Gas named after I.M.Gubkin
Moscow, USSR

The possibility of carrying out high temperature heterogeneous catalysis at 760-780°C in order to make the production of lower olefines more efficient was first established in the Moscow Institute of Oil and Gas in the late sixties [1]. Research work on catalytic pyrolysis has been considerably developed in the subsequent years.

The use of heterogeneous catalysts in pyrolysis of hydrocarbon feed makes it possible to influence the rate of conversion and selectivity. The ethylene yield for different feedstocks increases by 5-10% absolute at temperatures 70-100°C lower than those used in thermal pyrolysis, as shown in table 1 [2,3,4].

Pyrolysis catalysts may comprise transfer metal oxides and other compounds, as well as compounds of nontransfer metals.

Common for most of pyrolysis catalysts is to give similar product slates and product distributions. The selectivity does not change significantly, although the total amount of gaseous and liquid products may noticeably differ.

These facts considered, we believe that the process develops according to the free radical mechanism, characteristic of thermal pyrolysis, but proceeding with the formation of radicals in the free space between catalyst pellets and radical- like particles upon the surface of the catalyst.

Transfer metal catalysts display their activity in their lower or higher valencies, as represented in fig.1 for a vanadium and indium oxide catalysts. During the first 30 minutes of the run the catalysts are reduced by the propane feed to a lower valency, resulting in a change of their activity. After that conversion of propane remains constant.

It should be noted that the ethylene and propylene yields for different catalysts can be increased up to a certain limit. This may result from a limited possibility of hydrogen redistribution and/or the limit in concentration of active radicals when further increase in the amount of radicals no longer enhances the reaction velocity because of recombination processes.

TABLE 1. Catalytic Pyrolysis of Hydrocarbon Feedstocks on a Laboratory Unit

Feedstock	C_2H_6	C_3H_8	C_4H_{10}	Naphtha	Raffinate	Gasoil 235-350°C	Hydrocracked gasoil
Temperature, °C	820	810	800	780	780	770	780
Residence time, s	0.4	0.3	0.6	0.1	0.1	0.3	0.1
Product yields, % mass.							
H_2	4.2	2.7	1.2	1.3	1.2	1.0	1.4
CH_4	6.2	20.0	21.2	14.0	16.3	9.3	13.2
C_2H_4	60.8	45.0	38.0-40.0	38.0	35.5	30.7	33.3
C_3H_6	0.7	16.0	17.0	14.5	15.5	14.4	11.7
C_4H_6	1.9	3.5	2.9	5.0	4.8	4.4	4.5
C_4H_8	0.4	2.5	3.0	3.0	3.2	5.6	2.2
Uns. C_2-C_4	63.8	63.0	60.9	60.5	58.0	55.1	52.7
Conversion	74.6	97.0	95.1	-	-	-	-
Gaseous hydrocarbons	-	-	-	80.3	76.0	68.7	69.5
C_2H_4 selectivity	81.5	53.8	47.3	-	-	-	-

Besides the basic catalysts, there are compounds with specific properties responsible for a high level of feed conversion, a large yield of coke, hydrogen and carbon oxides as well as a smaller yield of ethylene. These compounds can be regarded as catalytic poisons. The coke yield is not only a function of the chemical nature of the catalyst, but also of the pore and structure characteristics of the catalytic system. Supports having small pores cannot be used for this process [5].

The selection of a catalyst support is one of the key problems of catalytic pyrolysis. We have developed a ceramic support with predetermined properties, which possesses large pores, small specific surface and high mechanical strength.

Modifiers only slightly affect the composition and yield of gaseous products and mostly influence the yield of coke and the liquid products distibution [6].

The kinetic parameters for different catalysts may vary considerably. For instance the activation energy for vanadium catalysts is 180-190 kJ/mole while for potassium chloride - 90-105 kJ/mole. Taking into account these differences it hardly seems possible to interpret catalytic pyrolysis for different catalysts by a single mechanism.

At the same time the process undoubtedly proceeds as a homogeneous-heterogeneous reaction since it takes place at temperatures characteristic of low severity pyrolysis (780-790° C) and one cannot neglect the homogeneous reaction running within the volume between the catalyst pellets. For this reason elucidation in details of the mechanism of catalytic pyrolysis is very complicated. Therefore it is expedient to examine some of its peculiarities: the influence of inhibitors, initiators and diluents for the case of utilizing a vanadium catalyst.

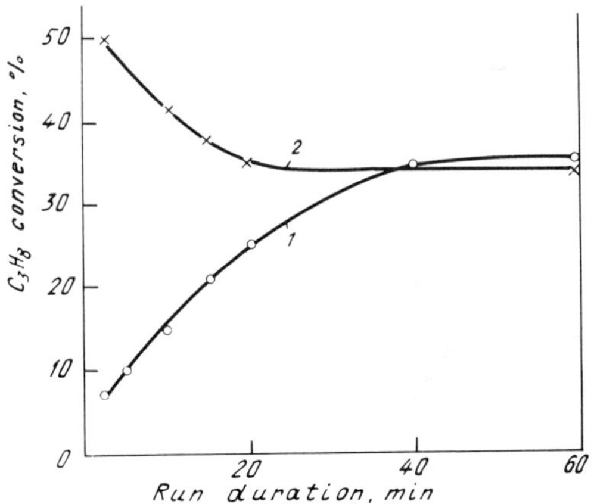

FIGURE 1. Propane conversion vs. run duration (min.) for catalytic pyrolysis on a vanadium (1) and indium oxide (2) catalysts at 760°C.

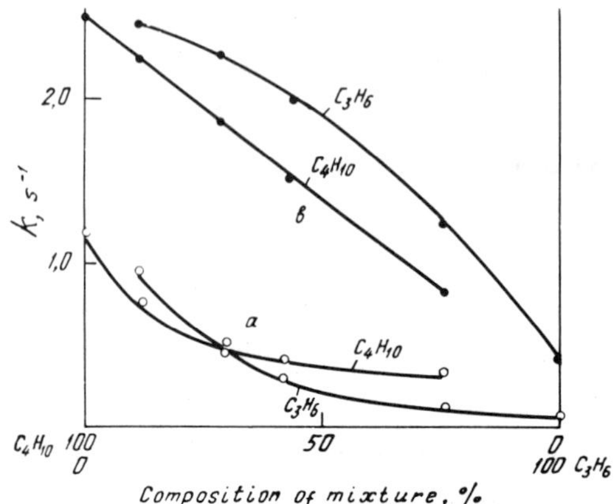

FIGURE 2. The specific rate constants of n-butane and propylene decomposition vs. composition of n-butane-propylene mixtures for thermal (a) and catalytic (b) pyrolysis.

The Influence of Inhibitors

As was shown earlier, propylene and isobutylene are typical inhibitors in the process of thermal pyrolysis [7].

We have found that in catalytic pyrolysis propylene, isobutylene and hexene-1 inhibit catalytic decomposition of ethane, propane, butanes and n-hexane, while the decomposition of the olefines is enhanced. However, the above-mentioned phenomena differ from those observed in thermal pyrolysis in that there is no

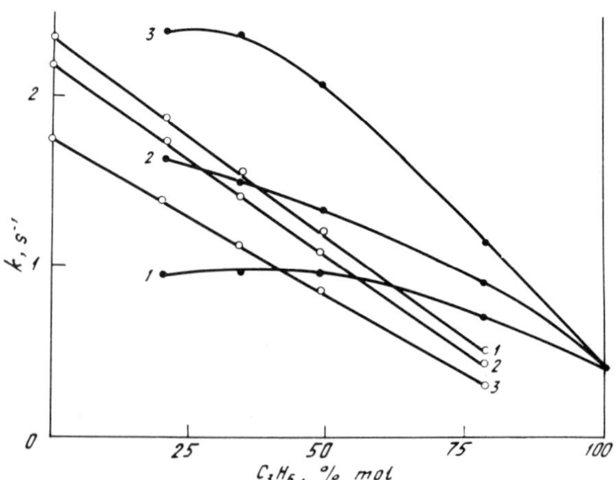

FIGURE 3. The specific rate constants of paraffins and propylene decomposition vs. propylene concentration for catalytic pyrolysis of propylene-ethane (1), propane (2) and n-butane (3) mixtures.

inhibition limit and the inhibition rate is almost a linear function of the olefine concentration, as can be seen in figure 2. In the case of catalytic pyrolysis the first order specific rate constants in all mixtures with maximum inhibition change about 3-4 times and are not practically influenced by the molecular mass of the paraffin (fig.3).

In thermal pyrolysis the selectivity of hydrogen formation in mixtures decreases and that of methane and ethylene increases in consistency with the inhibition mechanism [7].

$$\dot{H} + C_3H_6 \dashrightarrow C_2H_4 + \dot{C}H_3$$

In catalytic pyrolysis the selectivity of hydrogen formation is slightly higher and the selectivity of methane and ethylene lower than that calculated according to the additivity rule [8].

These data suggest that the paraffin decomposition is deterred by olefines mainly on the catalyst surface. Some of the data indicate that the inhibition may result from covering the catalyst surface by the olefine. On the other hand, the change in selectivity is indicative of a change in mechanism. This can be attributed to replacement of the active radical $\dot{C}H_3$ by the low active allyl radical on the surface as well as to the penetration of the $\dot{C}H_3$ radical into the propylene molecule with $\dot{H}$ radical being thrown out into the free volume [8].

$$CH_2=CH-CH_3 \xrightarrow{surface} CH_2=CH-CH_2 \cdots H \xleftarrow{\dot{C}H_3} \longrightarrow CH_2=CH-\dot{C}H_2 + CH_4$$

$$CH_2=CH-CH_3 \xrightarrow{surface} CH_2=CH-CH_2 \cdots H \xleftarrow{CH_3} \longrightarrow CH_2=CH-CH_2-CH_3 + \dot{H}$$

$$CH_2=CH-\dot{C}H_2 + C_3H_6 \longrightarrow \text{reaction products}$$

These presumptions lead to a conclusion that the methyl radical in catalytic pyrolysis plays a more important role than in thermal pyrolysis. The data obtained show that the catalyst is not just a generator of additional radicals and that chain propagation on the catalyst surface is of great importance.

The Influence of Initiators

The study of the influence of initiators upon catalytic pyrolysis provides additional information about the kinetics and mechanism of the process. Different organic and inorganic initiators were investigated and the most effective of them being found to be acetone [9].

The compounds studied are only slightly effective or ineffective in thermal pyrolysis at 700-720°C and totally ineffective at higher temperatures.

Small amounts of initiators significantly enhance the rate of catalytic pyrolysis. At the same time there exists an optimum concentration of initiators and when this concentration is exceeded, the rate of initiation begins to decline.

A study of the decomposition of pure initiators showed that the rate of decomposition increases in the presence of a pyrolysis catalyst. In acetone pyrolysis the stage of chain propagation produces a ketene which takes up water forming acetic acid. At the same time this ketene is a trap for $\dot{H}$ atoms.

$$CH_3-CO-CH_3 \xrightarrow{\dot{C}H_3} CH_3-CO-\dot{C}H_2 + CH_4$$
$$CH_3-CO-\dot{C}H_2 \longrightarrow CH_2{=}CO + \dot{C}H_3$$
$$CH_2{=}CO + H_2O \longrightarrow CH_3COOH$$
$$CH_2{=}CO + \dot{H} \longrightarrow CH_3\dot{C}O$$

The decrease in the initiation rate of the propane decomposition at initiators concentrations exceeding the optimum can be the result of an increase in the ketene formation and/or of recombination processes.

The rate of initiation of propane decomposition decreases with increasing temperatures and conversion (fig.4) as well as in the presence of an inhibitor (propylene).

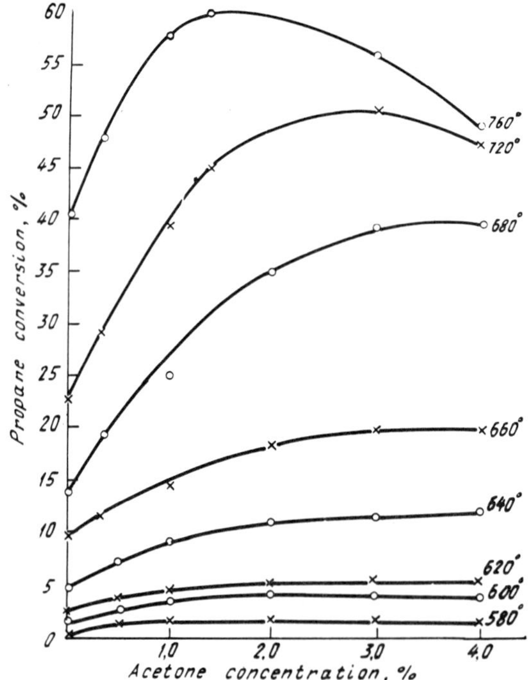

FIGURE 4. Propane conversion vs. acetone concentration for initiated catalytic pyrolysis of propane at different temperatures (°C).

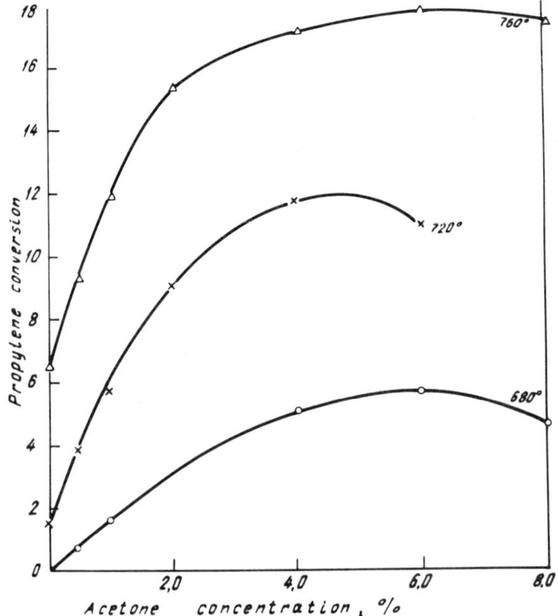

FIGURE 5. Propylene conversion vs. acetone concentration for initiated catalytic pyrolysis of propylene.

Pure propylene begins to decompose in the presence of acetone under the conditions when catalytic pyrolysis does not yet start (fig.5).

When small amounts of acetone are added the activation energy of n-butane decomposition decreases by 40 kJ/mole and the reaction order is 0.9 or near 1. This means that the methyl radical termination occurs linearly on the catalyst surface. The conclusions arrived at are.

The increasing effect of acetone in the presence of pyrolysis catalysts can be attributed to a higher concentrations of CH_3 radicals resulting from the initiator decomposition. The ethylene selectivity is lower and themethane selectivity is higher for catalytic pyrolysis of pure acetone. Thus, the catalyst surface has a certain effect on the reaction: acetone and intermediates interact with the surface enhancing the reaction rates. Such intermediates can be the methyl radical and ketene, since the decomposition of $CH_3\dot{C}O$ radical can occur only in the free space between the catalyst pellets and not on the catalyst surface because the lifetime of the radical is only 10^{-7} s.

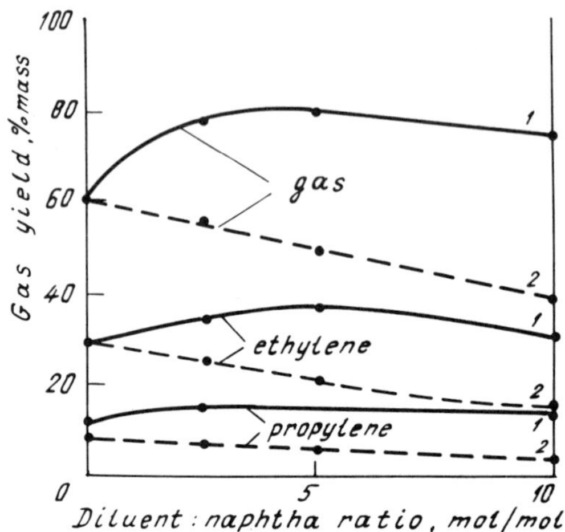

FIGURE 6. The influence of steam (1) and nitrogen (2) dilution on catalytic pyrolysis of naphtha.

These data confirm the importance of homogeneous initiation and of $\dot{C}H_3$ radicals in catalytic pyrolysis.

The Influence of Diluents

The rate of dilution and the nature of diluent are found to considerably influence catalytic pyrolysis of individual hydrocarbons and naphtha feed [9,10]. The regularities of diluent influence depend on the properties of catalyst support and active components. These regularities may differ for catalytic pyrolysis of various hydrocarbons.

In the process of catalytic pyrolysis of individual paraffin hydrocarbons dilution by steam up to a certain ratio does not affect the product yields; at higher dilution rates the hydrocarbon conversion and the yield of unsaturated gaseous hydrocarbons decreases. When pyrolysing naphtha dilution by steam up to a certain ratio enhances the yield of ethylene and gaseous hydrocarbons with the decrease in yield at higher dilution rates (fig.6).
Diminishing of conversion at high dilution rates can be attributed to a decrease in concentration of the reagents on the catalyst surface. When nitrogen is used as a diluent the conversion of the hydrocarbon and the yield of gaseous reaction products continually drop and are much lower than the values observed in the presence of steam. Thus steam is not an inert diluent and can enhance the reaction and even compensate for the decrease of the concentration of the reacting species.

The greater conversion and product yields can be ascribed to dissociative adsorption of steam on the catalyst surface with simultaneous formation of $\dot{H}$ and $\dot{O}H$ radicals and/or interaction between steam and low active radical-like species of allyl type which are present on the catalyst surface with the resulting of $\dot{O}H$ radical formation.

$$MeO + HOH \longrightarrow MeO\text{-}\dot{O}H + \dot{H}$$
$$MeO\text{-}\dot{O}H \longrightarrow MeO + \dot{O}H$$
$$\dot{R} + HOH \longrightarrow RH + \dot{O}H$$

The latter reaction also supresses formation of polymerization products and coke.

The Mechanism of Catalytic Pyrolysis

The mechanism of catalytic pyrolysis can be described by the conventional thermal pyrolysis reaction scheme supplemented with the stages occuring on the surface (s - surface, v - volume).

Initiation

$$C_3H_8 \rightleftharpoons \dot{C}H_3 + \dot{C}_2H_5 \quad (v)$$

We consider it more probable that on the catalyst surface initiation proceeds by a break-off of the more mobile H-atom rather than the C-C bonds cleavage

$$C_3H_8 \xrightarrow{surface} CH_3\text{-}CH_2\text{-}CH_2 \ldots H \longrightarrow \dot{H} + \dot{C}_3H_7$$

Chain propagation

$$C_3H_8 \xrightarrow{surface} CH_3\text{-}CH_2\text{-}CH_2 \ldots \overset{\dot{C}H_3\downarrow}{H} \longrightarrow n\text{-}\dot{C}_3H_7 + CH_4$$

$$C_3H_8 + \dot{C}H_3(\dot{H}) \rightleftharpoons n\text{-}\dot{C}_3H_7 + CH_4(H_2) \quad (v)$$
$$\rightleftharpoons iso\text{-}\dot{C}_3H_7 + CH_4(H_2) \quad (v)$$
$$n\text{-}\dot{C}_3H_7 \rightleftharpoons C_2H_4 + \dot{C}H_3 \quad (v,s)$$
$$iso\text{-}\dot{C}_3H_7 \rightleftharpoons C_3H_6 + \dot{H} \quad (v)$$

Chain termination

$$\dot{C}H_3 \xrightarrow{surface} termination$$
$$2\dot{C}H_3 \rightleftharpoons C_2H_6 \quad (v)$$
$$\dot{C}H_3 + \dot{C}_2H_5 \rightleftharpoons C_3H_8 \quad (v)$$
$$2\dot{C}_2H_5 \rightleftharpoons C_4H_{10} \quad (v)$$

Summary

Catalytic pyrolysis of naphtha and gaseous feedstock has been investigated on various units from laboratory to semicommercial ones. These investigations confirmed the process efficiency and extended runs without regeneration (1500 h).

Various feedstocks from ethane to hydrotreated vacuum gasoil can be pyrolysed.

At optimum conditions of naphtha catalytic pyrolysis - 780-790°C, space velocity of 3-4 h^{-1} and steam dilution of 50-70 wt.% - the average yields (% mass.) are as follows: ethylene - 34 -35%, propylene - 16-18%, butadiene - 4.5-5%, butenes - 6-7%. When pyrolysing gaseous hydrocarbons the yield of ethylene and the selectivity increases (table 2).

TABLE 2. Catalytic Pyrolysis on a Pilot Plant at Optimum Conditions.

Product yields, % mass.	Raffinate	Naphtha	Light hydrocarbons cut	Ethane
H_2	0.9	0.8	1.3	-
CH_4	16.0	15.0	21.5	-
C_2H_4	34.0	35.3	37.6	44.5
C_3H_6	16.8	16.5	16.5	-
C_4H_6	4.1	4.5	3.6	-
C_4H_8	5.2	4.8	4.7	-
Conversion	-	-	-	64.7
Selectivity	-	-	-	84.9

The process can be used for revamping existing ethylene plants and desingning grass-roots ones.

REFERENCES

1. Ya.M.Paushkin, A.G.Liakumovich, S.V.Adelson et al.Catalytic pyrolysis of cyclohexane and methylcyclohexane to produce butadiene. Khim. prom., No 11 pp. 811-813, 1968

2. S.V.Adelson, T.A.Vorontsova, C.A.Melnikova et al. Some peculiarities of catalytic pyrolysis in the presence of heterogeneous and homogeneous catalysts. Neftekhimiya, v 19, No.4, p.577-582, 1979.

3. S.V.Adelson, V.I.Nikonov, G.P.Kreynina et al.Intensification of olefine producing processes. Khimia i tekhnologiya topliv i masel, No.7,pp.19-22,1980.

4. S.V.Adelson, V.G.Sokolovskaya. Catalytic pyrolysis of n-butane.Neftekhimiya. vol.24, No.3, pp.371-375, 1984.

5. S.V.Adelson, G.P.Kreynina, B.A.Lipkind, T.N.Mukhina, S.P.Chernykh. Staightrun naphtha pyrolysis with KVO3 catalysts suspended on different supports. Neftekhimiya. No.4, pp.32-35, 1980.

6. S.V.Adelson, F.G.Zhagfarov, V.I.Nikonov. The influence of promoters on the properties of pyrolysis catalysts. Neftepererabotka i neftekhimiya, No.4, pp.25-27, 1983.

7. R.A.Kalinenko, K.P.Lavrovsky, L.V.Shevelkova et al. The mutnal influence of alkane and alkene hydrocarbons on their decomposition rate in high temperature cracking. Neftekhimiya, vol.9, No.4, pp.542-551, 1969.

8. S.V.Adelson, T.A.Vorontsova, O.V.Kuznetsova et al. The mechanism of the influence of initiators and inhibitors on the catalytic pyrolysis process. DAN USSR, vol.275, No.6, pp.1467-1470, 1984.

9. S.V.Adelson, V.G.Sokolovskaya, V.I.Nikonov et al. The influence of dilution upon catalytic pyrolysis of individual hydrocarbons. DAN USSR, vol.275, No.6, p.1467-1470, 1977.

10. S.V.Adelson, O.V.Kuklina. The influence of diluents on catalytic pyrolysis.Kinetika i katalis, vol.25, vip.1, pp.103-106, 1984.

11. S.V.Adelson, V.G.Sokolovskaya. The mechanism of catalytic pyrolysis of hydrocarbon feed. Kinetika i catalis, vol.22, No.2, pp.390-395, 1981.

12. S.P.Chernykh, S.V.Adelson, E.M.Rudyk et al. Catalytic pyrolysis of straight rum naphtha on a modified vanadium catalyst. Khim. prom., No.4, pp.10-12, 1983.

17

STEAMLESS PYROLYSIS OF ETHANE TO ETHYLENE

Y. Song, L.J. Velenyi, A.A. Leff, W.R. Kliewer and J.E. Metcalfe

BP Research
Warrensville Research Center
Cleveland, Ohio 44128

I. ABSTRACT

In order to circumvent the drawbacks of carbon deposition in commercial steel crackers, high temperature, short reaction time, steamless cracking of ethane to ethylene in an α-SiC reactor has been investigated. By optimizing temperature and reaction time and using a non-catalytic surface (α-SiC), high ethylene yields, exceeding significantly the best results of the state-of-the-art technologies, have been obtained with a very low concomitant coke deposition. Under our optimum condition, ethane conversion of 84% with an 83 wt% selectivity to ethylene is achieved. Kinetic modeling reveals that a global first order Arrhenius rate expression provides an excellent fit to the rate of ethane disappearance. The effects of addition of H_2S and C_3H_8 have also been studied. While carbon deposition increases slightly with increasing H_2S concentration, propane addition, at up to 12.5 vol%, does not adversely affect carbon deposition.

II. INTRODUCTION

While thermal conversion of ethane to ethylene has been a commercial technology for many years, research toward finding alternative ways or improving the process is still of great interest and can be very rewarding. The current state-of-the-art technologies and commercial ethane cracking processes include Lummus Crest's conventional furnace, Lummus Crest's Short Residence Time (SRT) cracking furnace, M. W. Kellogg's Millisecond Pyrolysis Furnace, and other pyrolysis heaters designed by Kinetic Technology International (KTI), Stone & Webster, Foster Wheeler, Mitsubishi, etc. A typical ethane cracking process requires a temperature range of 750 to 900°C and a reaction time of 100 to 600 milliseconds in order to achieve a high selectivity to olefins and diolefins and a low methane yield (1,2).

For all of the above technologies, metal alloys have been the only choice for tube material of the reactor coils because of a lack of material in the past that can satisfy the following design criteria for the reactor (3):

1. It should withstand the high temperature environment,
2. It should endure the severe stress-strain relationship, and
3. It should be resistant against carburization.

However, an inherent problem associated with the metal alloys is that they tend to promote or enhance carbon deposition on the tube surface. This is especially true for alloys that have a high Ni content (4). Furthermore, it has been shown that more coke is formed on some metal surfaces that have been subjected to frequent coking and decoking cycles as compared to new surfaces. This is a critical drawback because coke formation in a cracker unit is highly undesirable for several reasons:

1. It decreases yields of desired products,

2. It increases resistance of heat transfer to reactants, and

3. It leads to eventual plugging of the reactor and thus necessitates the shut-down of the cracker unit for decoking.

To reduce the amount of carbon being deposited on the tube walls, all of the above technologies add steam to the hydrocarbon feedstock and crack the hydrocarbon in the presence of steam. The steam-to-hydrocarbon weight ratios usually vary from 0.3 for ethane (i.e., 1/2 volume ratio steam to ethane) to as high as 1.0 for gas oils. Higher steam to hydrocarbon ratio is needed for higher hydrocarbons because of their higher tendencies for carbon deposition. Although steam is effective, the addition of steam has several disadvantages. First, it limits the throughput of hydrocarbon for a given size of reactor and operating pressure. Second, the energy efficiencies of reactant heating and product quenching suffer because of the presence of steam. Finally, the addition of steam in the front end necessitates the process of separating it from hydrocarbon products in the back end.

Besides the steam addition, a second characteristic of commercial ethane cracking processes is the requirement of ethane recycle. This is to increase ethane utilization efficiency and thus to improve ethane conversion from about 60% once-through to about 80% overall. Typical product yields from ethane cracking in commercial processes are summarized in Table 1 (2).

Kinetically, the decomposition of ethane is believed to be initiated by the scission of carbon-carbon bond in the ethane molecule. As the hydrogen radical (H·) and methyl radical ($CH_3\cdot$) concentrations increase, the abstraction of hydrogen atom from ethane molecule by H· or $CH_3\cdot$ to form ethyl radical ($C_2H_5\cdot$) becomes dominant reactions for ethane decomposition (5). Subsequent chain reactions involving $C_2H_5\cdot$ and other species lead to the formation of olefins (mainly, ethylene) and aromatic products. Two undesired products during ethane pyrolysis are

Table 1 *Typical product yields from ethane and propane cracking in commercial processes*

Feed	Ethane	Propane
Product	Yield, wt% (Once Through)	
H_2	3.3	1.5
CH_4	4.3	23.8
C_2H_2	0.2	0.7
C_2H_4	47.7	36.5
C_2H_6	40.0	3.2
C_3H_4	0.1	0.5
C_3H_6	1.1	14.7
C_3H_8	0.4	10.0
C_4H_6	1.4	2.7
C_4H_8	0.2	1.1
C_4H_{10}	0.2	0.1
C_5+	0.7	5.0

coke and methane. Coke is the consequence of degradation of higher molecular weight products while methane is formed by the recombination of $CH_3\cdot$ with $H\cdot$ or the hydrogen abstraction from C_2H_6 or other species by $CH_3\cdot$.

Based on the above kinetic consideration, we set as our goal high ethane conversion with high selectivity to ethylene. To reach this goal, we had to establish a delicate balance. The temperature must be high enough for the conversion of ethane into ethylene at acceptable rates, yet the contact time must be short enough so that only minimum amounts of degradation products, i.e., coke, and methane are produced. These considerations have led us to design our experimental approach of high temperature, short reaction time, steamless pyrolysis of ethane.

This paper presents the results of our study based on the experimental approach mentioned above. The focus of this paper is on the effects of temperature and reaction time on ethane conversion and major product distributions. The emphasis is the search for an optimum operating condition based on ethane conversion and ethylene yield. The coke deposition during ethane pyrolysis and its relationship with reactor material, surface, and additives etc. are subjects of another paper presented at the Nineteenth Biennial Conference on Carbon (6) and a paper submitted to Carbon for publication (7).

III. EXPERIMENTAL

The experimental apparatus used in this study consisted of a tubular flow reactor that was electrically heated by a Lindberg Model 54233 furnace. The reactors used during the course of this study included INCONEL 600 tubes (0.64 cm O.D. x 0.46 cm I.D. x 91 cm long) and α-SiC tubes (0.8 cm O.D. x 0.5 cm I.D. x 91 cm long). A schematic diagram of the vertical downflow reactor system used in the study is shown in Figure 1.

The furnace temperature was controlled and monitored by a platinum/platinum - 13% rhodium thermocouple (type R). The power to the silicon carbide heating elements was controlled by a Texas Instrument PM 550 Control Programmer which was connected to a Texas Instrument Workstation CVU 5000.

Pure ethane was fed to the reaction chamber from the top of the reactor. Its flow rate was varied according to the reaction temperature and the desired reaction time. Heat transfer calculations showed that heating rates of 10^3 to 10^4°C/sec were achieved. Gas temperature along the center axis of the reactor was measured by a type K thermocouple.

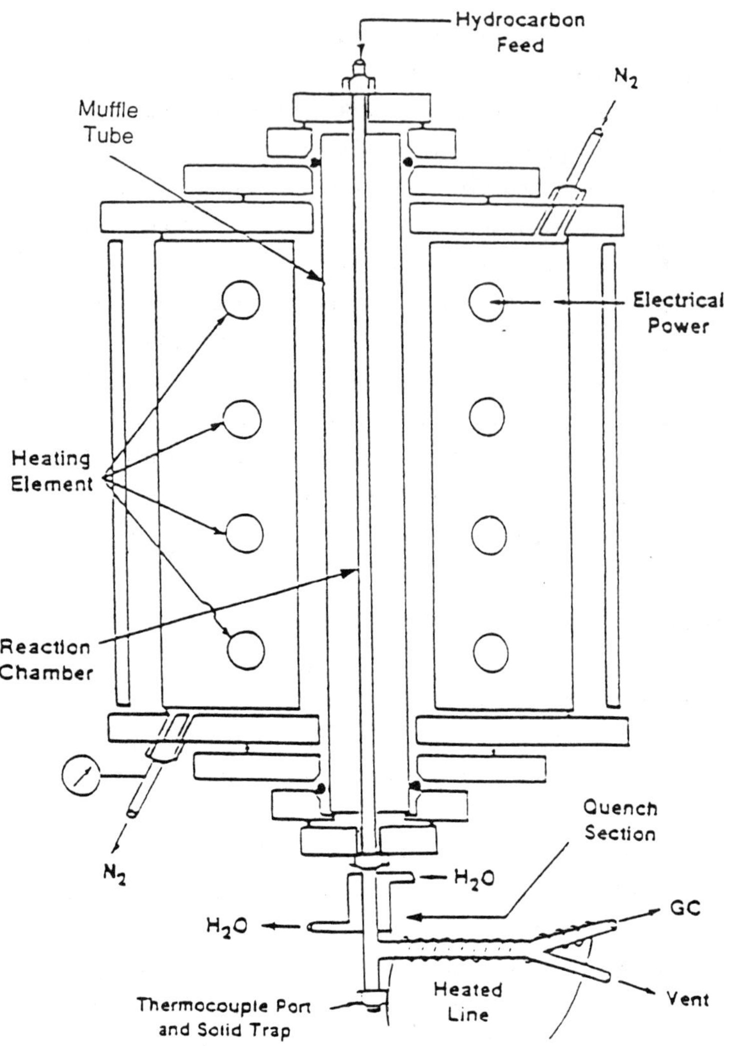

Fig. 1 *Schematic of experimental apparatus*

Reaction products were withdrawn and quenched from the bottom of the reactor. The extracted gas sample was sent to an on-line Varian 6000 GC, equipped with both thermal conductivity (TCD) and flame ionization detectors (FID). Poropak S packed columns were used for gas analyses, with good resolution of H_2 by TCD and unreacted ethane and hydrocarbon products by FID.

The carbon deposited on the reactor walls was measured by a burn-off experiment following each pyrolysis run. Burn-off gas products were collected in gas sampling bags. The volume of gas collected was measured during burn-off with a dry test meter. The carbon oxides formed during the burn-off cycle were analyzed on a Carle 311H GC. The amount of carbon deposited on the reactor walls was calculated based on the amount of CO_2 and CO measured. The hydrocarbon feed and the ensuing burn-off complete one cycle of the experiment.

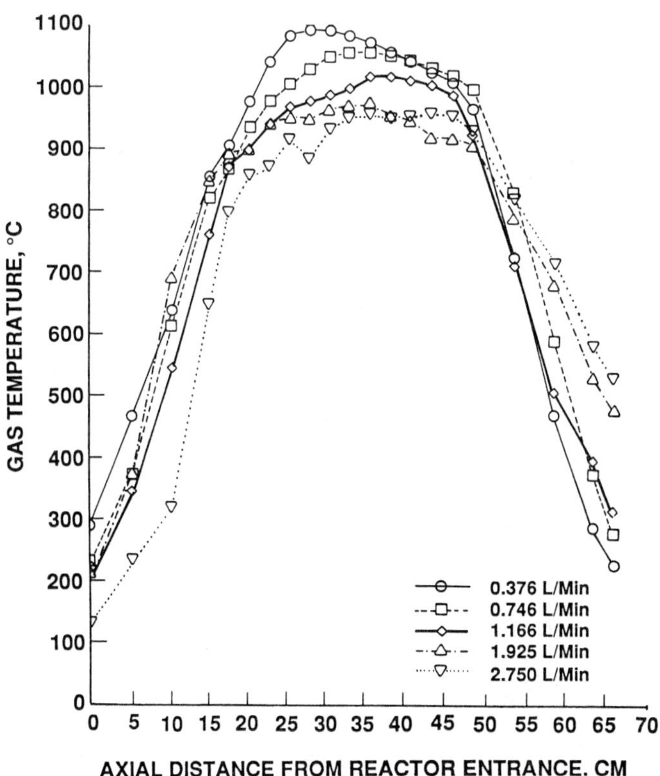

Fig. 2 *Temperature profiles of ethane pyrolysis at a furnace set-point of 1050°C*

IV. RESULTS AND DISCUSSION

The experimental program of this study consisted of investigations of the effects of reaction temperature and reaction time on ethane conversion and pyrolysis product distribution. The ultimate objective was to define an optimum operating condition for steamless ethane cracking to meet ethylene yield requirements. All the results reported here were obtained in replicate measurements at each set of conditions.

Table 2 *Effective reaction temperature and reaction time for ethane pyrolysis experiments*

Furnace Set-Point °C	Ethane Feed Rate l / min	Effective Temperature °C	Effective Reaction Time msec
850	0.858	836	69
	0.430	860	132
	0.300	860	201
1050	2.750	918	24
	1.925	926	35
	1.166	957	46
	0.740	979	78
	0.370	987	148
1100	2.750	930	23
	1.925	946	32
	1.650	949	37

A. *TEMPERATURE PROFILES AND EFFECTIVE TEMPERATURES*

Three furnace set-points were chosen for the experimental runs: 850°C, 1050°C, and 1100°C. Various ethane feed rates were used at each set-point. Figure 2 shows typical temperature profiles, as a function of axial distance from reactor entrance, of reacting gas mixtures corresponding to each feed rate at a furnace set-point of 1050°C.

The effective temperatures taking into account both thermal and chemical expansion of reacting gas were calculated using a one dimensional transport model for ethane pyrolysis. The one dimensional transport model calculated the temperature as a

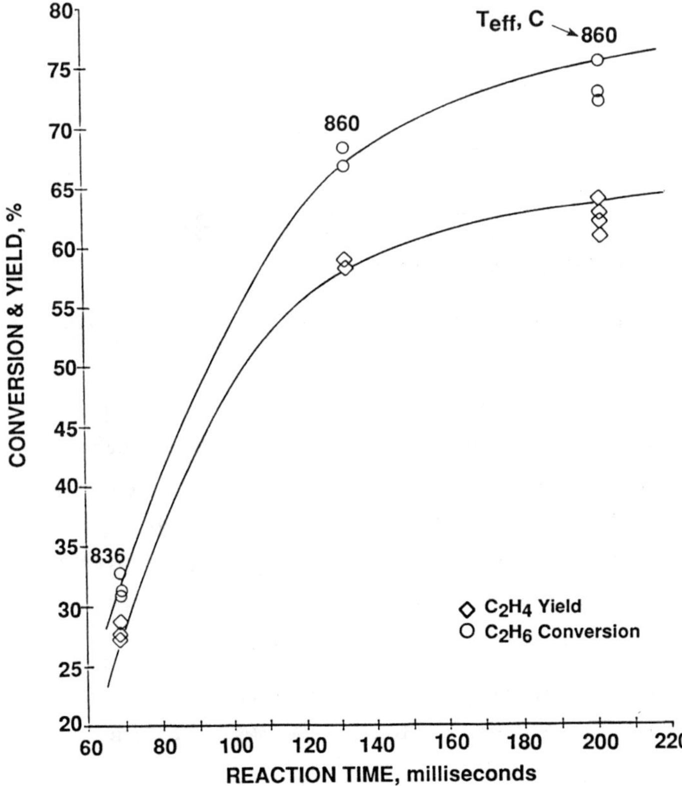

Fig. 3 *Ethane conversion and ethylene yield at a furnace set-point of 850°C*

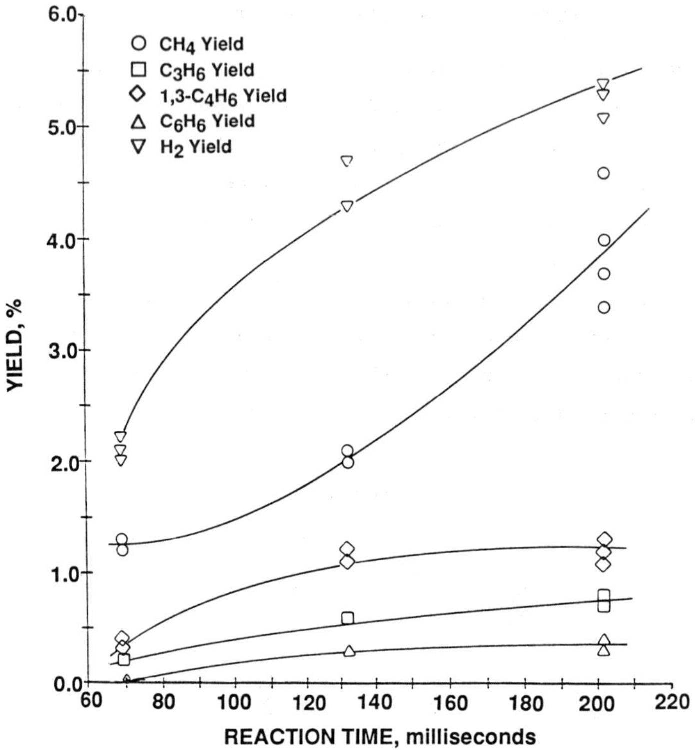

Fig. 4 *Pyrolysis product distribution at a furnace set-point of 850°C*

function of reaction time based on the actual measurements of the temperature as a function of distance along the reactor tube. The model then used the effective reaction time to calculate the time-averaged reaction temperature, i.e., the effective temperature, for each condition. The effective temperatures and reaction times are summarized in Table 2.

B. *ETHANE CONVERSION AND PRODUCT DISTRIBUTION*

The results of ethane conversion, ethylene yield, and distribution of other product species are depicted in Figures 3 to 8.

Figures 3 and 4 show the experimental results obtained at a furnace set-point of 850°C. Since the effective temperatures for

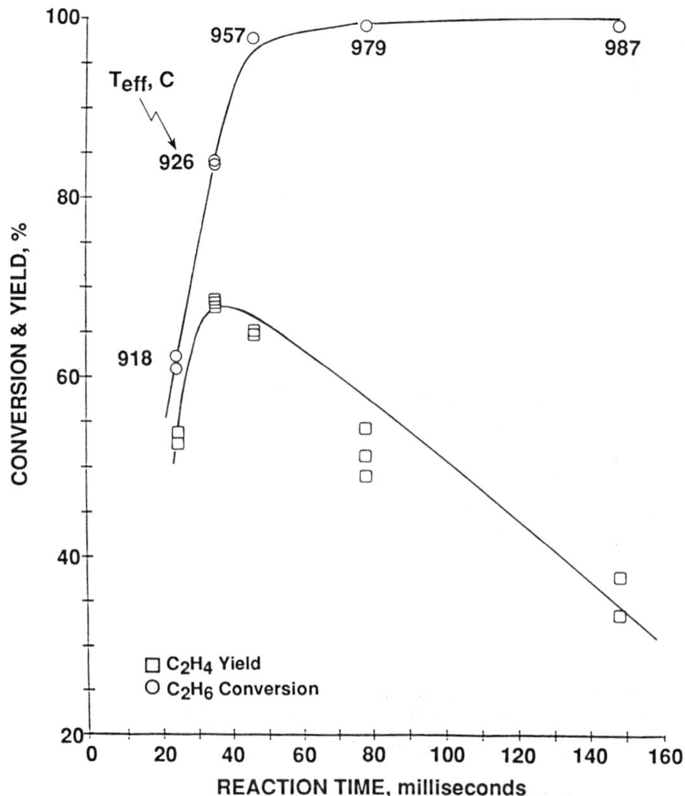

Fig. 5 *Ethane conversion and ethylene yield at a furnace set-point of 1050°C*

all the flow rates are very close, both Figures 3 and 4 can be viewed as showing the effects of reaction time on ethane pyrolysis at approximately 850°C. At 69 milliseconds, a 33% ethane conversion with a 28 wt% ethylene yield is obtained. As the reaction time increases, both ethane conversion and ethylene yield also increase. At t = 201 milliseconds, ethane conversion increases to 75% while ethylene yield increases to 62 wt%. Under the same conditions above, H_2 and CH_4 yields increase from 2 wt% to 5 wt% and from 1 wt% to 4 wt%, respectively. The product yields of other species such as propylene, 1,3-butadiene, and benzene, etc. are rather insignificant (i.e., below 1 wt%), as shown in Figure 4.

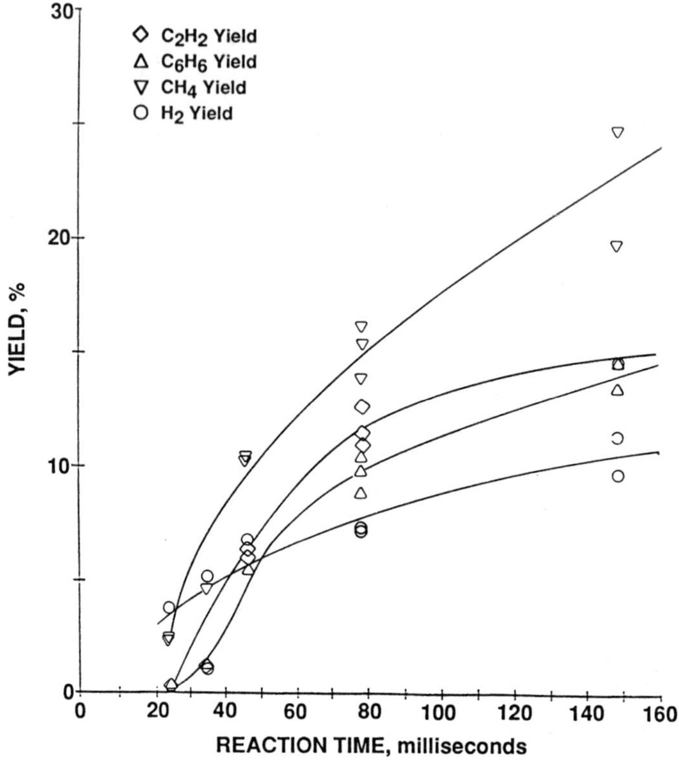

Fig. 6 *Pyrolysis product distribution at a furnace set-point of 1050°C*

Figures 5 and 6 present the results obtained at a furnace set-point of 1050°C. As shown in Figure 5, while ethane conversion increases from about 60% to almost 100% as reaction time increases, the ethylene yield curve peaks at T_{eff} = 926°C and t = 35 milliseconds. At the peak, a 70 wt% ethylene yield with an 83% ethane conversion, thus an 84 wt% selectivity to ethylene, is obtained. This defines the optimum condition for steamless ethane conversion to ethylene. The above ethylene yield exceeds significantly the best result of the current state-of-the-art technology which produces 45~50 wt% ethylene yield with about 60% ethane conversion on a per pass basis.

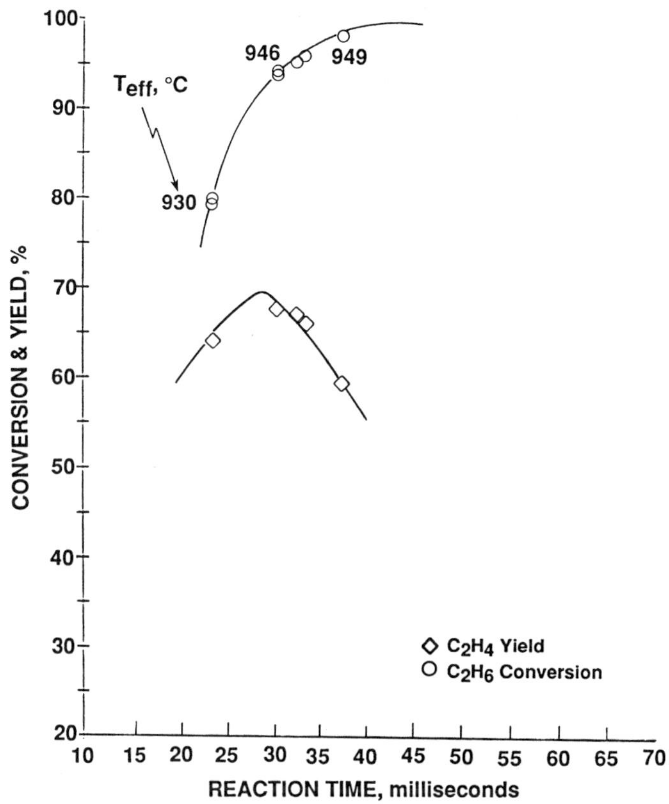

Fig. 7 *Ethane conversion and ethylene yield at a furnace set-point of 1100°C*

Ethane conversion reaches 100% when reaction time increases to 46 milliseconds. However, under this condition, ethylene yield decreases to 65 wt%. In fact, it keeps decreasing as reaction time increases. Ethylene yield reduces to 33 wt% when t = 148 milliseconds. Under these temperatures, other pyrolysis products increase very rapidly at the expense of ethylene. For example, as shown in Figure 6, CH_4 yield increases from 2 wt% at T_{eff} = 918°C and t = 24 milliseconds to 25 wt% at T_{eff} = 987°C and t = 148 milliseconds. In addition to methane, acetylene and benzene also increase rapidly and significantly under above conditions. They both increase from trace amounts to almost 15 wt%, respectively. Hydrogen increases from 4 wt% to 11 wt%.

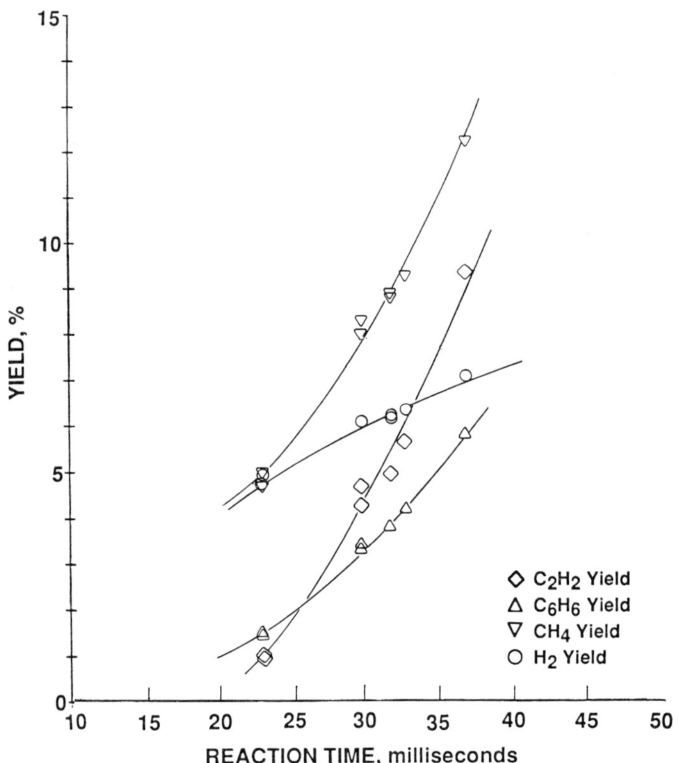

Fig. 8 *Pyrolysis product distribution at a furnace set-point of 1100°C*

The furnace set-point was raised to 1100°C to see whether ethylene yield could be further increased. The results are shown in Figures 7 and 8. As can be seen, the ethylene curve again exhibits a maximum. It peaks at T = 946°C and t = 32 milliseconds resulting in an ethylene yield of 67 wt% with a 95% ethane conversion, i.e., a 70.5 wt% ethylene selectivity. Apparently, this performance is not as good as that shown previously in Figure 5. Methane is still a major by-product. It increases from 5 wt% at T_{eff} = 930°C and t = 23 milliseconds to 12 wt% at T_{eff} = 949°C and t = 37 milliseconds. Both acetylene and benzene increase with increasing reaction time. However, acetylene exhibits a steeper upward slope than benzene, mainly because these high temperature conditions favor the formation of acetylene.

C. *EFFECTS OF REACTOR MATERIAL*

It is important to mention that all of the above results have been reproduced in both INCONEL 600 and α-SiC reactors. As far as ethane conversion and gaseous and liquid product distributions are concerned, there is no significant difference between these two reactor materials.

As regard to carbon deposition on reactor walls, there is no significant difference between α-SiC reactors and brand-new INCONEL 600 reactors. The carbon yield is in general less than 0.1 wt% under above conditions. However, carbon deposition becomes more and more significant in an INCONEL 600 reactor as the reactor tube is subjected to the coking and decoking cycle again and again and as the reactor tube ages. Up to 2 wt% carbon has been observed in pure ethane pyrolysis in INCONEL 600 reactors under above conditions. The details of the effects of reactor material on carbon deposition were presented elsewhere (6,7).

D. *EFFECTS OF H_2S AND H_2O*

Depicted in Figure 9 is the enhancement of coke deposition in an INCONEL 600 tube when ethane is pyrolyzed in a steamless environment with 25 and 100 ppmv of H_2S. Apparently, there is a synergistic effect of H_2S and surface roughness, composition, oxidation state, etc. Finally, when ethane is pyrolyzed in the presence of steam (0.3 to 1 H_2O to C_2H_6 weight ratio), the amount of carbon deposition in the INCONEL 600 reactor is reduced to a constant level regardless of the pyrolysis-decoking cycles. This amount coincides with that produced in an α-SiC reactor regardless of the age.

E. *EFFECTS OF PROPANE IN ETHANE FEED*

Industrial ethane feedstocks often contain various amounts of propane. Since propane pyrolyzes at lower temperature than ethane there is a concern that higher carbon formation may result

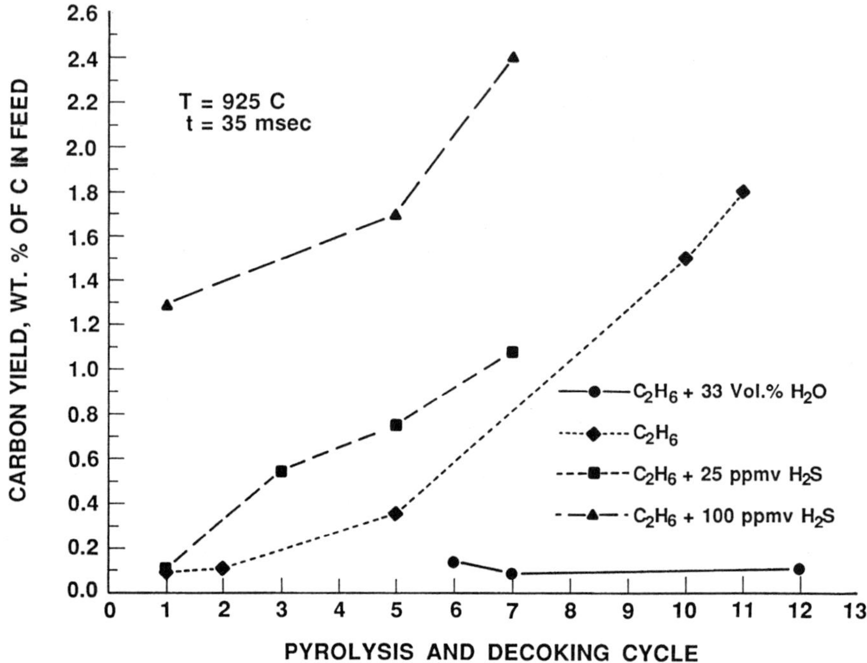

Fig. 9 *Effects of steam, hydrogen sulfide, pyrolysis-decoking cycle on carbon deposition*

from the pyrolysis of ethane-propane mixture at the usual ethane pyrolysis temperature. Removing propane from ethane feedstock is expensive and lowering the pyrolysis temperature will result in lower ethane conversion.

As shown in Table 3, using inert reactor surface (α-SiC) no increase in carbon formation was observed when the feed consisted of 12.5 vol% propane and 87.5 vol% ethane as compared to 100% ethane. With propane in the feed, the selectivity to methane increased while the selectivity to ethylene decreased. This is because for every ethylene produced a methane is formed. As expected, the selectivity to hydrogen is also lower because the major by-product is methane instead of hydrogen as in the pure ethane pyrolysis.

It is believed that the low carbon deposition on the reactor wall is related to propane having higher heat capacity than ethane. Thus propane has a cooling effect and the carbon precursors experience lower temperatures than that in pure ethane

Table 3 *Comparison of ethane pyrolysis with and without propane in feed*

Feed Composition, mol%		
C_2H_6	100	87.5
C_3H_8	0	12.5
Effective Temperature, °C	880	885
Conversion, wt%		
C_2H_6	83.5	85.2
C_3H_8	----	97.4
Total HC	83.5	87.4
Product Yield, wt%		
Liquid	2.81	5.63
Carbon	0.05	0.04
C_2H_4	61.93	55.49
Product Selectivity, wt%		
H_2	7.01	5.51
CH_4	7.66	13.57
C_2H_4	74.18	63.53
C_2H_2	1.88	3.68
C_3's	1.47	2.56
C_4's	4.00	4.41
C5+	3.37	6.45
Carbon	0.06	0.04

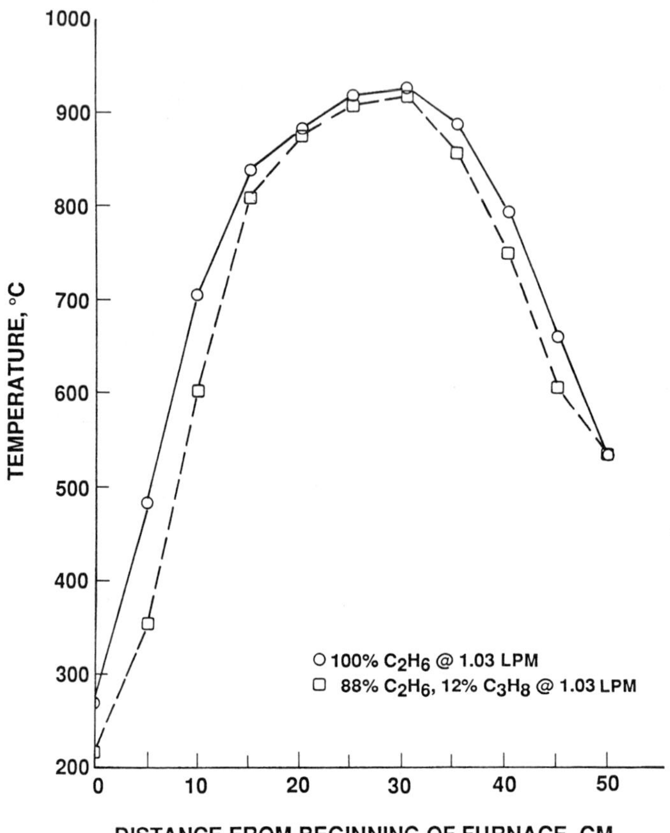

Fig. 10 *Comparison of temperature profiles of pure ethane and ethane with 12% propane in a 6 mm α-silicon carbide reactor at a furnace set-point of 985°C*

pyrolysis. Figure 10 shows the temperature profiles with a pure ethane feed and with a feed of ethane-propane mixture. Clearly, the latter achieved lower temperature at the same furnace setting.

F. *COMPARISON WITH MODEL PREDICTIONS*

The experimental results are compared with the one dimensional transport model predictions with a global first order ethane decomposition kinetics:

Table 4 *Comparison of experimental results and one-D transport model calculations*

Effective Temperature °C	Effective Reaction Time msec	Ethane Conversion, %	
		Experimental Results	One-D Model Calculations
836	69	23.1	23.5
860	132	64.0	64.2
854	201	74.3	74.4
918	24	59.6	55.5
926	35	82.7	73.6
957	46	97.7	98.7
979	78	98.9	99.7
987	148	99.1	100.0
930	23	78.5	68.5
946	32	95.4	94.6
949	37	98.4	98.5

$$k = 1 \times 10^{13} \exp(-65{,}500 \text{ cal-mole}^{-1}/RT) \text{ sec}^{-1}$$

The model provides an excellent fit to the experimental data. The comparison is shown in Table 4.

V. CONCLUSIONS

1. The optimum operating condition for steamless pyrolytic conversion of ethane to ethylene is identified as effective temperature = 925°C and reaction time = 35

milliseconds. Under this condition, a 70 wt% yield of ethylene with an 83% ethane conversion is obtained. This ethylene yield exceeds significantly the best result of the current state-of-the-art technology, which is in the range of 45~50 wt% ethylene yield with about 60% ethane conversion, on a per pass basis.

2. During ethane pyrolysis, yields of major by-products such as CH_4, C_2H_2, and C_6H_6 increase rapidly with increasing temperature and reaction time, at the expense of ethylene.
3. As far as ethane conversion and distribution of major gaseous and liquid products are concerned, there is no difference between INCONEL 600 and α-SiC reactors. However, carbon deposition is much more significant in INCONEL 600 reactors.
4. Propane addition to the ethane feed, up to 12.5 vol%, does not produce any adverse effect on carbon deposition.
5. The One-D transport model with a global first order ethane decomposition kinetics:

$$k = 1 \times 10^{13} \exp(-65{,}500 \text{ cal-mole}^{-1}/RT) \text{ sec}^{-1}$$

provides an excellent fit to our experimental data.

VI. REFERENCES

1. Hydrocarbon Processing, Special Report: 1987 Petrochemical Handbook, November 1987.

2. Kirk-Othmer Encyclopedia of Chemical Technology, Vols. 9 and 12, 1978

3. MacNab, A. J., Hydrocarbon Processing, December 1987.

4. Dunkleman, J. J., and Albright, L. F., Surface Effects During Pyrolysis of Ethane in Tubular Flow Reactors, in Industrial and Laboratory Pyrolysis, ACS Symp. Ser. No. 32, ed. by Albright, L. F., and Crynes, B. L. (1976).

5. McConnell, C. F., and Head, B. D. Pyrolysis of Ethane and Propane, in Pyrolysis, Theory and Industrial Practice, ed. by Albright, L. F., Crynes, B. L., and Corcoran, W. H., Academic Press, New York (1983).

6. Velenyi, L. J., Song, Y., and Metcalfe, J. E., 19th Biennial Conference on Carbon, University Park, PA (1989).

7. Velenyi, L. J., Song, Y., and Fagley, J. C., paper submitted to Carbon for publication (1990).

18

PRETREATMENT OF HIGH-ALLOY STEELS TO MINIMIZE COKING IN ETHYLENE FURNACES

GABOR SZECHY[1], TA-CHI LUAN, and L.F. ALBRIGHT*

School of Chemical Engineering
Purdue University
West Lafayette, IN 47907

Coke formation in the coils or transfer line exchangers (TLX's) of ethylene or vinyl chloride furnaces causes numerous detrimental effects including increased resistance to heat transfer, increased energy demands (and increased energy losses), increased temperatures in the coil walls resulting in reduced coil life, and frequent shut-downs of the unit for decoking purposes. There is hence considerable incentive to reduce, if not eliminate, coke formation or build up in the furnaces.

Coke is produced in ethylene furnaces by three mechanisms (1). In the first mechanism, coke is catalyzed by iron and nickel on the surface forming typically filamenteous coke. Such coke is particularly deleterious for at least three reasons. First nickel and iron are extracted from the surface resulting in corrosion of the steel surfaces. The nickel and iron becomes impregnated in the resulting coke. Second the metal-catalyzed coke acts as a collection site for the coke formed by the other two mechanisms, and hence promotes coke collection on the surface. Third, a porous type of coke is generally produced at and near the steel-coke interface. Such a porous

[1] Current address: Dept. of Chemical Technology, Technical University of Budapest, Hungary
* To whom correspondence should be addressed.

layer has poor thermal conductivity which reduces the overall heat transfer coefficients. The coke of the second mechanism is formed after tar or semi-tar droplets collect or are deposited on metal or coke surfaces. The coke of the third mechanism is produced when acetylene, olefins, and/or free radicals in the gas phase react with free radicals on the coke surface.

Techniques to reduce the nickel and iron contents on the surface of the high-alloy steels are highly desired. Aluminumized surfaces such as commercialized by The Alon Processing Company of Tarentum, Pennsylvania have been found to reduce drastically coke formation in both the laboratory (2,3,4) and in a commercial ethylene unit (5). Surfaces covered and impregnated with a thin layer of silica (or glass) also result in much lower levels of coking and in surfaces with little or no filamenteous coke (6,7). Coret tubes of the Toya Engineering Corp. are coated on the inner surfaces with a Fe-25Cr-9Mn-3Ni alloy. The high manganese content of this alloy is reported to be highly effective in reducing coke formation and in promoting long tube life.

In ethylene units, major changes occur in the surface composition of the high alloy steels after the coils are placed in service. With high-alloy steels such as Incoloy 800 or HK-40, the iron and nickel contents at the surface becomes greatly reduced with use and the chromium content becomes much enriched (8,9). Marek and Albright (4) found that the chromium content could also be increased on Incoloy 800 surfaces by heating it at 900°C in the presence of steam or oxygen for four hours. The titanium content on the surface also simultaneously increased from about 0.3 to 2-3%. The inner surfaces of the coils of select ethylene furnaces have nickel and iron contents that total 12-15 weight %. Yet it is significantly higher in other furnaces. The previous investigators did not, however, report the manganese contents on the surfaces of these coils.

Horsley and Cairns (10) pretreated a high alloy steel with hydrogen/steam mixtures at about 800°C for 4 hours; the nickel and iron contents at the surface of the steel were greatly reduced. In one case, they produced a surface containing 75% chromium and 25% manganese. The stabilities of the metal oxides at the pretreatment conditions explain these results. For example, when chromium and manganese diffuse to the surface, they are converted to stable oxides that have very low diffusivity values and hence remain at the surface. Nickel oxides and iron oxides are however not stable, and they are rather quickly replaced at the surface with chromium and manganese. Such pretreatments resulted in surfaces that had much reduced levels of coke formation. Pretreatments with gaseous mixtures of steam, hydrogen, and carbon oxides are also effective in reducing coke formation on the metal surfaces (11). These or modified treatments of the stainless steel surfaces could be adapted to ethylene furnaces. In addition, both steam and hydrogen are effective decoking gases (12). Steam forms metal oxides including chromium oxides. Metal carbides can also be formed when carbon oxides are present.

Methods of pretreating Incoloy 800 and HK-40 with mixtures of hydrogen, steam, and carbon oxides were investigated. Surfaces were found for which coke formation or collection were greatly reduced and for which the adhesion between the coke and metal was very poor.

EXPERIMENTAL DETAILS

The equipment and operating procedures reported by Albright and Marek (1) were essentially those employed in the present investigation. Various stainless steel coupons were positioned on the bottom of a 2 cm ID Vycor tube heated in a horizontal electrical furnace. The tube temperature was controlled at any desired level up to about 1000°C. Provisions were made to pass any of the following gases at controlled rates over the coupons in the tube: H_2, CO, acetylene, air or oxygen, helium, and/or nitrogen. Gaseous mixtures of about 2.6 mole % toluene in nitrogen were used at 860°C in coking experiments. Such a mixture was produced by bubbling nitrogen through liquid toluene in a flask at room conditions. For pretreating the coupons, gaseous mixtures of H_2/H_2O or CO/H_2O were used. Such mixtures were obtained by bubbling H_2 or CO through water in a flask usually maintained at room temperature.

The coupons were weighed with a Mettler balance before and after pretreatments, coking, or decoking experiments. The following were employed for analyses of the metal surface of the coupons, cross-sectional cuts of the coupons, and the coke produced: scanning-electron microscope (SEM) and attached EDAX; optical microscope; X-ray diffraction unit (XRD), and XPS unit with a sputtering auxiliary device attached. Surface roughness was measured using a Tencor profilometer.

RESULTS

Pretreatment of High Alloy Steels

Incoloy 800 and HK-40 coupons were pretreated with gaseous mixtures of hydrogen, carbon oxides, and steam at temperatures varying from about 660° to 960°C. Both unpolished (or as received) coupons and polished coupons were employed. The arithmetic average roughness of the surfaces of the unpolished Incoloy 800 coupons varied from about 2.2-2.7 μm, and that of polished coupons varied from 0.07-0.10 μm. Surface roughnesses of the polished HK-40 coupons varied from 0.01-0.015 μm. The polished coupons had surface compositions identical to those of the bulk steel. The surface compositions of the unpolished coupons were, however, slightly higher in both iron and nickel.

Several polished and several unpolished Incoloy 800 coupons were pretreated for four hours with the gas mixture containing approximately a 30:1 to 50:1 molar ratios of hydrogen to water vapor. Table 1 indicates the EDAX results for the surfaces of the polished coupons when the energy level of the electron beam was set at 15 kv. Large changes in the surface compositions occurred because of the pretreatments at 860° and especially 960°C. Much smaller changes occurred at 760°C and especially 660°C. At 960°C, and with four hours of pretreatment, the chromium, manganese, and titanium concentrations increased by factors of approximately 3, 30, and 10 respectively; simultaneously the iron and nickel values decreased by factors of about 16 and 22 respectively. The aluminum and silicon readings also increased significantly in some cases. Somewhat smaller changes of the surface composition were noted when unpolished Incoloy 800 coupons were pretreated with the H_2/steam mixture.

At 860°C, major changes of the surface composition occurred for pretreatments with H_2/steam mixtures within 1.5 hours, as indicated in Table 2. Changes in the next 6.5 hours were much smaller. During 8 hours of pretreatment, the chromium, manganese, and titanium contents had increased by factors of about 3.5, 20, and 11 respectively, whereas the surface changes for iron and nickel decreased by factors of about 13 and 21 respectively.

TABLE 1. Effect of Temperature on Surface Composition of Polished Incoloy 800 Surfaces after Four Hour Pretreatment Using H_2/Steam Mixture: EDAX setting of 15 kv

T°C	time, hr	wt, %						
		Al	Si	Ti	Cr	Mn	Fe	Ni
Untreated		0.41	0.96	0.24	21.67	0.82	42.90	32.28
		0.33	1.05	0.35	21.53	0.61	43.49	32.64
660	4	1.18	1.89	0.40	22.26	1.90	41.31	30.79
760	4	0.52	1.94	1.51	36.84	2.45	33.81	21.62
860	4	0.21	1.26	2.64	68.57	12.50	9.17	4.81
		0.92	1.64	2.63	66.67	11.10	10.25	6.03
960	4	0.26	0.24	2.81	65.72	26.07	3.29	0.96
		1.18	0.99	3.34	64.42	25.11	2.45	1.74
		0.15	0.54	2.69	68.55	23.78	2.42	1.61

TABLE 2. Effect of Time of Pretreatments on the Surface Composition of Polished Incoloy 800: EDAX setting of 15 kv

Pretreatment	T°C	Time, hr	wt. %						
			Al	Si	Ti	Cr	Mn	Fe	Ni
None	-	-	0.41	0.96	0.24	21.67	0.82	42.90	32.28
None	-	-	0.33	1.05	0.35	21.53	0.61	43.49	32.64
H_2/H_2O	860	1.5	0.77	1.11	2.68	62.35	9.63	14.18	8.05
		4	0.21	1.26	2.64	68.57	12.50	9.17	4.81
			0.92	1.64	2.63	66.67	11.10	10.25	6.03
		6	0.91	1.09	2.22	72.87	13.14	5.60	3.92
			0.97	1.09	2.23	70.86	12.72	7.65	3.15
		8	0.00	0.23	2.47	76.98	15.88	3.05	1.35
			0.32	0.28	2.93	77.46	15.87	2.26	0.87
CO	960	1	0.41	0.73	2.44	72.84	17.40	3.47	2.46
		3	0.27	0.34	2.13	71.29	20.42	3.14	1.47
		6	0.26	0.42	2.10	67.15	25.58	2.83	1.08
		24	-	0.24	2.78	63.77	31.60	1.57	0.04

EDAX readings, as presented here, are measurements of X-rays emitted from a layer of 1-2 microns thickness at the surface. The readings are not then necessarily the true composition at the surface itself. Several comparative readings were made using energy levels of the electron beam from the EDAX unit of 10, 15, and 25 kv. The beams with higher energy levels result in deeper penetrations below the surface. For untreated but polished coupons, the EDAX readings were almost identical at all energy levels, but were not for the pretreated coupons. At especially 25 kv, the iron and nickel readings of the pretreated coupons were usually significantly higher which suggests greater iron and nickel concentrations perhaps 1 to 2 microns below the surface. At 10 kv, high silicon and aluminum readings sometimes occurred. As will be indicated later when coke results are presented, at least the high silicon (or silica) readings are probably qualitatively correct. 10 kv may not, however, completely excite nickel, and hence 15 kv was used in most cases in this investigation.

The manganese values as reported here are considered to be essentially correct even though the manganese peaks coincide with chromium and iron peaks. Tests with pure chromium and essentially pure iron indicated no manganese. XPS analyses of several pretreated coupons gave manganese results consistent to those of EDAX. Several metal samples from the coils of industrial furnaces were re-analyzed using EDAX at 15 kv; these coupons had earlier been analyzed but not for manganese by Albright (8,9). Upon re-analysis, the inner surfaces of the coil of the first company were found to contain 12-21% manganese but those of the second company contained 8-10%.

The weight gains of the coupons, because of the pretreatments with hydrogen/steam mixtures, and the SEM results support the conclusion that the metal surfaces are oxidized as a result of the pretreatment. Photomicrographs of cross-sectional areas of the coupons indicate a white layer, as shown on Figure 1A. This layer as indicated by EDAX analysis had a relatively uniform composition and one that was similar to the surface itself. This layer was often 0.5-1.0 microns thick. After the average thickness was estimated, the volume of the white layer was estimated. The measured weight gain of the coupon suggests this layer is primarily Cr_2O_3 and manganese oxides. The silicon, aluminum, and titanium in the white layer are also undoubtedly oxidized.

Several polished Incoloy 800 coupons were pretreated with either pure CO or a 30:1 molar mixture of CO and steam. These pretreatments also resulted in large changes of the surface composition. Table 3 shows a comparison of the three pretreatments of 860°C for four hours. The changes in surface composition were similar in all three cases.

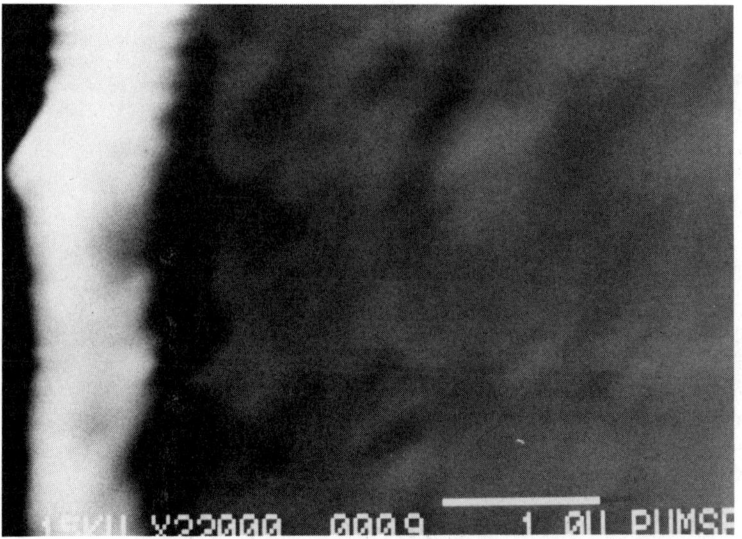

Figure 1A Cross-sectional View of Pretreated Incoloy 800 Coupon: Outer surface appears white and inner portion of metal is light gray.

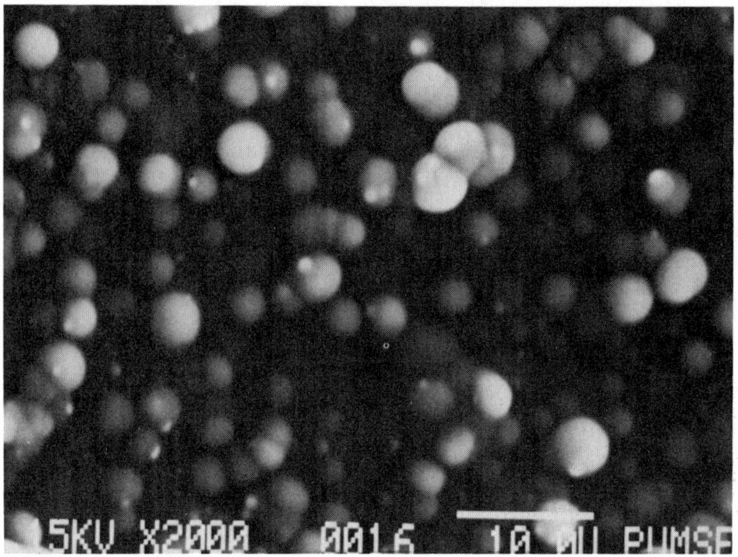

Figure 1B Globular Coke Deposited on Pretreated HK-40 Coupon

TABLE 3. Comparison of Surface Compositions for Three Pretreatments of Polished Incoloy 800 Coupons for Four Hours at 860°C; EDAX setting of 15 kv

	H_2/steam	CO/steam	CO
	wt %	wt %	wt %
Al	0.21	0.61	0.31
Si	1.26	1.27	1.13
Ti	2.64	2.43	2.64
Cr	68.57	69.22	69.67
Mn	12.50	11.25	11.32
Fe	9.17	9.65	9.76
Ni	4.81	5.57	5.18

Based especially on the results of pretreatments at 860° and 960°C, the changes in surface composition as determined by EDAX measurements at 15 kv can be divided into two time periods. In the first time period, the chromium content increases with time from 21-22% (for the untreated coupon) to about 75-77%; the manganese content increases to about 15-16%; the titanium content increases to about 2-3%; and the iron and nickel concentrations decrease to about 2-3% and 1-2% respectively. The results of the H_2/steam pretreatment at 860°C for up to 8 hours as reported in Table 2 are an example of the first time period. At 960°C, approximately 1.0 hour is needed to complete the first time period.

In the second time period, the chromium content at the surface decreases to values at least as low as 55%. The manganese content, however, increases to values that can be at least as high as 37%. During the second time period, the combined concentrations of chromium and manganese are about 89-92%. During this time period, the iron and nickel concentrations can decrease to values of at least <2% and <1% respectively. The results of the CO pretreatment at 960°C for pretreatments from 1 to 24 hours, as shown in Table 2, are an example of the second time period. The titanium concentration does not change appreciably during the last portion of the first time period or during the second time period.

Several HK-40 coupons were pretreated with H_2/steam, CO, and CO/steam. The changes in surface composition were smaller than those that occurred for Incoloy 800 as shown in Table 4. For the same pretreatment time, higher temperatures are required in order to obtain surface compositions for HK-40 similar to those obtained with Incoloy 800.

TABLE 4. Effect of Pretreatments on Polished HK-40 Coupons; EDAX setting of 15 kv

Pretreatment	T°C	Time, hr.	Surface wt. %				
			Si	Cr	Mn	Fe	Ni
Untreated	-	-	1.5	24.8	0.3	52.2	21.1
H_2/H_2O	860	4	2.5	55.8	2.8	30.8	7.6
H_2/H_2O	960	4	2.9	66.1	11.9	13.2	4.9
H_2/H_2O	960	12	2.4	71.6	13.6	6.9	1.9
CO	860	4	7.9	47.4	6.4	28.9	9.4
CO	960	4	5.5	61.1	8.6	19.1	5.3
CO/H_2O	960	4	2.9	63.9	10.0	20.5	2.4

TABLE 5. Coking Rates on Incoloy 800 Coupons (in grams/sq. meter, hr)

Coupon	Surface	Treatment	1st Rate	Decoking	2nd Rate	Decoking	3rd Rate
Incoloy 800	Unpolished	none	5.0-6.2	air	5.4	air	5.7
	Unpolished	H_2/H_2O	1.7-2.5	air	6.5	air	6.4
	Unpolished	none	5.0-6.2	H_2/H_2O	5.7	H_2H_2O	2.9 (P)
	Unpolished	H_2/H_2O	1.7-2.5	H_2/H_2O	3.1	H_2/H_2O	1.6 (P)
	Polished	none	1.7-2.0	air	3.7	air	5.6
	Polished	H_2/H_2O	1.8-2.8	air	4.9	air	5.6
	Polished	none	1.7-2.0	H_2/H_2O	1.8	H_2/H_2O	1.5 (P)
	Polished	H_2/H_2O	1.8-2.8	H_2/H_2O	2.3	H_2/H_2O	0.6 (P)
	Unpolished	CO	2.3 (P)	brushed	1.5 (P)		
	Polished	CO	1.4 (P)	air/H_2O	3.1		
	Polished	CO	2.7 (P)	brushed	1.0 (P)	brushed	1.5 (P)
	Polished	CO/H_2O	1.7 (P)	brushed	0 (P)	brushed	0.8 (P)
HK-40	Polished	none	4.4	H_2/H_2O	5.2	H_2/H_2O	4.0 (SP)
	Polished	H_2/H_2O	1.5 (P)	H_2/H_2O	0.5 (P)	H_2/H_2O	1.6 (P)
	Polished	CO	5.4 (P)	brushed	1.0 (P)	air/H_2O	3.0 (SP)

Notes: (P) in coking rate indicates easy peeling of the coke from the surface and (SP) indicates some peeling.

Coking Results:

Both Incoloy 800 and HK-40 coupons were coked using gaseous mixtures of 2.6 volume % toluene in nitrogen for four hours at 860°C. The results obtained with several Incoloy 800 coupons are summarized in Table 5. The coking rates are reported as grams coke/sq. meter, hr. In several cases, some of the coke apparently was either blown off or fell off the coupon before it was weighed which helps explain why the results of duplicate runs sometimes differed by 25-40%. In calculating the coking rates, the amount of coke on the coupon was assumed to be the difference in the weight of the coupon before and after coking. Better reproducibility was, in general, obtained on a relative basis when the rates were fairly high.

Unpolished Incoloy 800 coupons resulted in much higher rates of coking, 5.0-6.2 g/sq. meter, hr., as compared to those of the polished Incoloy 800 coupons, 1.7-2.0. Crynes and Crynes (13) had reported similar results. On unpolished Incoloy 800 coupons, both H_2/steam and CO pretreatments resulted in much reduced coking rates. The initial rate of coking on the polished Incoloy 800 coupons did not however change appreciably as a result of any of the three pretreatments.

The decoking technique used, however, had a major effect on the subsequent coking rate. Air decoking resulted in large increases in the rates of coking for all coupons. The rate at which the air was introduced to the coupon was maintained at a low level to prevent significant overheating. After two air decokings of Incoloy 800 coupons, the rates of coking were about 5.6-6.4. When, however, the H_2/steam mixture was passed over the coked coupons, decoking occurred resulting in the formation of carbon oxides and in the loss of adhesion between the coke and the metal. After two such decokings (following coking runs), the rate of coking on the Incoloy 800 coupons was much reduced, to values in the range of 0.6-1.6 g/m^2, hr. Similar low values were also noted for a HK-40 coupon that had been pretreated with H_2/H_2O or with CO and then brushed. In several cases for both Incoloy 800 and HK-40 coupons, the coke was removed from the coupon by either gentle brushing or gentle blowing. Part of the loss of adhesion may have occurred as the coupon was cooled and been caused by differences in the levels of thermal contraction of the coke and coupon. The lack of adhesion does not seem to be affected significantly by increased roughness of the surface resulting from the pretreatments or coking/decoking sequences, as shown in Table 6.

Tables 7 and 8 indicate the weights and weight changes of two Incoloy 800 coupons after various pretreatments, cokings, and decokings. Significant weight gains occurred for both coupons as a result of the pretreatment and/or the first coking-decoking sequence. No significant weight changes occurred subsequently which suggests that a stable and fairly durable surface layer had been produced. In addition, no differences in appearance or roughness were noted.

TABLE 6. Surface Roughness of Polished Incoloy 800 and HK-40 Coupons before and after Pretreatments

Coupon	Treatment			Roughness (arithmetic average) μm		
	Gas	T°C	Time, hr.	Before	After	After two cokings/decokings
Incoloy 800	H_2/H_2O	860	4	0.07-0.08	0.08-0.09	-
	CO	860	4	0.07-0.08	0.08-0.09	0.10-0.18
	CO	860	4	0.09-0.10	0.15-0.18	0.19-0.55
	CO/H_2O	860	4	0.08-0.09	0.13-0.18	0.16-0.26
HK-40	H_2/H_2O	860	4	0.01-0.015	0.08-0.09	
	CO	760	4	0.01-0.015	0.21-0.25	
	CO/H_2O	860	4	0.01-0.015	0.25-0.40	

Note: Decoking was made with H_2/H_2O mixtures at 860°C.

TABLE 7. Weight Change of Polished Incoloy 800 Coupon

Operation	Weight, g	Weight Change Between Cokings, mg	Weight Change Between Decokings, mg
None	3.3688	-	-
CO treatment	3.3714	-	-
After 1st coking	3.3763	-	-
After decoking and brushing	3.3740	-	2.6
After 2nd coking	3.3760	-0.3	-
After decoking and brushing	3.3738	-	-0.2
After 3rd decoking	3.3764	0.4	-
After decoking and brushing	3.3734	-	-0.4

Note: Decoking made with H_2/H_2O mixture for 4 hours. Coke peeled off surface.

TABLE 8. Weight Change of Unpolished Incoloy 800 Coupon

Operation	Weight g	Weight Change Between Cokings, mg	Weight Change Between Decoking, mg
None (initial coupon)	3.2191	-	-
After 1st coking	3.2327	-	-
After decoking and brushing	3.2296	-	10.5
After 2nd coking	3.2421	9.4	-
After decoking and brushing	3.2282	-	-1.4
After 3rd coking	3.2353	-6.8	-
After decoking and brushing	3.2283	-	0.1
After 4th coking	3.2332	-2.1	-
After decoking and brushing	3.2276	-	-0.7

Note: Decoking made with H_2/H_2O mixture for 4 hours.

TABLE 9. Surface Composition of Incoloy 800 Coupons after Coking-Decoking Sequences Leading to Poor Adhesion with Coke: EDAX Setting of 15 kv

	Al	Si	Ti	Cr	Mn	Fe	Ni
	as weight %						
No Pretreatment: 6 Cokings-Decokings	0.10	1.83	0.97	69.83	22.42	2.87	1.99
H_2/Steam Pretreatment	0.51	3.96	3.75	60.07	28.22	3.25	0.24
Plus 3 Coking-Decokings	0.20	3.33	2.92	59.91	28.42	3.37	2.03

Table 9 indicates the surface compositions, as measured by EDAX at 15 kv, for two Incoloy 800 coupons to which coke failed to adhere. The first coupon was coked and decoked several times. The first several decokings were accomplished using a H_2/steam-mixture, but brushing was effective for later decokings. The second coupon had initially been pretreated with the H_2/steam mixture, and then decoked several times by either H_2/steam or brushing. The chromium and manganese concentrations at the surfaces of these two coupons were in the ranges of 60-70% and 22-28% respectively; the combined totals for chromium and manganese were about 88-90%. The iron and nickel concentrations were about 3% and 2.0% respectively.

A coke sample that had flaked off an Incoloy 800 because of poor adhesion to the surface was analyzed using EDAX, as indicated in Table 10. On the side of coke that had been adjacent to the metal, chromium, manganese, and iron were detected in appreciable amounts. On the gas-side of the coke, chromium, iron, aluminum, and silicon were also detected in significant amounts. The X-ray count was lower by a factor of about three on the gas-side of the coke indicating less metal was present on that side. Some metals are apparently diffusing from the metal to the coke, and then through the coke. This finding seems to confirm that aluminum and especially silicon diffused in appreciable amounts to the surface of the stainless steel coupons.

TABLE 10. Composition of Metal in Coke that Failed to Adhere to Pretreated Incoloy 800: EDAX Setting of 15 kv

	Weight %						
Location	Al	Si	Ti	Cr	Mn	Fe	Ni
Metal Side	0.30	6.43	1.31	57.84	1.925	11.61	3.25
Metal Side	2.54	11.46	0.47	45.89	18.87	15.93	4.94
Metal Side	2.42	6.76	0.08	61.99	17.20	10.89	0.70
Gas Side	21.15	20.87	0.00	38.08	0.00	19.91	0.00

Note: The X-ray count was significantly higher on the metal side of the coke.

Filamenteous coke was never detected on Incoloy 800 or HK-40 surfaces that had been pretreated with H_2/H_2O, CO, or CO/H_2O at 860°C for four or more hours. All SEM microphotographs indicate a coke which is formed by deposition of tar or semi-tar droplets, as shown in Figure 1B. This particular coke was detected on an HK-40 coupon pretreated for four hours with H_2/H_2O at 860°C.

DISCUSSION OF RESULTS

The results of this investigation strongly suggest that suitable combinations of pretreatment and subsequent decoking of the inner surfaces of the coils in ethylene furnaces can minimize, if not completely eliminate, coke formation (or collection) in the coils. This conclusion is based on the assumption that the coke deposited on the surfaces adheres poorly when it is deposited (at high temperatures) as well as at room temperature (as demonstrated in this investigation). This assumption is supported by the low rates of coke formation on the pretreated coupons. The low rates probably indicate that some of the coke (or its precursors) were blown from the surface (because of poor adhesion) even though the velocities of the gas stream in the present investigation were very low.

At least three methods can be used to produce surfaces that result in both low rates of coke formation and in poor adhesion between the coke and the metal surface. These three methods are arranged in the probable order of easy application to ethylene coils.

(a) Pretreatment with CO/steam or CO. For subsequent decoking when required, a H_2/steam mixture is recommended.

(b) Pretreatment with H_2/steam followed at least one coking/decoking sequence. The decoking is accomplished using H_2/steam mixtures.

(c) No pretreatment, but two or three coking/decoking sequences are required in which a H_2/steam mixture is used for decoking.

Since decoking with H_2/steam mixtures result in the formation of CO and some CO_2, either CO or CO_2 is apparently needed to obtain the preferred metal surfaces. When only CO is used, some CO_2 is probably also formed on the heated metal surfaces. The preferred surface(s) is (are) apparently mixtures of metal oxides and metal carbides rich in chromium, manganese, plus possibly also silicon, aluminum and titanium. Large reductions of the iron and nickel concentrations obviously occur.

A 1991 patent application (15) reports a hydrogen/stream pretreatment for heat-resistant steels such as HK-40 and Incoloy 800. It was specified that the pretreatment should be at temperatures greater than 1800°F (982°C) in order to produce surfaces containing less than 3% nickel but having increased amounts of chromium or chromium oxides. These surfaces are reported to minimize coke formation. Of interest, surfaces with less than 3% nickel are currently found in at least some commercial ethylene furnaces (8). In the present investigation, such surfaces were, however, often produced at much lower temperatures, as shown in Tables 2, 3, and 4. The results of this investigation also strongly suggest the following. First, to obtain both reduced coking and probably improved stability, improved surfaces can be produced by procedures developed in this investigation. Second, the pretreatment techniques developed here are easier and cheaper to apply. Third, the criterion of a low nickel content plus a high chromium content on the surface is not sufficient for developing the preferred surfaces. The patent application further does not consider improved decoking procedures as outlined here.

The lack of filamenteous coke on pretreated surfaces explains in part at least why there was poor adhesion between the coke and the steel surface since filamenteous coke adheres strongly to the surface. The high surface concentrations of manganese and/or manganese oxides in combination with chromium and/or chromium oxides may, however, also provide other benefits including poor adhesion in their own right and desired catalytic reactions. Eastman, Kolts and Kimble (14) in Chapter 2 of this book report that supported manganese oxides are catalysts for oxidative dehydrogenation of ethane to produce ethylene; at pyrolysis conditions, steam acts as an oxidizing agent to produce metal oxides (12), such as manganese oxides.

Only limited information has yet been obtained on the durability or longevity of the pretreated surfaces while in service. Some surfaces may be brittle or have poor adhesion to the metal substrate; and certain surfaces may "wear" or flake off relatively rapidly. There is hence a need to use a pretreatment-decoking combination that also results in a stable and durable inner surface.

Although pretreating-decoking techniques such as reported here will greatly reduce coke collection on the inner surfaces of the coil, increased levels of coke formation may occur in the transfer-line exchanger (TLX) or the U-bends of the coil. With the proposed techniques, increased amounts of coke or coke precursors will likely become entrained in the high velocity gas stream passing through the straight sections of the coil, the U-bends in the coil, the transfer line, and finally the TLX. To obtain the projected benefits including longer run lengths between decokings or no decokings at all, the TLX and U-bends may need to be re-designed. Claims have been made that certain TLX's result in relatively small rates of coke formation. Or possibly the operation of the TLX will need to be modified.

Work is currently in progress to quantify better how different pretreatments-decokings and different operating conditions affect the surface compositions. The compositions of the pretreating gases are being expanded, and carbon dioxide will be tested in these mixtures. There is also a need to determine better how the metal composition changes as a function of depth (or distance below the surface). The chromium, manganese, and titanium concentrations for perhaps the 2-10 micron range below the surface are presumably reduced after a pretreatment. The iron and nickel concentrations at such depths are, however, probably increased. Such information will be useful in finding surfaces that will be durable in ethylene or other furnaces.

ACKNOWLEDGMENT

Shell Development Co. of Westhollow, TX. provided financial support.

REFERENCES

1. L.F. Albright and J.C. Marek, Ind. Eng. Chem. Research 27, 755 (1988).

2. L.F. Albright, C.F. McConnell, and K. Welther in Thermal Hydrocarbon Chemistry, Advances in Chemistry Series #183, Chapt. 11, pp. 193-203, American Chemical Soc., Washington, D.C. (1979).

3. M.J. Graff and L.F. Albright, Carbon 20, #4, 319 (1982).

4. J.C. Marek and L.F. Albright in Coke Formation on Metal Surfaces, eds. L.F. Albright and R.T.K. Baker, ACS Symposium Series #202, Chapt. 8, pp. 151-176, American Chemical Soc., Washington, D.C., 1982.

5. L.F. Albright and W.A. McGill, Oil and Gas Journal, Aug. 31, 1987, pp. 46-50.

6. R.T.K. Baker and J.J. Chludzinski, J. Catalysis 64, 464 (1980).

7. D.E. Brown, J.T.K. Clark, A.I. Foster, J.J. McCarroll, and M.L. Sims in Coke Formation on Metal Surfaces, eds. L.F. Albright and R.T.K. Baker, ACS Symposium Series #202, American Chemical Soc., Washington, D.C., 1982.

8. L.F. Albright, Oil and Gas Journal, Aug. 15, 1988, pp. 69-75.

9. L.F. Albright, Oil and Gas Journal, Sept. 15, 1988, pp. 90-96.

10. G.W. Horsley and J.A. Cairns, Application of Surface Science 18, 273 (1984).

11. British Patent #656,910 (1948).

12. C.H. Tsai and L.F. Albright in Industrial and Laboratory Pyrolyses, eds. L.F. Albright and B.L. Crynes, ACS Symposium Series #32, Chapt. 16, pp. 274-295, American Chemical Soc., Washington, D.C., 1976.

13. L.L. Crynes and B.L. Crynes, Ind. Eng. Chem. Research 26, 2139 (1987).

14. A.D. Eastman, J.H. Kolts, and J.B. Kimble, This Book, Chapter 2, 1991.

15. Shell Internationale Research Maatschappij, B.V., UK Patent Application #2,234,530, Feb. 6, 1991.

19

COUPLED METHANOL/HYDROCARBON CRACKING (CMHC)–A NEW ROUTE TO LOWER OLEFINS FROM METHANOL

S. Nowak, A. Martin and H. Günschel

Academy of Sciences
Central Institute of Organic Chemistry
Rudower Chaussee 5
1199 Berlin

I. INTRODUCTION

For the last ten years, the conversion of methanol into hydrocarbons has been an important research topic in C_1-chemistry. The synthesis of light olefins (MTO process) has been one of the most important subjects (1-4). Thus, methanol becomes an increasingly attractive raw material for the chemical industry. The development of ZSM-catalysts by the Mobil Oil Company has opened the way for producing low olefins, aromatics and high octane gasoline (5-7). Despite the impressive scientific results and technological advances including the MTG-plant in New Zealand (8), the conversion of methanol to hydrocarbons is not free of some shortcomings. They are the unsatisfactory degree of olefin selectivity, the high degree of product dilution, low space velocities and a low degree in methanol conversion. A fundamental problem for the conversion of methanol is the heat control.

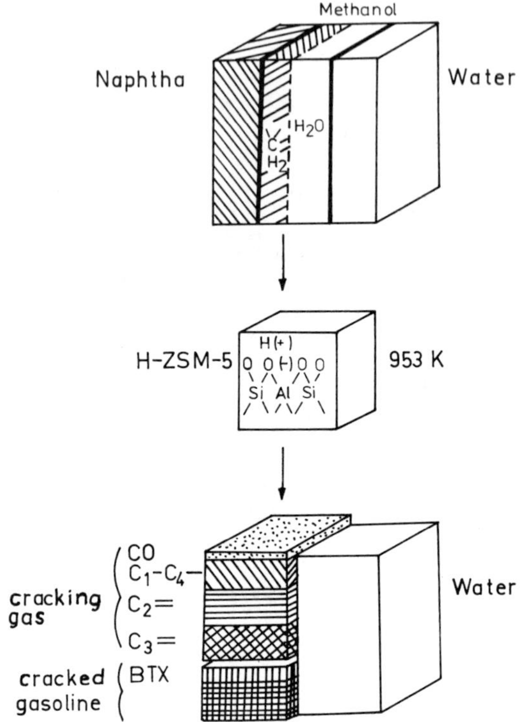

Figure 1 Scheme of CMHC reaction principle

A new reaction principle was introduced by the Coupled Methanol/Hydrocarbon Cracking (CMHC) - in German: GEMEKS - Gekoppelte Methanol/Kohlenwasserstoff Spaltung - carried out under the same reaction conditions (9-13) on H-ZSM-5 type zeolites. At high temperatures, up to 953 K, the addition of hydrocarbon to the methanol feed in a defined ratio, calculated by adding up the standard reaction enthalpies of the feed components (law of HESS), leads to the conversion of both methanol and hydrocarbon into lower olefins, gasoline and aromatics in a nearly thermoneutral reaction in the presence of H-ZSM-5 catalysts (Fig. 1). The process requires the development of a thermostable catalyst sim-

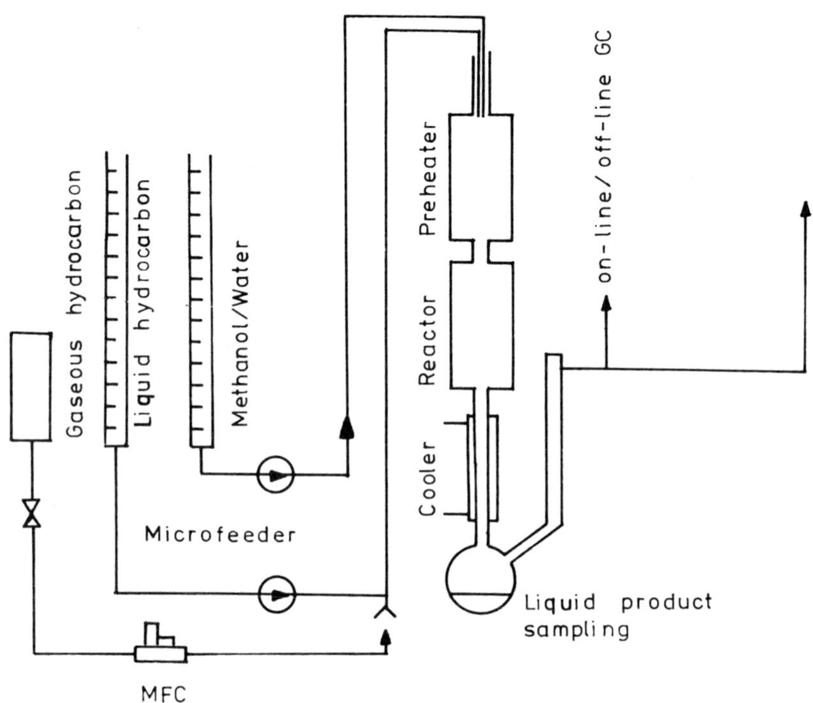

Figure 2 Laboratory equipment

ultaneously converting methanol and hydrocarbons with high selectivity into lower olefins in high space-time yields.

II. EXPERIMENTAL

A. Apparatus and reaction procedure

The CMHC experiments were carried out in bench-scale reactors of quartz glass with catalyst volumes up to 30 ml. Figure 2 demonstrates a principle view of a CMHC laboratory equipment. Methanol, hydrocarbons and water were pumped and vaporized separately. Gaseous hydrocarbons were fed directly into the preheater. After preheating the mixed streams were passed over the catalyst bed. The effluents were quenched to separate gaseous and liquid products.

The same reaction has also been studied on technical scale using two stainless steel reactors with catalyst volumes of 250 ml each. This apparatus operated in a swing regime, one reactor worked under stream, the other one under decoking. In this way, operation times up to 80 hours on-stream were realized.

The reaction products were analyzed by on-line or off-line gas chromatography and the results were stored in a computer data base.

Pure hydrocarbons, e.g. n-decane, n-octane, n-butane or isobutene, and technical fractions, e.g. naphtha (boiling range 331 - 449 K) or the isobutene/n-butene fraction from pyrolysis gases, methanol as well as mixtures of methanol and hydrocarbons were employed in the investigations. All conversion data and product yields described in the tables and figures have been calculated

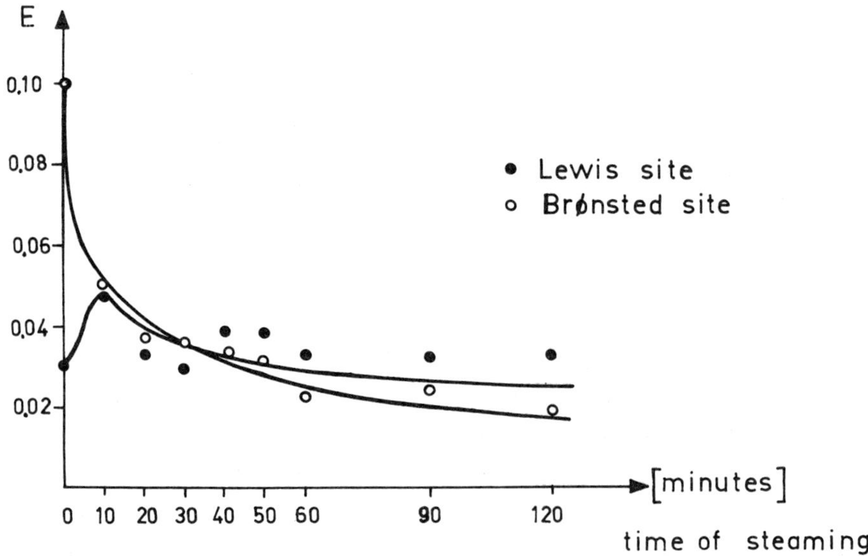

Figure 3 Bronsted- and Lewis-site extinction depending on steaming time at 953 K (from IR spectra using the pyridine chemisorption technique)

on the CH_2 part in the feed. After the reaction the zeolite catalysts could be regenerated by air or other oxygen-containing gases at ca. 873 - 923 K.

B. Catalysts

As standard catalyst an industrial H-ZSM-5 zeolite (template-free synthesis) with a Si-to-Al ratio of 16 was used. Furthermore, pretreated ZSM-5 samples were carried out as catalysts. These samples were dealuminated by steam treatment with subsequent acid leaching to rearrange the extra-framework Al formed. During such pretreatments, the number of acidic sites (Bronsted) was drastically reduced (Fig. 3) as shown by IR spectroscopy pyridine chemisorption method (14). Other catalysts applied were Si/Fe- and Si/Fe/Al-containing ZSM-5 type zeolites (13). Recently in the CMHC investigations also P-containing ZSM-5 samples were included to lower the Bronsted acidity and to suppress the dealumination phenomenon.

C. Characterization methods

IR spectroscopic characterization of fresh and used, pretreated and regenerated zeolites was carried out with a ZEISS UR 20 spectrometer using the pyridine adsorption technique to determine the acidity of both Bronsted and Lewis sites

By means of ^{1}H, ^{27}Al and ^{29}Si MAS NMR the content of OH-groups in silanol and bridging (Bronsted) sites as well as the ratios of framework Si-to-Al were determined in the samples mentioned. The measurements were performed on HSS-270 pulse spectrometer and MSL-300 (Bruker) in the laboratory of Prof. H. Pfeifer at the Leipzig university.

EPR spectra of the applied Fe-containing zeolites were recorded on a Varian E 4 spectrometer operating at X-band microwave frequency. The recorded spectra allow to distinguish between Fe^{3+} in tetrahedral coordination in the zeolite framework and Fe^{3+} in occluded oxides or hydroxides and in OH symmetry in cationic sites of the hydrated zeolite.

X-ray methods were applied to determine the crystallinity of fresh and used samples. Powder patterns were performed with a Guinier equipment with CuK_{α} - radiation.

Furthermore, number and strength of acidic sites of some selected samples were determined by the TPD of ammonia. These experiments were performed by gas-chromatographic measurement of the amount of ammonia desorbed upon heating ammonia-treated samples (treated with ammonia at 393 K followed by desorption of physically adsorbed ammonia at the same temperature for 30 min) at a rate of 12 deg/min.

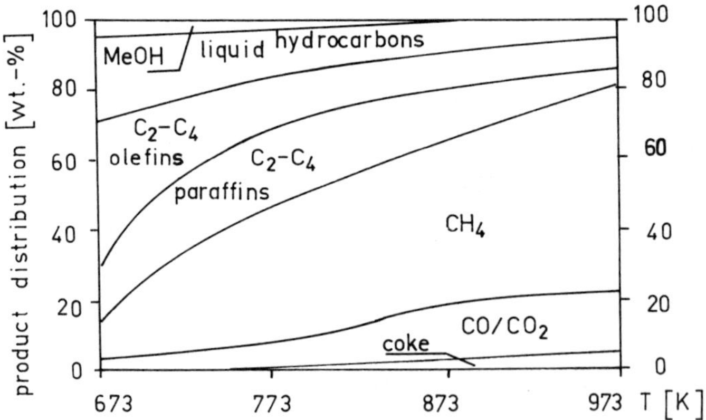

Figure 4 Product distribution obtained during methanol conversion at reaction temperatures up to 973 K on H-ZSM-5 (WHSV = 1.5/h, atmospheric pressure)

III. RESULTS AND DISCUSSION

A. CMHC investigations on H-ZSM-5 (Si/Al)

Typical results of the methanol-to-hydrocarbon reaction on H-ZSM-5 zeolite (Si-to-Al synthesis ratio = 16) are shown in Fig. 4. The conversion is completed at ca. 873 K. As expected, above this temperature the formation of methanol dehydrogenation products (CO and H_2) and other undesired by-products (CH_4) and also coke is drastically increased.

The cracking activity has been investigated on the same catalytic system and under the same reaction conditions using industrial n-butane containing ca. 5 - 7 % isobutane as feed (Fig. 5). Beside a low n-butane conversion, at lower temperatures the generation of lower paraffins predominated, whereas at higher temperatures the formation of lower olefins increased. At ca. 873 K the highest yield of aromatic compounds was obtained. This amount declined with rising temperature up to 973 K and the coke formation nearly doubled.

In analogy to the conversion of the individual compounds the coupled reaction of methanol and n-butane has been investigated. The methanol-to-n-butane ratio was ca. 3 : 1 to obtain an almost thermoneutral reaction. Figure 6 mirrors the temperature dependence of this reaction. The combined conversion leads to high yields of ethylene and propylene. The major part of the C_2 - C_4 paraffin region is attributed to unreacted n-butane. The formation of the by-products CO and CH_4 was drastically suppressed, the methanol dehydrogenation is inhibited by admixed hydrocarbon feed as shown additionally by ^{14}C - tracer experiments with labelled methanol.

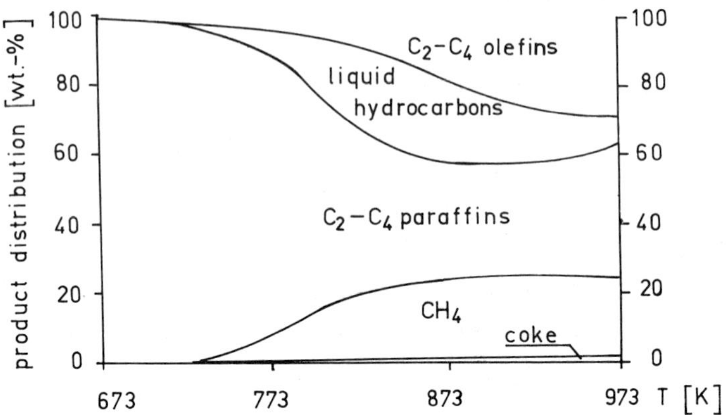

Figure 5 Product distribution obtained during n-butane cracking at reaction temperatures up to 973 K on H-ZSM-5 (WHSV = 1.5/h, atmospheric pressure)

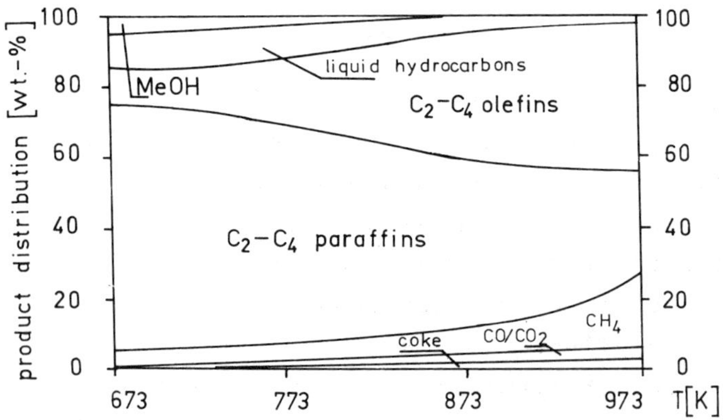

Figure 6 Product distribution obtained during CMHC of a methanol/n-butane feed (ratio = 3 : 1) at reaction temperatures up to 973 K on H-ZSM-5 (WHSV = 3/h, atmospheric pressure)

Table 1 underlines these observations of suppressing the by-product formation and rising the lower olefin yield by admixing hydrocarbon to the methanol feed. The table demonstrates results of methanol and methanol/naphtha conversion under the same reaction conditions and on the same catalyst as used in the experiments described above.

During our investigations using different types of model hydrocarbons, high yields and space-time yields of lower olefins were obtained. Table 2 shows the achieved space-time yields of lower olefins (C_2, C_3) and aromatics.

When the hydrocarbon feed is a straight - run naphtha, high space-time yields of C_2 - C_4 olefins and aromatics are obtained. No durene is found in the product stream. Compared with the naphtha fed, the liquid product has a 20 - 30 points higher octane number (ca. 80 - 90).

TABLE 1 Conversion of methanol and a methanol/naphtha mixture at 953 K on H-ZSM-5 and atmospheric pressure

Reaction conditions		without naphtha	with naphtha
Methanol	(g/g*h)	1.5	1.7
Naphtha	(g/g*h)	–	0.9
Water	(g/g*h)	1.3	0.4
Conversion	(wt.-%)		
Methanol		97.9	99.3
Yield	(wt.-%)		
Methane		20.3	10.3
Ethylene		18.0	21.0
Propylene		17.2	19.2
Butenes		6.6	4.5
Benzene		1.1	3.1
Toluene		5.8	3.4
Xylenes		11.7	12.1
Coke		2.8	1.2
C in CO_x		8.1	4.8

Long-time tests in CMHC reaction with naphtha as feed showed a decline in catalyst activity in spite of decoking cycles. Figure 7 shows this decrease in naphtha conversion related to the development of the product space-time yields during time on-stream. The reason for for the decline in conversion is the rapid dealumination of the zeolite framework, caused by the severe reaction conditions applied.

Another indication of dealumination is given by the propylene-to-propane ratio dependent on time on-stream. Starting at a ratio of ca. 4, we obtained a ratio of ca. 25 after 30 hours on-stream. The withdrawal of framework aluminium connected with a loss of Bronsted acid sites leads to a catalyst with less but stronger acid sites, which are applied especially for higher space-time yields of lower olefins and suppressed paraffin formation. The fact of dealumination is underlined also in Table 3 by some results of chemical and physical characterization of fresh and used H-ZSM-5. In partic-

TABLE 2 Space-time yields of lower olefins and aromatics in the CMHC reaction with different hydrocarbons as feed

Feed	Space-time yields (g/kg*h)	
	C_2/C_3-Olefins	Aromatics
n-Butane	530	39
Isobutene	1100	158
Butene-Mixture*	960	166
Naphtha	1000	250-400

* 45% isobutene, 27% n-butene, 15% cis-/trans-butene, 13% saturated C_4-hydrocarbons; (industrial fraction)

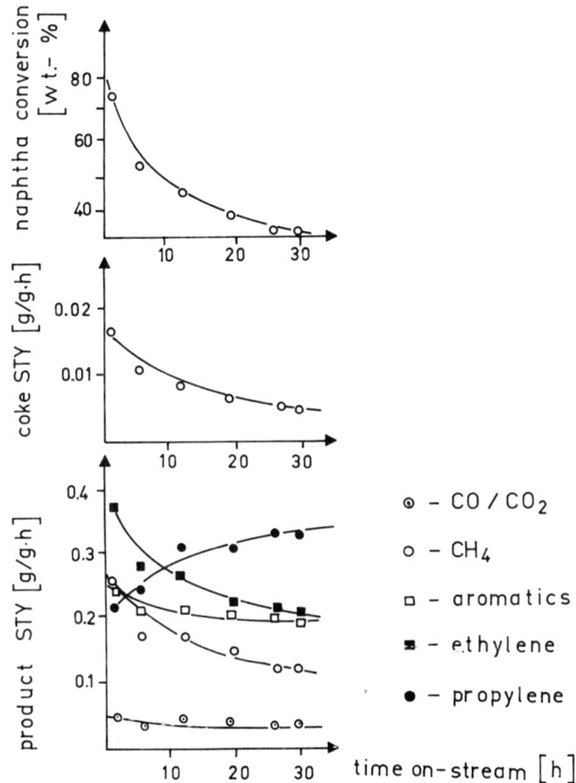

Figure 7 Naphtha conversion, coke and product space-time yields obtained during the first 30 hours of CMHC reaction (Methanol-to-naphtha ratio = 2 : 1, WHSV = 2.5/h, atmospheric pressure)

ular, MAS NMR data clearly show the dealumination of the used zeolite sample during time on-stream.

B. Acid reactivation of H-ZSM-5 (Si/Al)

The reaction conditions during CMHC correspond to a strong hydrothermal treatment and cause a rapid dealumination. This results in the proposed condensation of the extra-framework aluminium-species formed and their migration to the outer surface connected with a slight blockage of the pores. Dealumination and migration lead to a decrease in activity and declined yields of lower olefins.

Acid reactivation of such deactvated zeolites leads to an increased conversion (15). Furthermore, by steam pretreatment and acid leaching of fresh zeolite samples we obtained a more stable zeolite activity and constant product distribution during time on-stream.

Table 4 demonstrates, for example, reaction results obtained with pretreated H-ZSM-5 (synthesis Si-to-Al ratio = 16). The sample was hydrothermally treated for 2.5 hours at 953 K. After steaming the acidity was checked showing a drastic decline. For further treat-

TABLE 3 Chemical and physical characteristic parameters of applied fresh and used H-ZSM-5

H-ZSM-5		fresh	used
SiO_2	(mg/g)	905	940
Al_2O_3	(mg/g)	41	40
Si-to-Al ratio	(mol/mol)	19	20
Acidity	(meq/g)	0.79	0.09
^1H MAS NMR			
SiOH	(OH/g)	$27*10^{19}$	$17*10^{19}$
SiOHAl	(OH/g)	$38*10^{19}$	0
^{29}Si MAS NMR			
Si-to-Al (framework)		18	*
^{27}Al MAS NMR			
Si-to-Al (framework)		19	≥500
Al^{IV} (per channel crossing)		4.5	*

* no signal for $Si(OAl)(OSi)_3$ or framework Al in tetrahedral coordination

ment, the samples were leached with 0.2 N oxalic acid or with an equimolecular solution of the corresponding ammonium salt. After this acid treatment an increase in acidity could be observed again.

After the first hours on-stream the data obtained clearly show a stable yield of the main products, connected with a constant conversion of naphtha. Other pretreatment agents, such as hydrochloric acid and lactic acid or their ammonium salt solutions, have also been checked and they have proved to be useful, too.

Figure 8 demonstrates investigations using naphtha as feed and an above-described pretreated catalyst. No remarkable changes in the activity or in the selectivity could be detected.

TABLE 4 Main product distribution during CMHC with naphtha as hydrocarbon feed on H-ZSM-5 at 953 K pretreated hydrothermally with subsequent leaching by oxalic-acid or ammonium-oxalate solution (untreated sample C for comparison)

Parameter	Catalyst sample A-$(COOH)_2$			B-$(COONH_4)_2$			C-untreated		
Acidity (meq/g)									
- fresh	800			800			800		
- hydrotherm. tr.	76			76			-		
- acidic tr.	180			178			-		
CMHC-cycle*	1	2	3	1	2	3	1	2	3
Ethylene	25	21	19	23	22	21	25	19	12
Propylene	22	21	22	23	21	21	23	20	15
Aromatics	15	15	15	14	14	14	17	15	12
Acidity (meq/g)									
- after test 3	80			80			90		

* reaction cycle (1 hour with subsequent decoking)

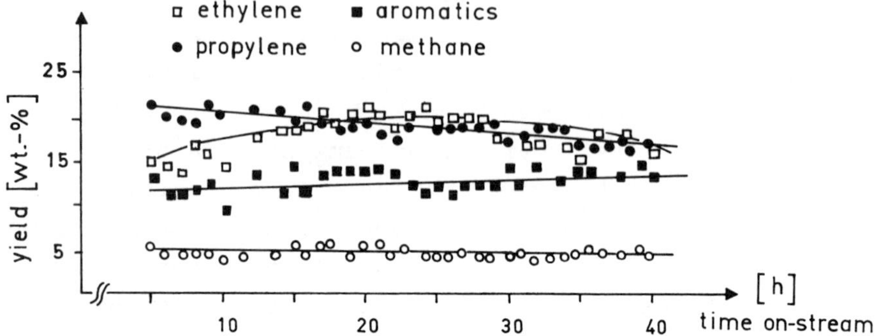

Figure 8 CMHC with naphtha as feed on hydrothermally and oxalic-acid pretreated H-ZSM-5 at 953 K (Methanol-to-naphtha ratio = 2 : 1, WHSV = 2.5/h, atmospheric pressure)

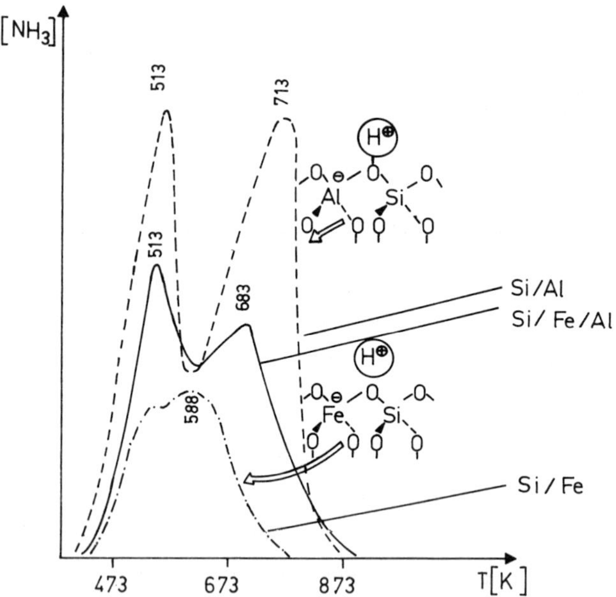

Figure 9 TPDA plot of Si/Al-, Si/Fe- and Si/Fe/Al-containing ZSM-5 samples

Such preactivation is aimed at lowering the zeolite acidity to obtain a stable product distribution with high olefin yields. Further possibilities of lowering the acidity are the isomorphous substitution of aluminium by iron during zeolite synthesis steps or the treatment of only Si/Al-containing zeolites with P-containing compounds.

C. Fe-containing zeolites used as catalysts during CMHC

The replacement of the Al ion in the ZSM-5 matrix by various kinds of metal ions could modify the behaviour of active sites along with the acidic properties. In comparison with Si/Al zeolites, the yields of desired olefins could be increased mainly by ZSM-5 type zeolites with different amounts of iron incorporated. Furthermore, owing to the lower acidity of such molecular sieves, a decline in the deposition of coke has been observed recently. A TPD plot (Fig. 9, TPD of ammonia)

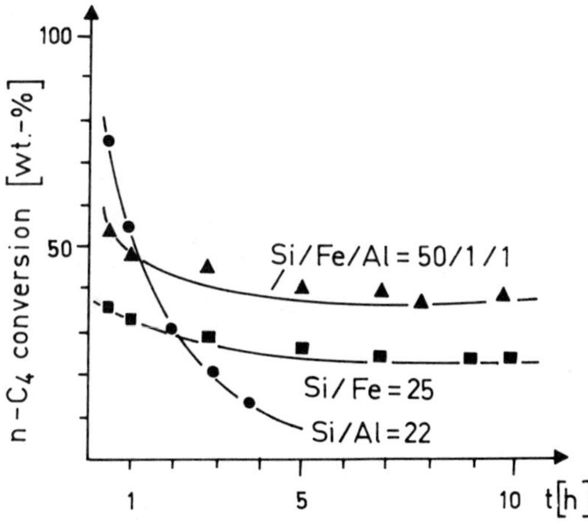

Figure 10 n-Butane conversion during CMHC at 953 K on Si/Al-, Si/Fe- and Si/Fe/Al-containing ZSM-5 samples depending on time on-stream (WHSV = 3/h, methanol-to-n-butane ratio = 3 : 1, atmospheric pressure)

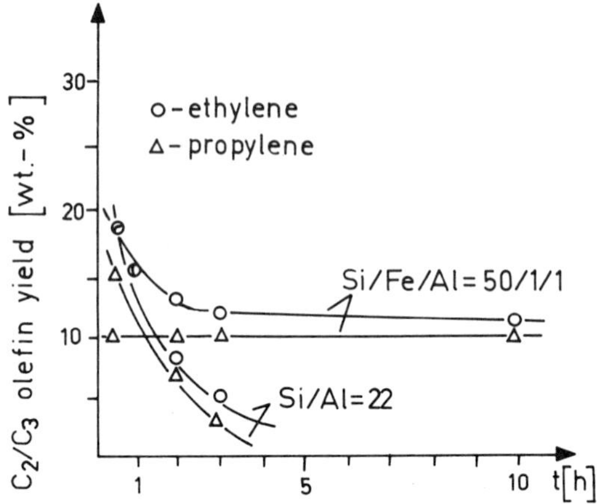

Figure 11 Olefin yields obtained during CMHC at 953 K on Si/Al- and Si/Fe/Al-containing ZSM-5 samples depending on time on-stream (WHSV = 3/h, methanol-to-n-butane = 3 : 1, atmospheric pressure)

shows different strengths of acidity for three samples. Framework Al results in two peaks at ca. 500 and 710 K. Framework Fe with the weaker acidity shows one broad peak at 585 K. Finally, framework Al and Fe in one sample result in the same peaks known from framework Al, but the high temperature peak indicating the strong Bronsted acidity is decreased.

The use of zeolites containing both iron and aluminium effected the supposed stabilization of the hydrocarbon conversion level caused by the lowered acidity (Bronsted). Figure 10 depicts these considerations during CMHC with n-butane as hydrocarbon feed.

Already after 10 minutes time on-stream the n-butane conversion on Si/Fe/Al zeolite was higher than on solely aluminium-containing ZSM-5. After four hours a significant decline in hydrocarbon conversion on Si/Al zeolite was observed, whereas the decrease on Si/Fe/Al was very low. The decrease in n-butane conversion led also to a decline in the yield of olefins formed, as shown in Fig. 11, but the olefin yields obtained on the Si/Fe/Al sample were nearly constant after three hours time on-stream.

In Table 5, the results obtained with iron-containing zeolites in the CMHC reaction are compared with the results observed with only aluminium-containing zeolite. These data reflect the more stable hydrocarbon conversion on iron-containing zeolites and the higher level of hydrocarbon conversion and olefin production on the Si/Fe/Al zeolite as well. The prevention of catalyst deactivation by incorporation of iron results in the prolongation of catalyst lifetime.

IV. ECONOMIC EVALUATION OF CMHC

The economy of CMHC firstly depends on the prices of the feedstocks methanol and the hydrocarbons used. The prices fluctuate according to the availability on the global market. Therefore, a consideration of relative prices is of advantage.

The following example demonstrates an evaluation of CMHC with a straight-run naphtha applied as hydrocarbon. The naphtha cracking in conventional olefin plants pro-

TABLE 5 Products obtained by CMHC reaction using Si/Al-, Si/Fe- and Si/Fe/Al-containing ZSM-5 zeolites as catalysts (WHSV = 3/h, methanol-to-n-butane ratio = 3 : 1, atmospheric pressure, reaction temperature = 953 K)

Catalyst		Si/Al		Si/Fe		Si/Fe/Al	
Reaction time	(h)	1	3	1	3	1	3
Conversion	(wt.-%)						
Methanol		99	98	99	99	100	100
n-Butane		53	30	40	40	63	61
Yield	(wt.-%)						
C_2-C_4 olefins		28	14	16	15	32	28

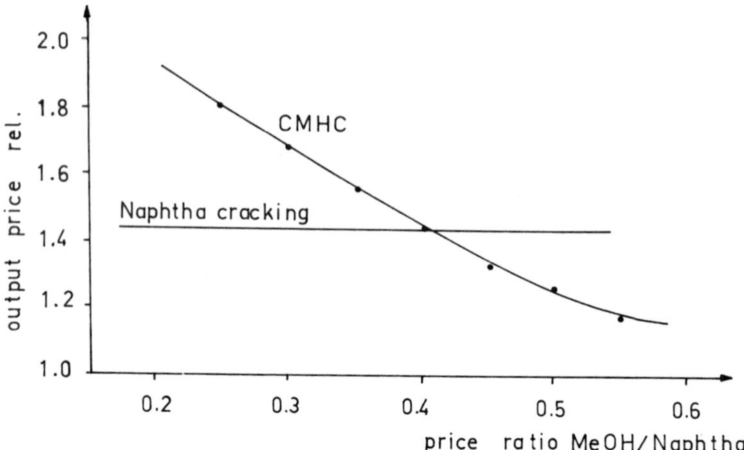

Figure 12 Uplift depending on the price methanol-to-naphtha ratio during CMHC with a methanol-to-naphtha feed ratio = 2 : 1

duces chemicals with a relative overall value of about 1.45 based on a feed value = 1. A corresponding value of 1.45 and an uplift in CMHC is possible only at a price methanol-to-naphtha ratio ≥ 0.4 implying a feed methanol-to-naphtha ratio of ca. 2 : 1.

These considerations concerning the output price in dependence on the price methanol-to-naphtha ratio are demonstrated in Fig. 12.

The feed ratio is the next important economic criterion. At the mentioned feed ratio of 2 : 1 the balance of the summarized standard enthalpies of all partial reactions is compensated. That means, no energy has to be introduced or discharged for maintaining temperature constancy and feed conversion in the catalyst bed. Another important economic parameter is the lifetime of the used ZSM-5 catalysts. Based on latest results in a laboratory unit, lifetimes of about 1000 hours may be forecasted.

V. CONCLUSIONS

Although both the knowledge of the specific mode of action of the zeolite catalyst and the detailed design of the technological process are still incomplete, the CMHC principle may open up an interesting route to the generation of lower olefins. The advantages include the realization of an almost thermoneutral reaction, the high space velocity of the feed products and high space-time yields of C_2 - C_4 olefins.

Especially in the field of catalyst development more stable catalysts for CMHC were obtained by different pretreatment methods or isomorphous substitution of aluminium in the zeolite framework. One reason for the better catalytic behaviour could be the lower acidity of these catalyst samples compared to freshly synthesized H-ZSM-5. Thus, the rapid dealumination connected with structural rearrangement in the zeolite crystallite could be suppressed and, furthermore, a lower deposit of coke could be observed.

VI. REFERENCES

1. S. L. Meisel, J. P. McCullough, C. H. Lechthaler and P. B. Weisz, CHEMTECH 6: 86 (1976).
2. T. Inui and Y. Takegami, Hydrocarbon Process. : 117 (1982/11).
3. C. D. Chang, Catal. Rev., Sci. Eng. 25: 1 (1983).
4. Y. C. Hu, Hydrocarbon Process. : 88 (1983/5).
5. R. J. Argauer and G. R. Landolt, US-Patent 3.702.886 (1972).
6. N. Y. Chen and W. E. Garwood, Catal. Rev., Sci. Eng. 28: 185 (1986).
7. D. H. Olson, G. T. Kokotailo, S. L. Lawton and W. M. Meier, J. Phys. Chem. 85: 2238 (1981)
8. C. J. Maiden in D. M. Bibby, C. D. Chang, R. F. Howe and S. Yurchak (Eds.), Studies in Surface Science and Catalysis, Vol. 36, Elsevier, Amsterdam,p. 1 (1988).

9. S. Nowak, H. Günschel, A. Martin, K. Anders and B. Lücke in M. J. Phillips and M. Ternan (Eds.), Proc. IXth Int. Congr. Catal., Calgary, Canada, 1988, Vol. 4, The Chemical Institute of Canada, Ottawa, p. 1735 (1988).

10. S. Nowak, Freiberger Forschungsh. A 763: 116 (1987).

11. S. Nowak, H. Günschel, K. Anders, A. Martin and B. Lücke, Erdöl & Kohle, Erdgas, Petrochemie 43: 57 (1990).

12. A. Martin, S. Nowak, B. Lücke and H. Günschel, Appl. Catal. 50: 149 (1989).

13. A. Martin, S. Nowak, B. Lücke, W. Wieker and B. Fahlke, Appl. Catal. 57: 203 (1990).

14. A. Martin, S. Nowak, U. Wolf and B. Lücke, Zeolites, in press.

15. H. Günschel, A. Martin and S. Nowak, Appl. Catal., in press.

20

SELECTIVE PRODUCTION OF LIGHT OLEFINS BY ZEOLITE-CATALYZED METHANOL CONVERSION

Bernard JUGUIN,

IFP, 1 et 4 Avenue de Bois Préau
BP. 311, F-92506 RUEIL MALMAISON FRANCE

François HUGUES

IFP CEDI,
BP. N° 3, F-69390 VERNAISON FRANCE

Christian HAMON

STE CHIMIQUE DE LA GRANDE PAROISSE
BP.2, F-44550 MONTOIR DE BRETAGNE FRANCE

I. INTRODUCTION

The production of light olefins from methanol is currently a very important goal. It would be highly desirable to obtain selectively a narrow range of C_2 - C_4 olefins, most particularly ethylene and/or propylene. Propylene is a target of choice, due to an expected shortage in the future of this important petrochemical intermediate.

By modifying the operating conditions of the methanol to gasoline process (MTG) or by using new catalysts, it has been possible to increase the share of light olefins (1). The main difficulty has been to reduce the amount of C_4^+ hydrocarbons.

IFP and Société Chimique de la Grande Paroisse have developed a new catalyst operating in a fluidised bed reactor in the temperature range of 500 - 600 °C and at low contact time leading to a C_2 - C_4 olefins yield up to 90 %, with propylene yield of up to 60 % at complete methanol conversion.

II. EXPERIMENTAL

The catalyst used is a proprietary zeolite based catalyst. The feed used is methanol diluted with nitrogen and/or water.

Catalytic tests are performed in a micro fluid bed reactor (see Fig. 1). From 0.1 to 1.0 g of powdered catalyst (particle size range 40 - 100 µm) are placed on the fritted quartz disk.

The liquid feed, metered by a motor-driven syringe, is vaporized in the chamber and diluted with nitrogen.

The gaseous mixture, heated to the reaction temperature, travels through the catalyst bed and maintain it in a fluidized state. In the fluid bed so produced, a good contact of the catalyst with the reactant gas and a continuous mixing of the catalyst particles provide an almost isothermal bed.

A thermocouple immersed in the fluid bed gives the real reaction temperature. To avoid entrainment of catalyst particles by the gaseous product the upper part of the micro-reactor is enlarged and contains a little cone in the middle of it.

The gaseous product is cooled to 0 °C to condense the liquids.

Both gaseous and liquid products are analysed by G.C.

The catalyst can be regenerated by switching to an air-nitrogen mixture.

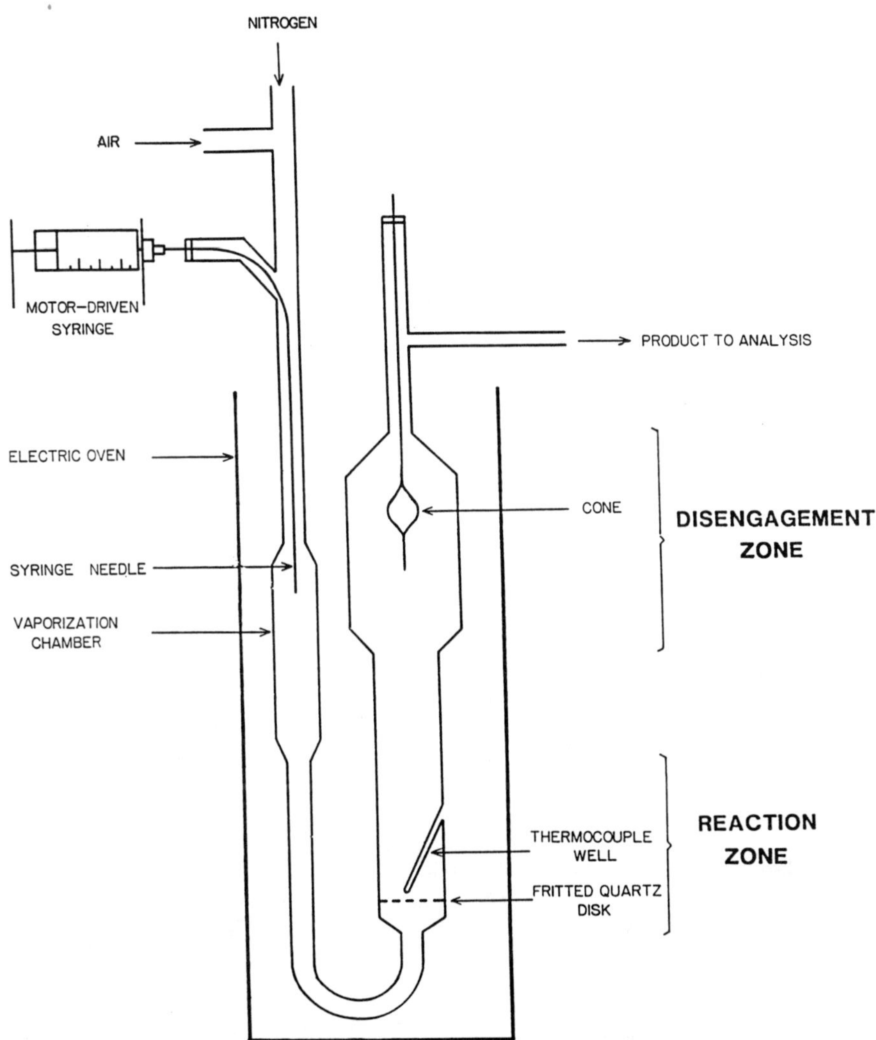

Fig. 1 *Micro fluid bed reactor*

Throughout the present paper the following terms will be used :

$$\text{Contact time} = \frac{1}{\text{GHSV}}$$ usually expressed in seconds.

$$\text{Selectivity} = \frac{\text{moles of carbon in the product}}{\text{moles of methanol converted}} \times 100$$

(for a particular product or a range of products)

$$\text{Yield} = \frac{\text{moles of carbon in the product}}{\text{moles of methanol introduced}} \times 100$$

III. RESULTS AND DISCUSSION

Fig. 2 shows the evolution of methanol conversion with time on stream.

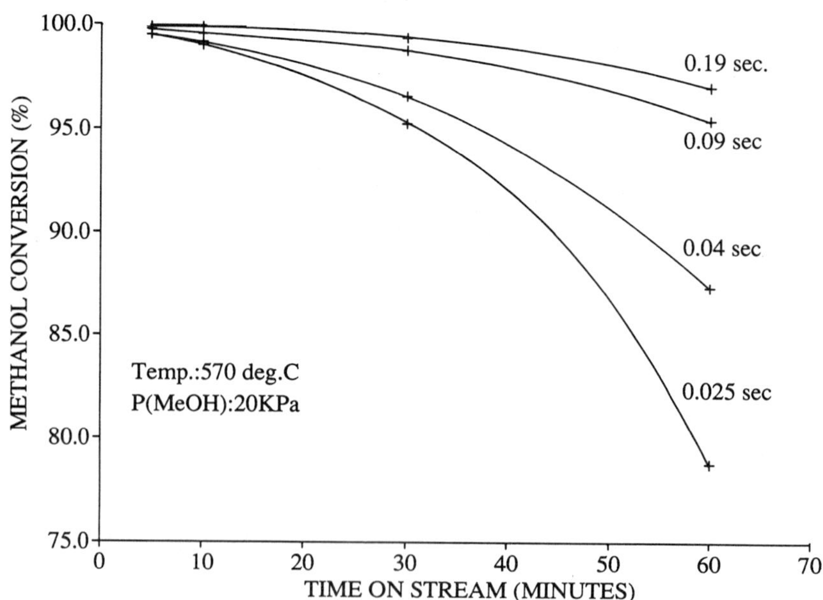

Fig. 2 *Methanol conversion as a function of time on stream and contact time. Temp. : 570 °C, P (MeOH) : 20 KPa*

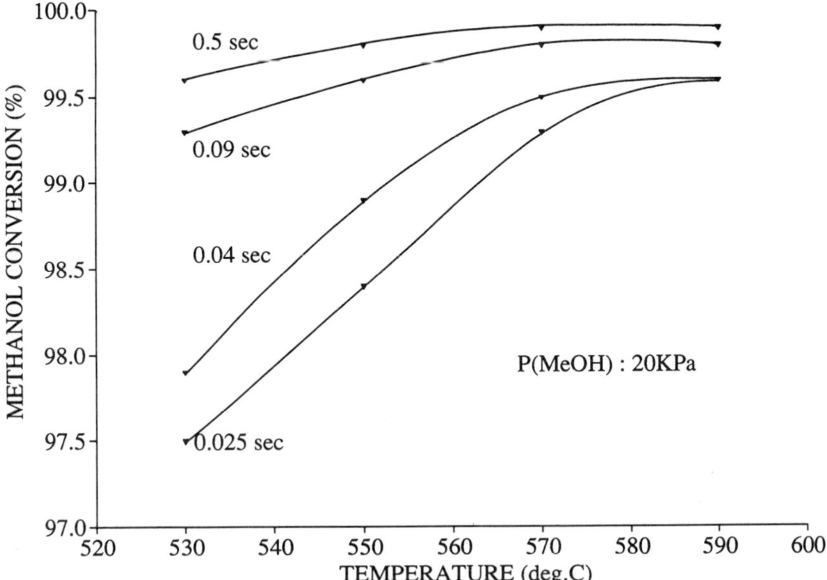

Fig. 3 *Methanol conversion as a function of temperature and contact time. P (MeOH) : 20 KPa*

It is apparent that the catalyst activity decrease with time is faster, the shorter the contact time. However the activity is completely restored by heating the catalyst in a stream of air at 500 - 600 °C.

A fairly rapid catalyst deactivation is expected at the reaction temperature used, but the complete regeneration by air treatment is consistent with coke deposition being responsible for loss of activity.

Therefore, in the following, the catalyst performances have been recorded after 5 min. on stream for each set of conditions chosen.

The effect of temperature on methanol conversion, shown in Fig. 3, indicates that almost complete conversion of the reactant is observed, even at short contact time, underlining the high activity of the catalyst.

Table 1 ***Typical product distribution for experimental conditions favouring either ethylene or propylene***

		Methanol conversion to	
		Ethylene + propylene	Propylene
OPERATING CONDITIONS			
Temperature	*°C*	*590*	*590*
Contact time	*Sec*	*0.19*	*0.025*
MeOH partial pressure	*KPa*	*8.6*	*77*
Time on stream	*min*	*5*	*5*
Methanol conversion	%	99.9	99.9
SELECTIVITIES	%		
Olefins			
Ethylene		*35.8*	*13.6*
Propylene		*39.7*	*63.3*
Butenes		*8.6*	*14.2*
Pentenes		*0.8*	*0.9*
Hexenes		*0.3*	*0.4*
By products			
Dimethylether		*0.0*	*0.1*
Methane		*3.6*	*3.0*
Ethane		*0.3*	*0.4*
Propane		*4.8*	*1.4*
N-butane		*5.0*	*2.1*
Isobutane		*1.1*	*0.5*
OVERALL SELECTIVITY			
Ethylene + propylene		*75.5*	*76.9*
C2 - C4 olefins		*84.1*	*91.1*

Typical product distributions are shown in Table 1, for two sets of conditions, one maximizing propylene, the other favoring both ethylene and propylene.

It is interesting to note the low amount of C_5 and C_6 hydrocarbons and furthermore, the absence of hydrocarbons higher than C_6.

The C_2 - C_4 olefin selectivity amounts to 90 %, the balance being paraffinic hydrocarbons.

The olefin distribution is dependent on temperature, contact time and methanol partial pressure (see Fig. 4, 5, 6).
Whereas propylene selectivity is almost unaffected by temperature, butene selectivity decreases and ethylene selectivity increases with temperature.

As far as contact time is concerned, there is a dramatic decrease of propylene and to a lesser extent of butene with

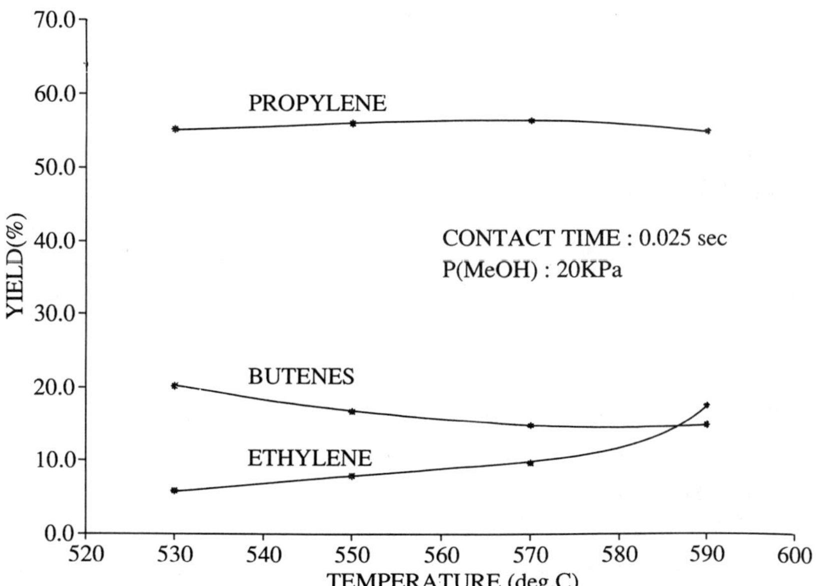

Fig. 4 *Distribution of light olefins as a function of temperature. Contact time : 0.025 sec. P (MeOH) : 20 KPa*

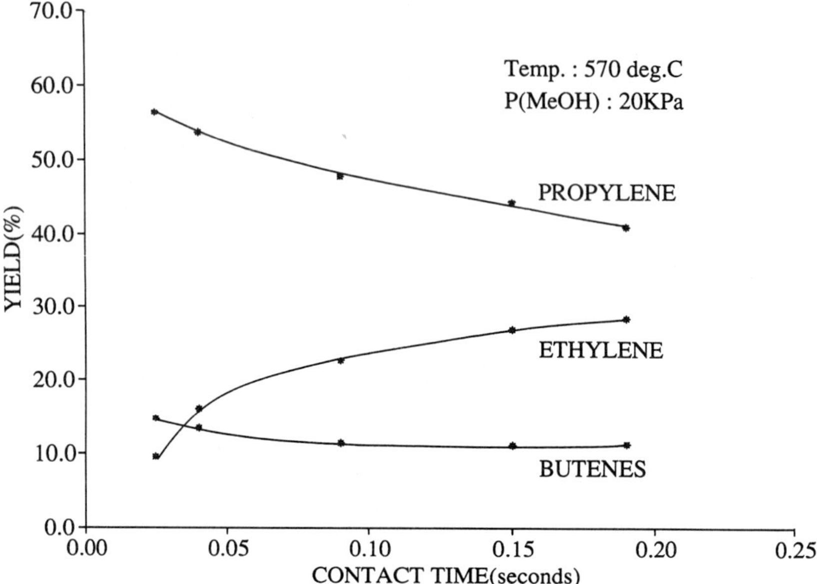

Fig. 5 *Distribution of light olefins as a function of contact time. Temp. : 570 °C, P (MeOH) : 20 KPa*

contact time with a corresponding increase of ethylene, the effect being more pronounced the shorter the contact time.

IV. CONCLUSION

The combination of low contact time and high temperature and the use of a novel zeolite catalyst operating in a fluidized bed has allowed us to obtain high selectivities toward light olefins especially propylene.

The incentive to produce more propylene in the near future calls for further development of the process.

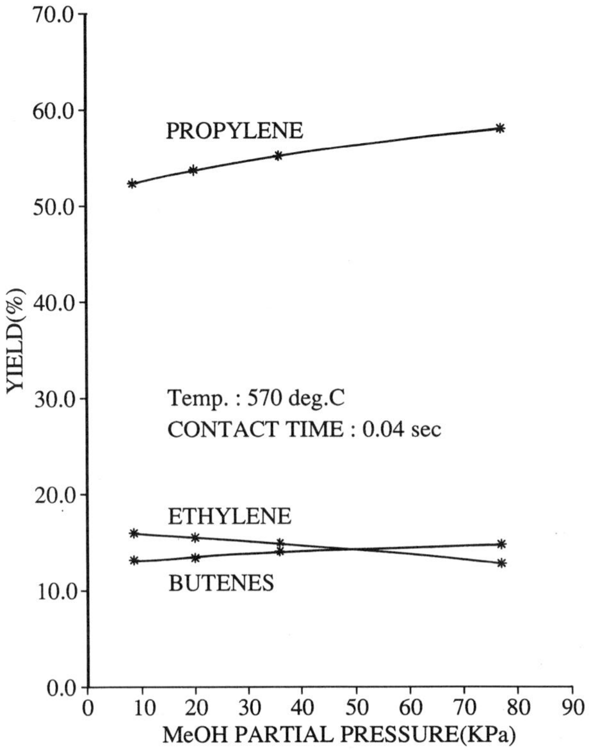

Fig. 6 *Distribution of light olefins as a function of MeOH partial pressure. Temp. : 570 °C, contact time : 0.04 sec*

REFERENCE

(1) C.D. CHANG Catal. Rev. 26, 323 (1984)

21

FACTORS INFLUENCING THE LIGHT OLEFIN SELECTIVITY DURING THE CONVERSION OF METHANOL OVER ZSM-5

M.G. Howden*, J.J.C. Botha and M.S. Scurrell

Catalysis Programme, Division of Energy Technology,
CSIR, P O Box 395, Pretoria 0001, South Africa

ABSTRACT

Due to market demands, the envisaged methanol conversion plants in South Africa will have to produce a large proportion of light olefins. The olefin selectivity depends on catalyst characteristics and on certain parameters used during the process, and consequentially this organisation has over the past few years gathered information about most of the key factors. This paper reviews our results about reaction conditions and composition of the ZSM-5 catalyst. It was found that the most important factors influencing olefin selectivity are the temperature of the reaction and the silica to alumina ratio of the ZSM-5. The work includes a distinction between the selectivity for ethene and the remaining light olefins, and shows that under optimum conditions the product can contain up to 65 % light olefins, of which about 15 % is ethene.

---

*To whom all correspondence should be addressed.

INTRODUCTION

While investigating methods to synthesise highly siliceous zeolites around 1970, Mobil discovered the now well-known product ZSM-5 [1]. Shortly afterwards it was found that ZSM-5 could catalytically convert methanol selectively to hydrocarbons, rich in aromatics and yet not containing much material with more than ten carbon atoms per molecule [2]. Such a product was obviously a good candidate as a synthetic petrol. These discoveries led to a commercial plant being built in New Zealand where natural gas from below the ocean bed is piped onshore, transformed first to methanol and then converted to petrol [3].

In South Africa, AECI Ltd and partners are evaluating the prospects of developing and operating a similar process. Certain differences have to be incorporated to suit local circumstances. One of them is that coal would be used to manufacture the methanol. Another is that to satisfy our fuel requirements, the conversion of the methanol should not be as complete as is practised in New Zealand. This means that the product spectrum would give besides petrol, a sizeable proportion of the intermediate product, namely light olefins needed for the subsequent manufacture of diesel. Further, additional olefins are required as chemical feedstock for other manufacturing processes.

To meet these requirements it is obviously important to be able to control the type and amounts of olefins that are obtained. The product spectrum depends on variations in conditions during the reaction, and the CSIR has, over the past few years, been researching the subject. The relevant information has been accumulated from both published literature [4] and our own research [5-7]. This paper is a summary of our current observations.

FACTORS STUDIED

We have classified the reaction variables or parameters into four categories; viz:

A. Invariable parameters; these are conditions which, although influencing the olefin selectivity, cannot be altered significantly due to other reasons.

B. Variations in reaction conditions.

C. Factors regarding the composition and structure of the ZSM-5 catalyst.

D. Parameters which, although influencing the olefin selectivity, are not recommended for practical application.

INVARIABLE PARAMETERS

1. MASS HOURLY SPACE VELOCITY (MHSV)

During the initial investigations, this parameter was studied by Mobil [2]; their results are reproduced in Fig. 1. It can be seen

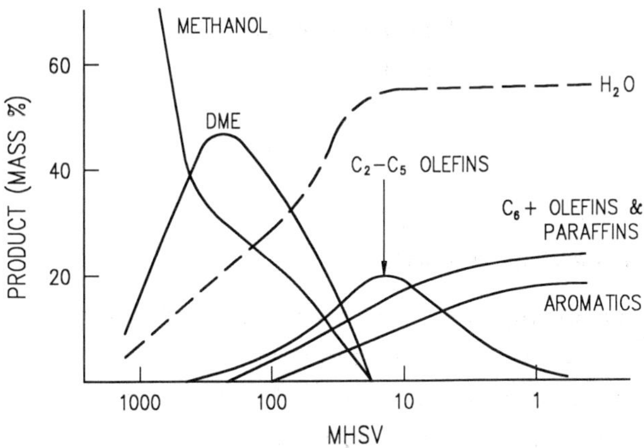

FIG. 1. *Reaction path for methanol conversion to hydrocarbons. Reaction temperature was 371 °C.*

that approximately half the hydrocarbon product can be in the form of light olefins if the MHSV is around 20. However, it is not practicable to use such a high space velocity since it would either give incomplete conversions or in the event that complete conversions could be achieved it would lead to temperature runaways since the reactions are very exothermic. Normally the MHSV is around one.

2. BINDER

The catalyst matrix is usually formed by extruding the ZSM-5 with pseudo-boehmite followed by calcination. In comparison with samples of pure ZSM-5 we found that the extrudates gave a slight increase in ethene selectivity. Regarding the remaining light olefins no simple pattern could be found; increases, decreases and no changes were recorded. However, these variations were also relatively small.

From world-wide sources we chose eight different brands of pseudo-boehmite and used them to extrude a selected sample of ZSM-5. Each of the extrudates were examined for their conversion of methanol and it was found that all of them resulted in similar olefin selectivity results. Therefore, although its presence does impart a small effect on the olefin selectivity, the choice of pseudo-boehmite for any given catalyst would depend rather on extrusion properties or physical strength.

REACTION CONDITIONS

1. REDUCED PARTIAL PRESSURE OF METHANOL

The first success in increasing the light olefin selectivity was reported by Chang *et al* [8], who reduced the partial pressure of the methanol feed. This technique would form the basis for the plant being planned by AECI Ltd. Dilution of the feed would be effected by using water recycled from the product [9].

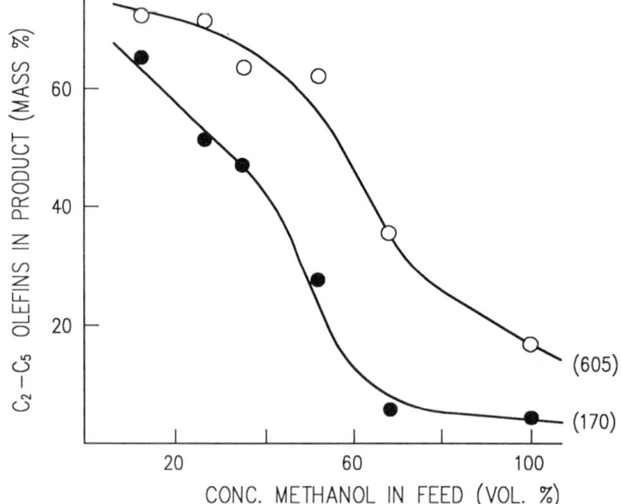

FIG. 2. *Light olefin selectivity in relation to methanol concentration. Methanol was diluted in nitrogen and complete conversion to hydrocarbons was attained using reaction temperature of 500 °C and MHSV of 0,63. Silica to alumina mole ratio of catalysts are given in brackets.*

Some of our results obtained with this system are given in Fig. 2, where the methanol was diluted with nitrogen. The methanol feed obviously has to have a relatively low partial pressure in order to obtain a good olefin selectivity. There are two sets of results which were obtained from samples of ZSM-5 having silica to alumina mole ratios of 170:1 and 605:1. The material with the higher silica to alumina ratio produced the higher olefin selectivity. This is due to it having a lower concentration of active sites, with a reduced chance of the olefins reacting further to form heavier products.

2. *TEMPERATURE OF REACTION*

Raising the reaction temperature is known to increase the proportion of light olefins obtained [2,10]. In our work [7], the olefins were divided into two separate fractions; (a) ethene and

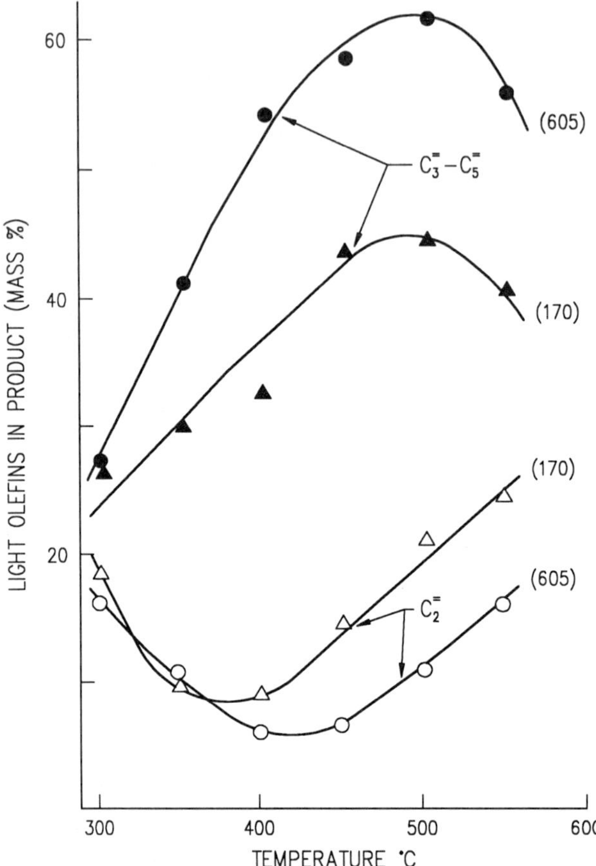

FIG. 3. *Light olefin selectivity in relation to reaction temperature. Methanol was diluted to 12,3 vol. % in nitrogen and MHSV was 0,48. Silica to alumina mole ratios of catalysts are given in brackets.*

(b) propene to pentene. Results are shown in Fig. 3: samples with different silica to alumina mole ratios were examined and complete conversion was obtained with a feed consisting of 12,3 vol % methanol in nitrogen. Although the overall result confirms an increase of olefins with increasing temperature, the individual components behave differently. The $C_3^=$-$C_5^=$ fraction has a maximum just below 500 °C and the ethene fraction actually has a minimum around 400 °C.

The likely explanation for the product distribution can be derived from the following sequence of reactions. The primary product from the dehydration of methanol/dimethyl ether is ethene [11,12], and therefore at the mild (lower) temperatures a large fraction of ethene is obtained. Higher olefins are formed through the alkylation of ethene, and as indicated by Fig. 3 by raising the temperature there is an increase in C_3-C_5 olefin distribution and a concomitant decrease in ethene. Above 400 °C the cracking of propene/butene to ethene [13,14] becomes significant; the higher the temperature, the faster the cracking. Therefore, as the reaction temperature increases the concentration of ethene in the product begins to rise, and the proportion of $C_3^=$-$C_5^=$, while initially increasing due to faster alkylation, will eventually start to decrease as cracking predominates. Finally, however, even the ethene itself and any other compounds are also cracked, and around 600 °C the only products are methane, hydrogen and carbon monoxide.

COMPOSITION AND STRUCTURE OF ZSM-5 CATALYST

1. SILICA TO ALUMINA RATIO OF THE ZSM-5

Chang *et al* [10] reported that the olefin selectivity can be improved by using ZSM-5 which has been synthesised with higher silica to alumina mole ratios. We repeated their work and our results are given in Fig. 4, which confirmed the sharp increase in olefin selectivity as the silica to alumina ratio is raised up to 500:1. With higher ratios, the olefin selectivity began to decrease and dimethyl ether was detected in the product spectrum. Incomplete conversion is obviously due to insufficient residence time for the reactant under the prevailing conditions where the catalyst has a limited density of active sites.

Our work expanded on that of Chang *et al* by distinguishing the ethene yield from that of the other light olefins, and we found that the ethene selectivity actually decreased [7]. When the

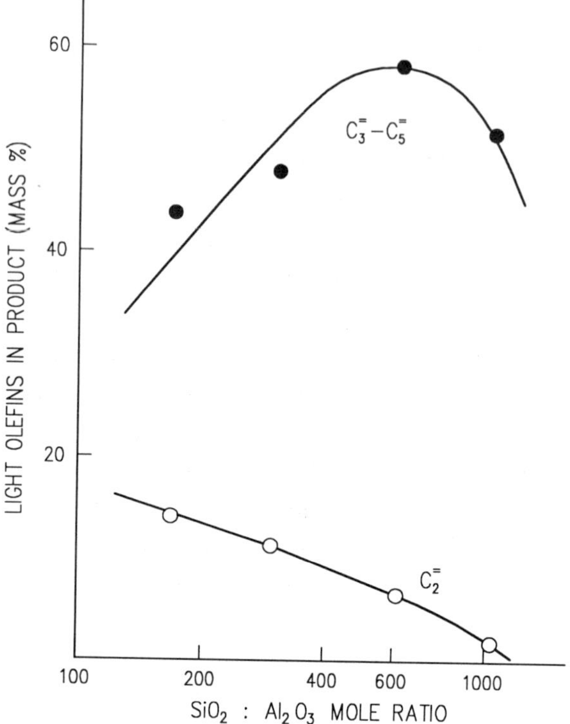

FIG. 4. *Light olefin selectivity in relation to silica to alumina mole ratio of the ZSM-5 catalyst. Reaction temperature was 450 °C, methanol was diluted to 12,3 vol. % in nitrogen and MHSV was 0,48.*

silica to alumina ratio of the ZSM-5 is reduced, there are fewer active sites and consequentially there is less likelihood of the C_3-C_5 olefins being cracked.

2. *STEAM DEACTIVATION*

In the preceding section we have seen the advantages of a ZSM-5 with a low concentration of aluminium. Besides direct synthesis, another method of obtaining a similar catalyst is through dealumination of ZSM-5 having an average aluminium content. This is usually effected by calcining the material in the presence of steam, where the amount of dealumination is dependent on the partial pressure of the steam, calcination temperature and time.

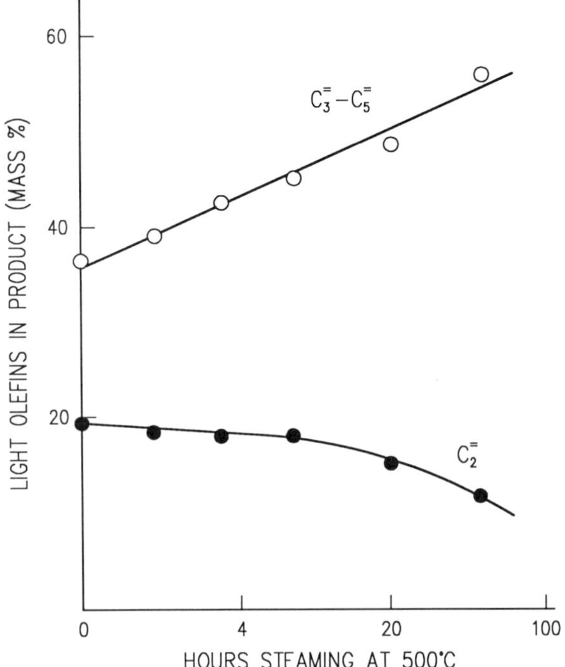

FIG. 5. *Light olefin selectivity in relation to duration of steaming the ZSM-5 catalyst at 500 °C. Reaction temperature was 450 °C, methanol was diluted to 24,0 vol % in nitrogen and MHSV was 0,66.*

With a sample of ZSM-5 that had an initial silica to alumina mole ratio of approximately 150:1, we calcined extrudates at 500 °C in an atmosphere of saturated steam for various periods of time. These preparations were used to convert methanol and the results are illustrated in Fig. 5. Dealumination through steaming has the same effect on the olefin selectivity (increasing the $C_3^=$-$C_5^=$ while decreasing the ethene fractions) as was found when decreasing the aluminium content of the ZSM-5 via direct synthesis.

Since the rate of dealumination is also temperature dependent [15], samples of the same preparation were dealuminated in steam for 50 hours at 400 °C and 450 °C. Their olefin selectivities from the conversion of methanol, as well as that of

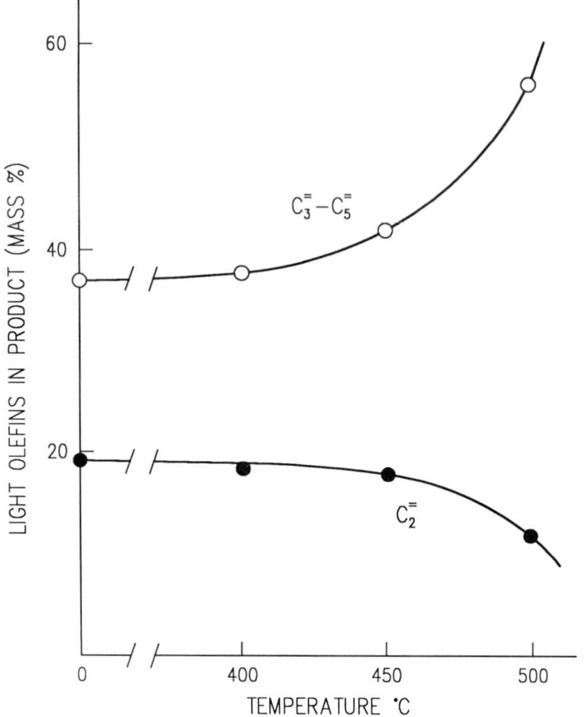

FIG. 6. *Light olefin selectivity in relation to temperature at which the ZSM-5 catalyst was steamed for 50 hours. Reaction temperature was 450 °C, methanol was diluted to 24,0 vol % in nitrogen and MHSV was 0,66.*

the sample steamed at 500 °C, are given in Fig. 6. The shape of the selectivity curve indicates that the rate of dealumination is rapid when the pre-steaming temperature exceeds 450 °C.

In practice, the feed will consist of a steam-rich mixture passing into the reactor and this obviously will cause some dealumination of the ZSM-5. The same extrudate mentioned above was reacted with a methanol/water mixture instead of using nitrogen gas as the diluent, and the results are shown in Fig. 7. As can be seen, there was only a small increase in the olefin selectivity. After testing, the samples that were reacted at 480 °C and 510 °C

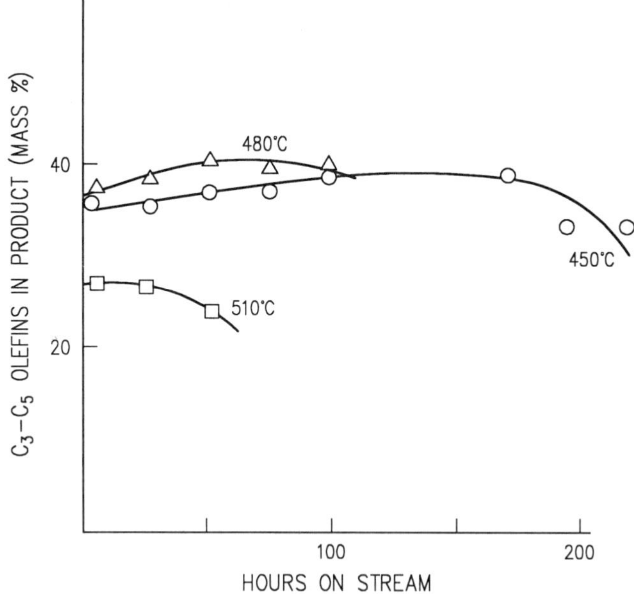

FIG. 7. *Light olefin selectivity over a long period, using a methanol/water feed and different reaction temperatures. Methanol concentration was 20,0 vol % in steam and MHSV was 0,62.*

were regenerated and re-examined under the same conditions. The product selectivities were virtually identical to those originally obtained, which indicates that hardly any dealumination occurred at these temperatures; changes in selectivity were caused by the build-up of coke. This means that the rate of dealumination was less severe than when saturated steam was used.

3. COKE DEACTIVATION

The samples for which the results are reported in Fig. 7 were left in the reactor until the C_3-C_5 olefin yield began to decrease significantly. At this point it was felt that the catalysts had been deactivated because of the deposition of coke. It was found that the amount of coke deposited on the samples was about 5 % by mass. Obviously the rate of coke deposition and period of reaction before regeneration is necessary, is adversely affected by using

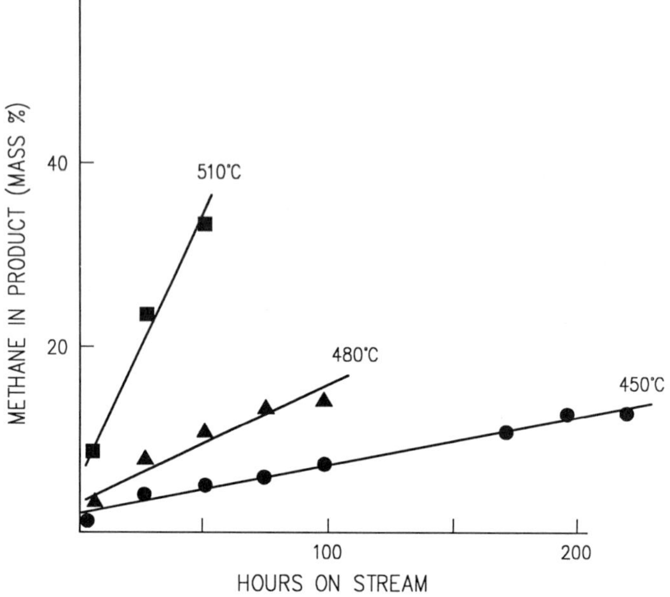

FIG. 8. *Changes in methane selectivity with time, using a methanol/water feed and different reaction temperatures. Methanol concentration was 20,0 vol % in steam and MHSV was 0,62.*

the higher reaction temperatures. This phenomenon criterion appears to set the maximum temperature at which this type of reaction could take place.

Although the percentage of C_3-C_5 olefins showed only a small variation for the duration of the reaction, there were large differences in the remainder of the product spectrum. The most significant was the amount of methane formed, which is illustrated in Fig. 8. This parameter also sets the same upper limit at which it is advisable to conduct the reaction; this temperature is around 450 °C.

4. CRYSTALLITE SIZE

From all our preparations of ZSM-5, we selected three pairs of samples which were similar except for the size of their crystallites. Their olefins selectivities are shown in Table 1.

TABLE 1. *Effect of crystallite size on light olefin selectivity.*

Ref.	Si/Al_2 mole ratio	Cryst. size µm	React. temp. °C	$C_2^=$ %	$C_3^=-C_5^=$ %
1-D	525	25	450	5,9	70,1
5-D	395	3		11,0	64,9
#-5	90	30	350	1,6	19,0
5-A	105	10		4,6	26,1
F22	250	5,9	450	17,5	56,6
F23	250	3,4		15,4	57,0

The first two sets of results indicate that by reducing the crystallites from 30 to 3 µm, the ethene selectivity is increased. In the final set of samples, a similar selectivity was obtained in spite of a slight difference in crystallite size. Thus, there is some enhancement in ethene selectivity provided that the catalyst contains ZSM-5 with significantly smaller crystallites.

Commenting on the $C_3^=-C_5^=$ selectivity is only possible when there is low overall olefin selectivity. Thus, as shown by the second pair of samples, the selectivity for this fraction of olefins appears to have improved because of decreased crystallite size.

A qualitative explanation for this increased olefin selectivity is relatively simple: small crystallites allow the light products to diffuse quicker from the catalyst reducing the opportunity for further reactions. However, we are not yet in a position to indicate how small the crystallites should be.

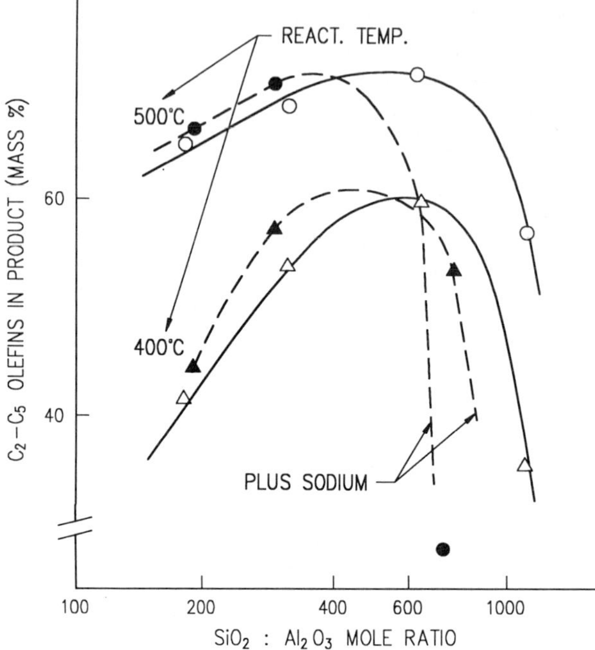

FIG. 9. *Comparison of product from catalyst partially poisoned by sodium and that using higher silica to alumina ratios. Methanol was diluted to 12,3 vol. % in nitrogen and MHSV was 0,48.*

NON-RECOMMENDED PARAMETERS

1. POISONING OF THE CATALYST

Another method of reducing the number of active sites is to poison some of them. For example, this can be done by cation exchanging with phosphorus compounds or sodium. We partially exchanged a sample with sodium, calculated the silica to alumina mole ratio from the remaining (non-poisoned) sites and used the resulting catalyst for the conversion of methanol; the results are shown in Fig. 9. For comparison, details from samples that were synthesised directly with higher silica to alumina mole ratios, as was discussed earlier, are also included. The findings confirm that this technique of poisoning does indeed increase the olefin selectivity.

However, the problem with adding sodium cations to the catalysts was that they tended to coke rather rapidly, particularly at higher temperatures. The use of poisons to improve olefin selectivity seems therefore restricted to low levels of conversion and low reaction temperatures. This approach is therefore unsuitable in industrial practice.

Besides having exchangeable cations, ZSM-5 can also contain sodium which cannot be removed through exchanging. This sodium is presumably located in amorphous phases formed together with ZSM-5 during synthesis. The influence of this non-exchangeable sodium is rather different to that mentioned above, and the presence of even small amounts actually reduces the olefin selectivity [7]. Unfortunately we were unable to give an explanation for this behaviour; nevertheless we concluded that for the best olefin selectivity, the ZSM-5 should contain little residual (non-exchangeable) sodium after synthesis.

2. ISOMORPHIC SUBSTITUTION FOR ALUMINIUM

Besides reducing the number of active sites, lowering the acidity of the existing ones has also been proposed for improving olefin selectivity. This can be achieved by synthesising ZSM-5 with different trivalent cations in the place of the normal aluminium. Examples are boron, chromium, iron and gallium. The boron analogue has been studied quite extensively. Unfortunately, it was found to have a very low acidity and as a result was not even capable of converting methanol to dimethyl ether [5].

The other analogues are more active, but further research on them has subsided due to their lack of thermal stability. It has already been demonstrated that ZSM-5 can undergo dealumination through hydrothermal treatment. This is generally even a bigger problem when the other trivalent elements are substituted for aluminium in ZSM-5.

CONCLUSION

Enough major factors have been examined that makes it possible to significantly raise the light olefin selectivity from the conversion of methanol over zeolite ZSM-5. However, we would like to caution that this information is based on small experiments in laboratories, and confirmation on larger scales and for longer periods of time is advisable to substantiate the effects of the parameters that have been discussed.

Nevertheless, based on our results outlined in this paper, it appears feasible to achieve a light olefin selectivity around 65 %, of which about 15 % would be ethene, by adopting the following parameters:

partial press. of methanol	±	0,2 atm.
reaction temperature	±	450 °C
$SiO_2:Al_2O_3$ mole ratio of ZSM-5	±	250:1
crystallite size	<	5 µm

The combined use of the above parameters would probably give a product spectrum that would satisfy or even exceed our present needs.

REFERENCES

1. R.J. Argauer and G.R. Landolt, *U.S. Patent* 3 702 886 (1972).
2. C.D. Chang and A.J. Silvestri, *J. Catal.* **47**: 249 (1977).
3. S.L. Meisel, *Chemtech,* January 1988, p 32.
4. M.G. Howden, CSIR Report CENG 654, 1987.
5. M.G. Howden, *Zeolites* **5**: 334 (1985).
6. M.G. Howden and J.J.C. Botha, CSIR Report ENER 89001, 1989.
7. M.G. Howden and J.J.C. Botha, CSIR Report ENER 89002, 1989.
8. C.D. Chang, W.H. Lang and R.L. Smith, *J. Catal.* **56**: 169 (1979).
9. G. Hutchings, *New Scientist*, 3 July 1986, p 35.
10. C.D. Chang, C.T-W. Chu and R.F. Socha, *J. Catal.* **86**: 289 (1984).

11. W.O. Haag, R.M. Lago and P.G. Rodewald, *J. Mol. Catal.* **17**: 161 (1982).

12. C.T-W. Chu and C.D. Chang, *J. Catal.* **86**: 297 (1984).

13. R.M. Dessau and R.M. LaPierre, *J. Catal.* **78**: 136 (1982).

14. R.M. Dessau, *J. Catal.* **99**: 111 (1986).

15. G.V. Echevskii, K.G. Ione, G.N. Nosyreva and G.S. Litvak, *Appl. Catal.* **43**: 85 (1988).

22

THE BIOETHANOL-TO-ETHYLENE (BETE) PROCESS: PRODUCTION OF ETHYLENE AND BTX AROMATICS FROM BIOMASS

R. LE VAN MAO(*), D. Ly and J. Yao

Department of Chemistry and Biochemistry,
Concordia University,
1455 De Maisonneuve Bld. W.,
Montréal, (Québec), H3G 1M8 Canada

ABSTRACT

Practically pure ethylene is produced directly from sugar fermentation broths which contain from 2 to 15 wt % of ethanol. Several ZSM-5 zeolite-based catalysts have been studied, including pure silica-synthesized ZSM-5, the asbestos-derived ZSM-5 and the steam-treated ZSM-5. Such "moderately" acidic zeolites are effective only at reaction temperatures above 275°C. With trifluoromethane sulfonic acid (TFA) bearing ZSM-5 zeolite, the reaction can be carried out at temperatures as low as 180°C. Data on kinetics, reaction network and catalyst stability are reported here. In addition, BTX aromatics can be produced by sending the gaseous effluents of the dehydration of aqueous ethanol over an aromatization hybrid catalyst, developed according to the concept of Hydrogen (Back) Spillover.

(*) To whom correspondence should be addressed

INTRODUCTION

Ethanol is currently produced by hydration of ethylene, which itself is derived mainly by steam-cracking of petroleum or natural gas feedstocks. Ethanol can also be obtained by the anaerobic fermentation of sugars which in turn are the products of acid or enzymatic hydrolysis of cellulosic compounds (biomass materials). Fermentation ethanol is recovered from the aqueous broth by distillation. Due to today's low cost of petroleum-derived ethanol and the relatively high production cost of fermentation ethanol, the competitiveness of the latter is not very obvious.

In our laboratory, zeolite catalysts which allow the production of practically pure ethylene from aqueous ethanol broth (concentration of ethanol ranging from 2 wt % to 15 wt %) have been developed. These include acidic ZSM-5 zeolite and related materials (steam-treated zeolite and asbestos-derived zeolite). Incorporating an organic superacid, trifluoromethane sulfonic or triflic acid (TFA), leads to lower reaction temperatures (less than 200°C with respect to temperatures much higher than 250°C when the previously mentioned zeolites are used).

This paper reviews the progress achieved in the development of the catalysts and the related "laboratory scale" process, the so-called: Bioethanol-to-Ethylene (B.E.T.E.) process [1-6].

The production of aromatics (and particularly BTX type aromatics) from the B.E.T.E. ethylene containing gaseous effluents is reported. The latter reaction is carried out by sending such gaseous effluents over special catalysts, namely Hydrogen Back Spillover Hybrid Catalysts [7-10], also developed in our laboratory. This B.E.T.E. process has to be seen as a portion of a larger (integrated) process for the

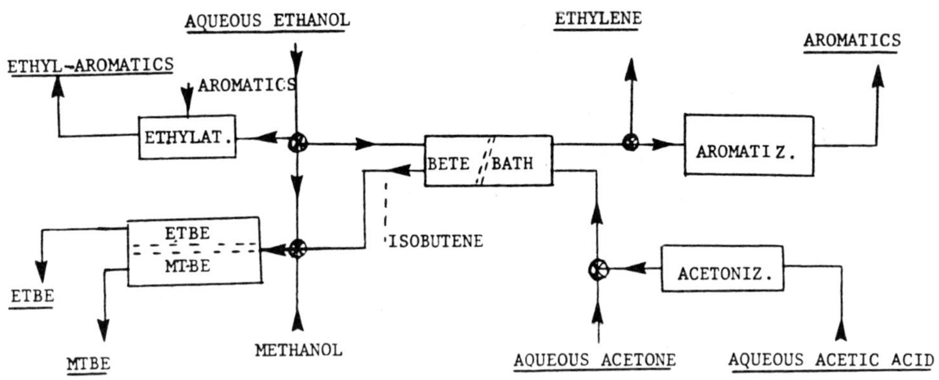

FIG. 1 *Integrated process for the synthesis of industrial chemicals from biomass and organic wastes.*

synthesis of industrial chemicals from biomass and organic wastes; such integrated process is in the development stage in our laboratory (See Figure 1).

EXPERIMENTAL

The preparation and the characterization of the catalysts used in this work are fully described elsewhere [1-6]. ZSM-5 zeolites are synthesized according to the conventional method [11].

ZSM-5 chryso-zeolites (or asbestos-derived zeolites) and steam-treated ZSM-5 zeolites are prepared according to the procedures described in references 5 and 6, whereas the TFA bearing ZSM-5 zeolites are obtained by the method of references 2, 3, 4 and 12. Hybrid catalysts used for the aromatization step are prepared according to the methods described in references 6 to 10.

The runs are performed in the experimental set-up shown in Figure 2. Part 1 is used when the dehydration of aqueous ethanol (5 to 15 wt %) is carried out alone. In this case, the reaction products are condensed and the gaseous phase is separated from the liquid one. The gaseous phase is analysed by gas chromatography (GC), and the liquids by GC and GC mass spectrometry (GC-MS), as already reported [2,5]. The ethanol conversion to ethylene is calculated by:

$$C_{et}\ (\%) = \frac{(Et\ OH)_i - (EtOH)_f}{(Et\ OH)_i} \times 100$$

where $(Et\ OH)_i$ is the number of carbon atoms (C-atoms) of ethanol in the feed, $(EtOH)_f$ is the C-atoms of ethanol recovered in the products of ethanol dehydration. The distribution of the reaction products (Selectivities based on C-atoms) is also determined (see references 5 and 6).

The entire set-up is used when the aromatization is coupled with the dehydration of ethanol. The gaseous effluents of the dehydration reaction are thus sent over the aromatization catalyst. Products from the aromatization reactor are condensed and then analysed by GC (gases), or GC and GC-MS as previously mentioned.

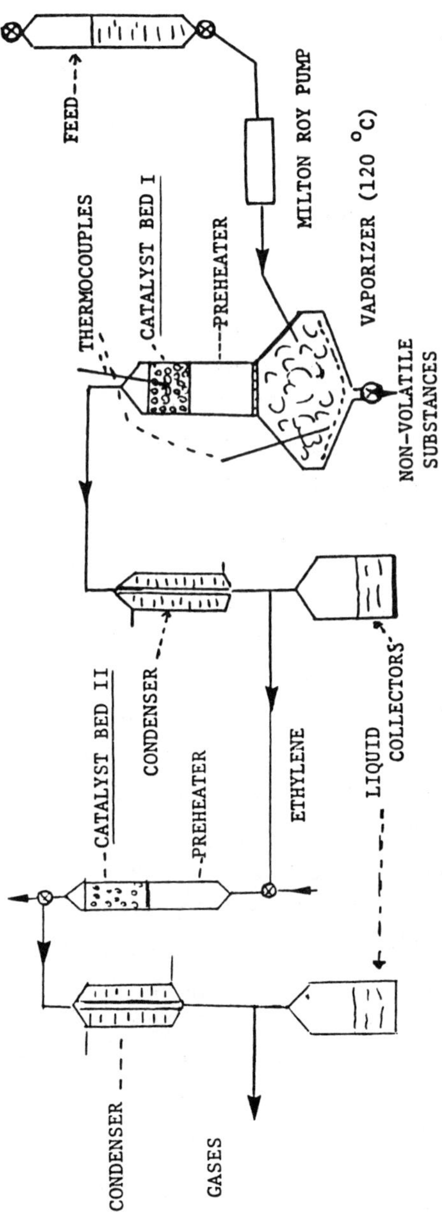

FIG. 2 *Laboratory reaction system for the production of ethylene and/or aromatics from aqueous ethanol.*

The conversion of total aromatics is calculated by:

$$C_{AR}\ (\%) = \frac{(C - AP)}{(C - EI)} \times 100$$

where (C - AP) and (C - EI) are the numbers of carbon atoms of aromatic products and of injected ethanol, respectively.

RESULTS AND DISCUSSIONS

Nearly complete conversion of aqueous ethanol to ethylene is obtained with steam-treated or asbestos-derived ZSM-5 zeolites only when the reaction temperature is higher than 275°C. The moderate surface acid density of these catalysts ensures a high selectivity towards ethylene and at the same time, prevents the formation of higher hydrocarbons (Tables 1 and 2) [5,6]. Parent ZSM-5 zeolites are also active and selective in the ethanol dehydration. However, such zeolites must have very precise acidic characteristics, i.e. a Si/Al ratio ranging from 35 to 55 or a surface acid density ranging from 2.4 to 3.7. 10^{-4} mol/g (Figure 3 and Table 2) [6].

The incorporation of an organic superacid, triflic acid (TFA) or trifluoromethanesulfonic acid, into the ZSM-5 zeolite, provides a very efficient ethanol dehydration catalyst which can totally convert the aqueous ethanol into ethylene at 180°C. The ethylene yield depends on the TFA loading (Figure 4) [2] and this behaviour is the result of the combined effects of surface acidity and pore size modification upon TFA loading [13]. The TFA incorporated onto the ZSM-5 zeolite is stable up to 240 - 245°C. This is presumably due to the interactions between the zeolite OH groups and the chemisorbed TFA species [2].

Studies of the reaction network for the conversion of aqueous ethanol over steam-treated ZSM-5 and Asb-

TABLE 1 *Effect of MLD of asbestos-derived ZSM zeolites on the conversion of aqueous ethanol (10 wt %) at various temperatures and at WHSV:3.2h^{-1}. (Permission of Appl. Catal.).*

Catalyst	Reaction temperature (°C)	Conversion (%)		Product selectivity (%)	
		Total C	HC	Ethylene	Other hydrocarbons
Asb-ZSM(26/99)	175	14.4	0.0	100.0	0.0
	200	27.0	0.5	100.0	0.0
	225	35.0	10.3	100.0	0.0
	250	63.6	42.7	99.9	0.1
	275	97.9	96.6	99.6	0.4
	300	99.8	99.8	99.5	0.5
	325	99.9	99.4	99.4	0.6
Asb-/ZSM(21/83)	225	17.1	0.9	100.0	0.0
	250	30.5	11.8	100.0	0.0
	275	55.5	52.6	99.4	0.6
	300	86.0	85.5	99.4	0.6
	325	99.2	99.2	99.4	0.6

TABLE 2 *Catalytic performances of the parent and steam-treated ZSM-zeolites. (Permission of Appl. Catal.). *Hydrocarbons heavier than ethylene, with some aromatics.*

Catalyst (extrudates)	Si/Al (of zeolites)	Conversion HC (%)	Selectivity (%)	
			Ethylene	*Other hydrocarbons*
ZSM(21)	21	*98.2*	*80.6*	*19.4**
ZSM(45)	45	*98.4*	*99.3*	*0.7*
ZSM(75)	75	*22.4*	*99.9*	*0.1*
ZSM(124)	124	*14.8*	*100.0*	*0.0*
ZSM(21.St)	21	*91.0*	*99.3*	*0.7*

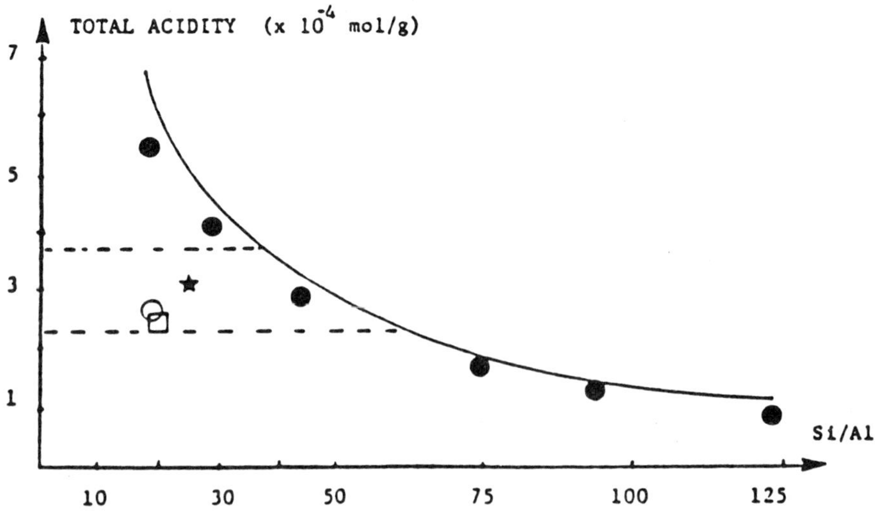

FIG. 3 *Acid density versus the Si/Al atomic ratio. (●) parent zeolite, (O) ZSM (21,St), (★) Asb-ZSM (26/99), (□) Asb-ZSM (21/83). Curve = theoretical values of total acidity versus the Si/Al atomic values (Permission of Appl. Catal.,).*

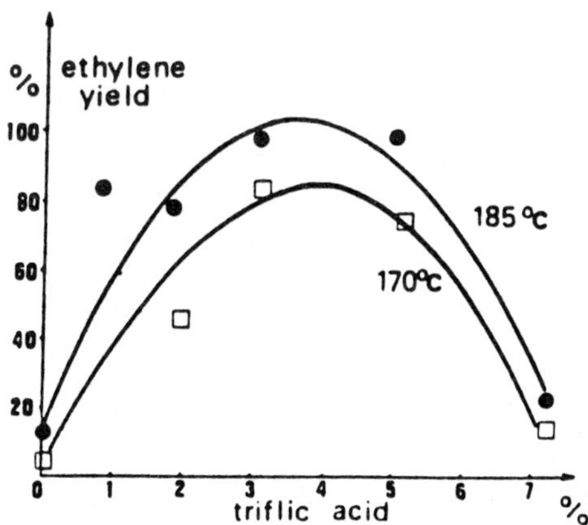

FIG. 4 *Effect of TFA loading on the ethylene yield (Permission of Appl. Catal.,).*

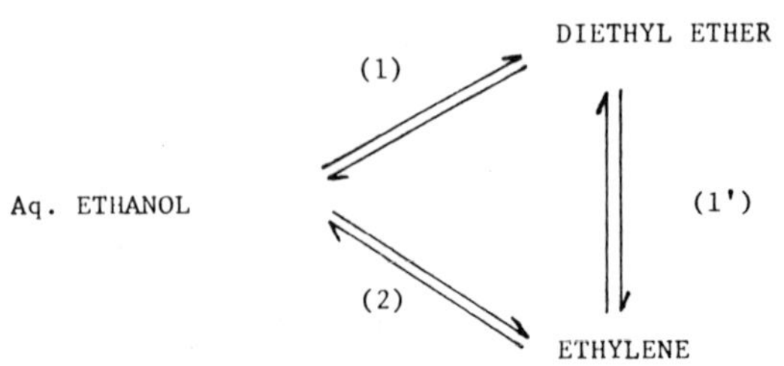

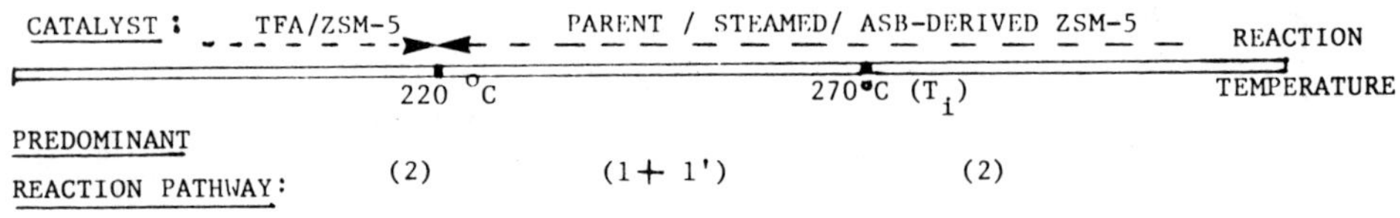

FIG. 5 *Predominant reaction path way in the conversion of aqueous ethanol to ethylene.*

ZSM-5 zeolites show that several pathways may exist. At a reaction temperature below T_i (270°C), the two-step pathway into diethyl ether as the intermediate coexists with the direct conversion-to-ethylene route. The latter appears to be favoured at temperatures above

TABLE 3 *Values of apparent activation energies (Permission of Appl. Catal.) *Pure reagents; catalyst ZSM-5 (Si/Al:25).*

Catalyst	Reaction	Temperature range (K)	Apparent Activation energy (kJ/mol)	
			Calculated	From ref.14*
ZSM(21,St)	*Aq.Ethanol-to-diethyl ether*	*448-498*	*56*	-
	Aq.Diethyl ether-to-ethylene/ethanol	*523-573*	*79*	*92*
	Aq.Ethanol-to-ethylene	*523-598*	*119*	*122*
Asb-*ZSM(26)*	*Aq.Ethanol-to-diethyl ether*	*448-498*	*56*	-
	Aq.Diethyl ether-ethylene/ethanol	*523-573*	*80*	*92*
	Aq.Ethanol-to-ethylene	*523-623*	*123*	*122*

T_i (Figure 5) [5]. Table 3 reports the values of apparent activation energies for different reaction steps [5]. Figures 6 and 7 show the stability tests with the steam-treated ZSM-5, the Asb-ZSM-5 and the ZSM-5/TFA zeolites. Particularly with the TFA loaded

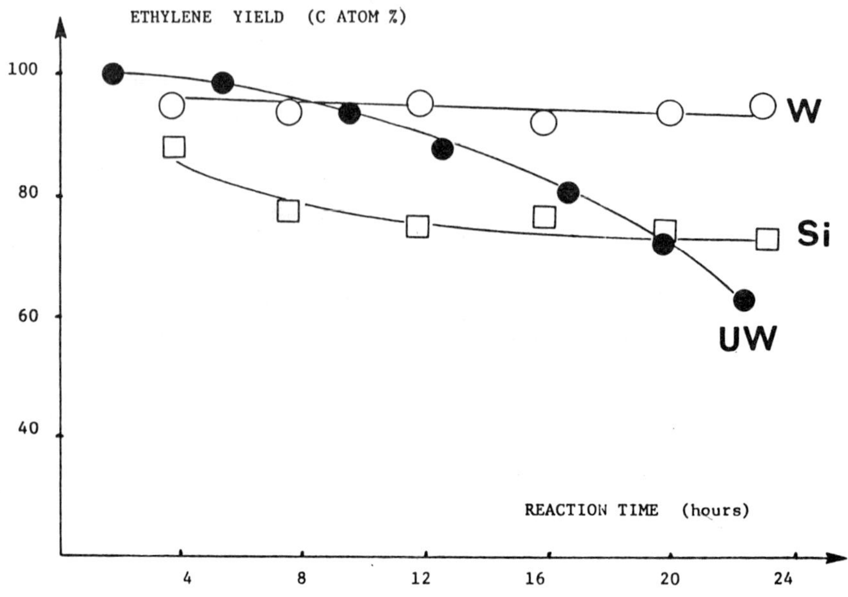

FIG. 6 *Tests of stability with the TFA loaded (4 wt %) ZSM-5 zeolites (T = 200°C).*

ZSM-5 catalysts, the sample (W) which has been washed with a small amount of acetone after the TFA impregnation operation shows a higher stability with respect to the unwashed catalyst (UW). This may be due to the elimination of the TFA external (to the zeolite particles) coating.

Table 4 reports the results obtained when the gaseous effluents of the dehydration reactor are sent over various aromatization catalysts developed in our laboratory [7-10]. Conversion to aromatics reaching the level of 60 (C atom) % is obtained with our HBS-hybrid catalyst whose co-catalyst is prepared by co-evaporation of a Ga salt with a colloidal silica [9].

Finally, Figure 8 gives an estimated figure of product yields of the process.

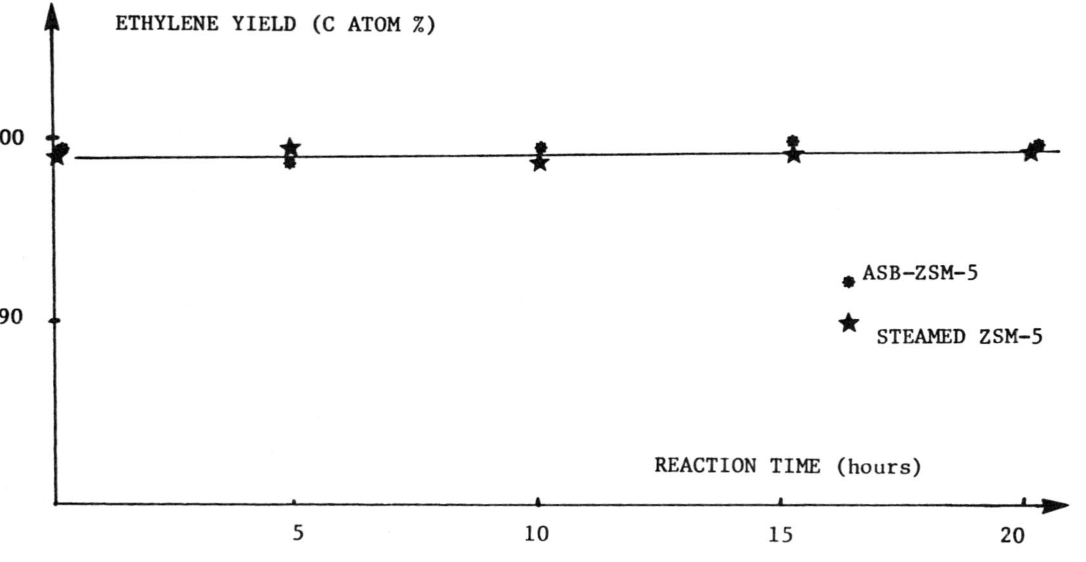

FIG. 7 *Tests of stability with the Asb-derived and steam-treated ZSM-t zeolites (T = 325°C).*

TABLE 4 *AROMATIZATION OF THE ETHYLENE CONTAINING GASEOUS EFFLUENTS*

CATALYST I	ZSM-5/TFA (3)	ZSM-5/TFA (3)	ZSM-5 (P:42)
OPERATING CONDITIONS	T: 200°C: W.H.S.V. : 3 h^{-1}	T: 200°C: W.H.S.V. : 3 h^{-1}	T: 370°C: W.H.S.V. : 3 h^{-1}
PRODUCT DISTRIBUTION (C atom %)			
Ethylene	99.2	99.8	95.7
Propylene	0.3	0.5	1.7
Butenes	0.2	0.2	1.5
C_1-C_4 Paraffins	0.1	0.1	0.4
C_5^+	0.3	0.4	0.7
Ethanol conversion (C atom %)	89.9	90.9	92.3
CATALYST II	ZSM-5 /// ZnO,Al_2O_3	ZSM-5/Ga_2O_3	ZSM-5 /// Ga_2O_3/SiO_2-L
OPERATING CONDITIONS	T: 540°C	T: 540°C	T: 540°C
PRODUCT DISTRIBUTION (C atom %)			
Ethylene	27.7	12.9	3.7
Propylene	4.4	4.1	2.9
Butenes	1.3	1.3	0.2
C_1-C_4 Paraffins	25.3	31.0	18.5
C_5^+ aliphatics	0.4	0.6	0.6
Total aromatics	40.9	50.1	74.1
CONVERSION TO AROMATICS (C atom %)	31.3	44.9	58.1
COMPOSITION OF THE AROMATICS			
Benzene	30.2	39.4	40.5
Toluene	34.2	33.2	35.6
Xylenes	15.7	11.3	9.3
Other C_8	4.5	2.8	1.1
C_9^+	15.4	13.3	13.3
BTX/Total aromatics	84	86	86

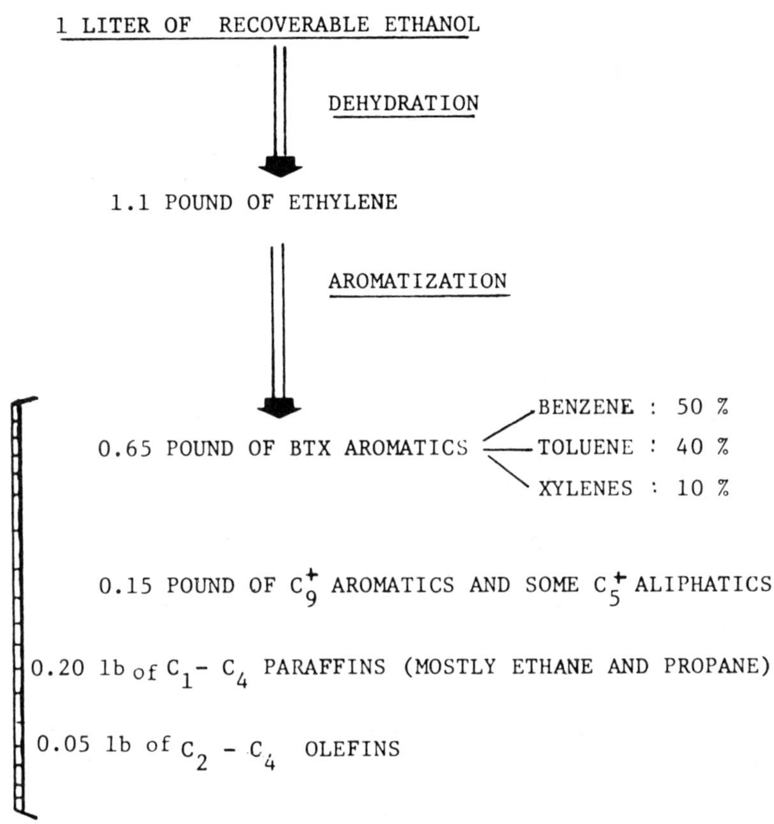

FIG. 8 *Estimated product yield of the two-step process.*

CONCLUSION

It is shown that ethylene can be produced by reaction of an ethanol fermentation broth with a ZSM-5 zeolite catalyst. Several formulations are proposed which include steam-treated, Asb-derived and TFA-loaded ZSM-5 zeolites. Ethylene in the gaseous effluents undergoes aromatization with very high yield over HBS-hybrid catalysts.

ACKNOWLEDGMENTS

The authors thank the Natural Science and Engineering Research Council of Canada (NSERC) and the Québec Actions Structurantes Program for financial support.

REFERENCES

1. R. Le Van Mao, P. Levesque, G. McLaughlin and L.H. Dao, *Appl. Catal.* 34 (1987), 163-179.
2. R. Le Van Mao, T.M. Nguyen and G.P. McLaughlin, *Appl. Catal.* 48, (1989), 267-277.
3. R. Le Van Mao and T.M. Nguyen, U.S. Patent 4 847 223 (July 11, 1989).
4. R. Le Van Mao, U.S. Patent 4 873 392 (Oct. 10, 1989).
5. T.M. Nguyen and R. Le Van Mao, *Appl. Catal.* 58 (1990), 119-129.
6. R. Le Van Mao, T.M. Nguyen and J. Yao, *Appl. Catal.* 61 (1990), 161-173.
7. R. Le Van Mao and L. Dufresne, *Appl. Catal.* 52 (1989), 1-18.
8. R. Le Van Mao, L. Dufresne and J. Yao, *Appl. Catal.*, 65 (1990), 143-157.
9. J. Yao, R. Le Van Mao and L. Dufresne, *Appl. Catal.*, 65 (1990), 175-188.
10. R. Le Van Mao and J. Yao and B. Sjiariel, *Catal. Lett.*, 6 (1990), 23-32.
11. R.J. Argauer and G.R. Landolt, U.S. Patent 3702 886 (Nov. 14, 1972).
12. a) R. Le Van Mao, R. Carli, H.A. Masheye and V. Ragaini, Preprints, *11th Canadian Symposium on Catalysis, Halifax (Canada)*, July 15-18 (1990); b) R. Le Van Mao, R. Carli, H. Ahlafi and V. Ragaini, *Catal. Lett.*, 6 (1990), 321 - 330.
13. R. Le Van Mao and L. Huang, *ACS Spring Natl Meeting, Boston (U.S.A.)*, April 1990.
14. J.H.C. van Hooft, J.P. van den Berg, J.P. Wolthuisen and A. Volmer, Proc. *6th Int. Zeolite Conf.*, Olson and Bisio (Editors), Butterworth, London 1984, p. 489.

23

THE BIOACIDS/BIOACETONE-TO-HYDROCARBONS (BATH) PROCESS

R. LE VAN MAO (*) and L. HUANG
Department of Chemistry and Biochemistry,
Concordia University, *1455 De Maisonneuve Bld. W.*,
Montréal, (Québec), H3G 1M8 Canada

ABSTRACT
Acetic acid in dilute aqueous solution, as obtained by fermentation of organic wastes, is coverted into hydrocarbons in a reactor which contains two catalyst beds set in series. Acetone is produced over CaO-based catalysts at 400 - 500 ° C, and then transformed into hydrocarbons over a trifluoromethane sulfonic acid (TFA) modified ZSM-5 zeolite at 210 ° C. High yields of isobutene, other light olefins and liquid hydrocarbons are achieved with a TFA loading of 2 wt %. Such products are probably derived from the olefinic intermediates formed by condensation reaction of acetone on acidic sites.

(*) To whom correspondence should be addressed.

Results from TGA/DTA, Solid-State NMR and FT-IR techniques show the strong interactions between the zeolite Al sites and the incorporated organic superacid.

INTRODUCTION

Carboxylic acids such as acetic, formic and propionic acids are obtained from various sources, in particular from organic wastes. In the latter case, the aqueous fermentation broths contain very low amounts of such acids (a few wt %). ZSM-5 zeolites can convert (pure) acetic acid into hydrocarbons [1]. However, although the selectivity towards isobutene is interesting, the total conversion to other hydrocarbons is relatively low at fairly high reaction temperatures. The same trends are observed with pure acetone [1].

In our laboratory, a two-step process to convert aqueous acetic acid into hydrocarbons has been devised [2]. Step I consists of using a CaO - based catalyst to convert aqueous acetic acid into acetone at a temperature ranging from 400 ° C to 450 ° C. The effluents of reactor I are sent over a triflic acid (TFA or trifluoromethane sulfonic acid) bearing ZSM-5 zeolite catalyst whose role is to convert the product acetone into hydrocarbons [3]. Among the light olefins which are produced by this process, isobutene is the most interesting because it is one of the two precursors (the other methanol) of MTBE (methyl terbutyl ether), used as octane enhancer for gasoline.

EXPERIMENTAL

Preparation of catalysts

The ZSM-5 zeolites were synthesized according to the known method of Argauer and Landolt [4]. The acid form of the ZSM-5 zeolites were prepared using the method as described elsewhere [5].

The triflic acid loading (2 wt %, as an example) was done according to the following procedure:

Circa 0.2 g of triflic acid (TFA from Fluka Chemie AG) were dissolved in 15 ml of pure acetone. This solution was then slowly added to 10 g of zeolite (powder, acid form). The resulting suspension was allowed to settle and dry in air at room temperature. The obtained solid was washed quickly with 5 ml of acetone and then heated at 120 ° C - 130 ° C in the air for 12 hours. The only difference with the previously described procedure [6,7] was that the TFA impregnated zeolite was washed with acetone prior to its heating at 120 ° C. It is believed that washing decreases significantly the amount of TFA simply chemisorbed on the zeolite surface, mostly on the external surface of the zeolite particles [8].

Several catalysts based on CaO were prepared. These include CaO (bentonite as binder) and the Ca-Y. The Ca-Y zeolite was obtained by ion-exchanging some protons of the zeolite acid form with Ca ions.

Zeolite characterization

The zeolite powders (parent or acid form, and the TFA modified ZSM-5 zeolites) were characterized by atomic absorption (chemical composition), X-ray powder diffraction (structure identification and degree of crystallinity [5]), nitrogen adsorption / desorption (BET/Langmuir specific surface area and pore size distribution using an automatic Micromeretics ASAP 2000 apparatus), water and n-hexane adsorption degree of hydrophobicity or RAI, relative affinity index, according to the method of Le Van Mao et al [9],

Thermogravimetric (TGA) and Differential Thermal (DTA) analyses (Setaram B70 TGA/DTA apparatus, thermal stability of the incorporated TFA surface species), Reflectance FT-Infra-Red (Bomen M 102 Model, for OH groups study) and ^{27}Al and ^{1}H MAS-NMR (Varian XR-300 spectrometer).

Catalyst testing

The final catalysts were obtained by extruding the zeolite or CaO-based material with bentonite following the procedure described elsewhere [6].

Figure 1 shows the experimental set-up which comprises two catalysts beds (I and II), separated by a temperature adjusting section. Bed I is used for the conversion of aqueous acetic acid into acetone whereas bed II is used for the production of hydrocarbons from aqueous acetone. The B.A.T.H. process involves the simultaneous operation of the two beds when an aqueous solution of acetic acid is fed to Bed I.

The reaction products are cooled and the two phases separated. The gaseous phase is analysed by gas chromatography (HP-FID Model GC equipped with a 5 m column packed with 20 wt % Squalane on Chromosorb P). For the GC analysis of the liquid phase, a 50 m PONA capillary column is used. Identification and quantification of O-compounds in the liquid phase are made with a GC-MSD using a capillary column of OV-101. The total conversion of acetic acid is calculated by:

$$C_{AC}\ (\%) = \frac{(Ac.ac.)_f - (Ac.ac.)_p}{(Ac.ac.)_f} \times 100$$

where $(Ac.\ ac.)_f$ and $(Ac.ac.)_p$ are the numbers of carbon atoms (C- atoms) of acetic acid in the feed and in the product stream, respectively.

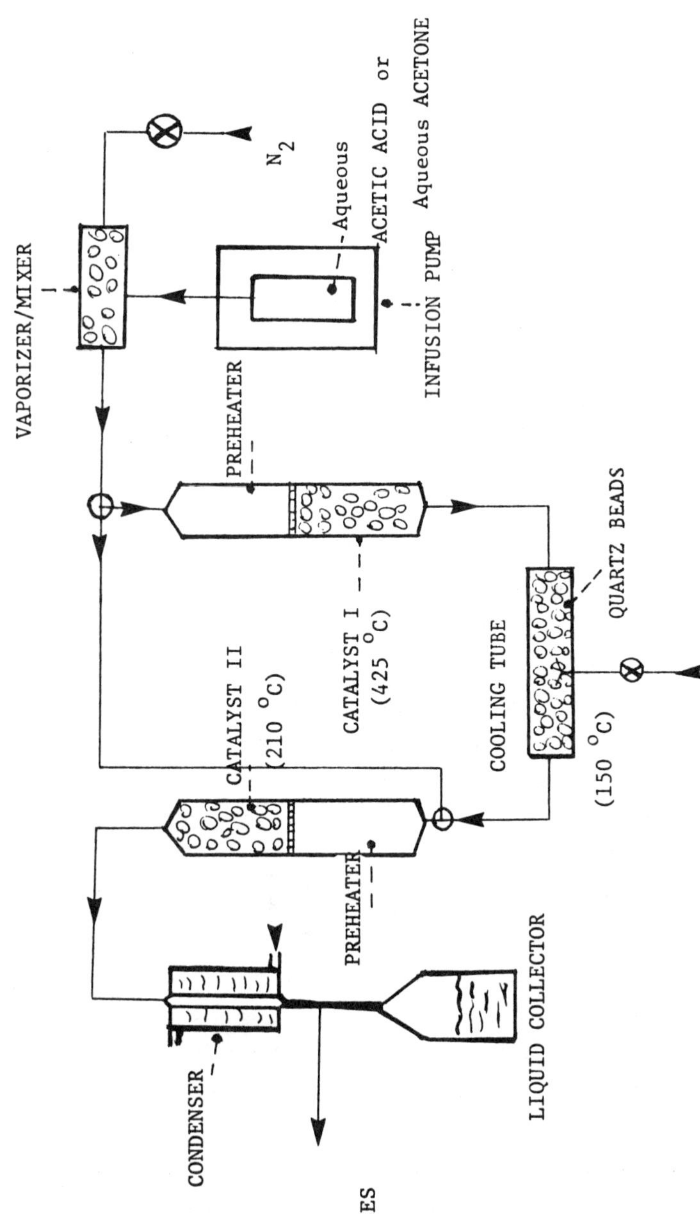

FIG. 1 *Experimental set-up for the conversion of aqueous acetic acid or aqueous acetone.*

The total conversion of acetone is defined as follows:

$$C_{Acet}\ (\%) = \frac{(Acet)_f - (Acet)_p}{(Acet)_f} \times 100$$

where $(Acet)_f$ and $(Acet)_p$ are the numbers of carbon atoms of acetone in the feed and in the product stream, respectively.

The yield in isobutene in the conversion of acetone is given by:

$$Y_{ISOB}\ (\%) = S_{ISOB}\ \%\ C_{Acet}\ /\ 100$$

where S_{ISOB} is the product selectivity in isobutene which is also expressed in C-atoms.

The contact time used in this work is the reciprocal of the weight hourly space velocity which is calculated based on the flow rate of the acetone feed.

RESULTS AND DISCUSSION

Figure 2 shows the adsorption trends (water, n-hexane, ethanol and acetone) of our catalysts versus the TFA loading. While the water adsorption does not change with increasing TFA loading, the effects of the gradual zeolite pore restriction upon TFA loading is visible with bulkier molecules such as n-hexane, ethanol and acetone.

Figure 3 shows the following results obtained with the TGA/DTA performed on the TFA bearing ZSM-5 zeolite:

i) The triflic acid, which normally boils at 161°C, is quantitatively removed only at a temperature above 240°C. This means that the TFA chemisorbed on the zeolite surface is strongly bound to the zeolite surface;

ii) the DTA spectra obtained in the presence of air exhibit two exothermic peaks with maximums located at 275°C (peak A) and 310°C (peak B). This indicates that

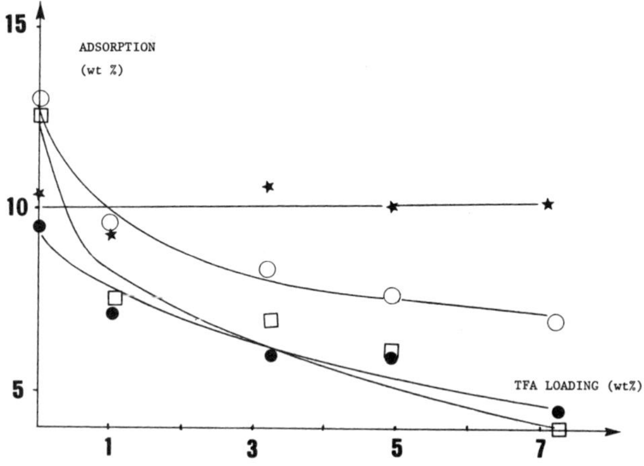

FIG. 2 *Adsorption versus the TFA loading.*
(★) = water: (●) = n-hexane: (O) = ethanol:
(□) = acetone.

there are two species of TFA (A and B) chemisorbed onto the zeolite surface. In the presence of air, these two TFA species undergo oxidation type reaction rather than decomposition or desorption. However, DTA spectra in inert atmosphere (argon) show only one exothermic peak (B). This means that the lesser bound TFA can react with the surface silanol groups (or lattice oxygens) when the sample is heated in inert ambient at 275°C. The more firmly bound TFA might be sorbed on loci which are close to the Al sites. Therefore, the higher thermal stability of species B may be derived from the stabilizing interactions of its fluorine atoms with OH groups associated with framework Al atoms (see Figure 4).

Washing (rapidly) the zeolite particles removes mainly species A which is normally deposited at the external surface of these particles [8].

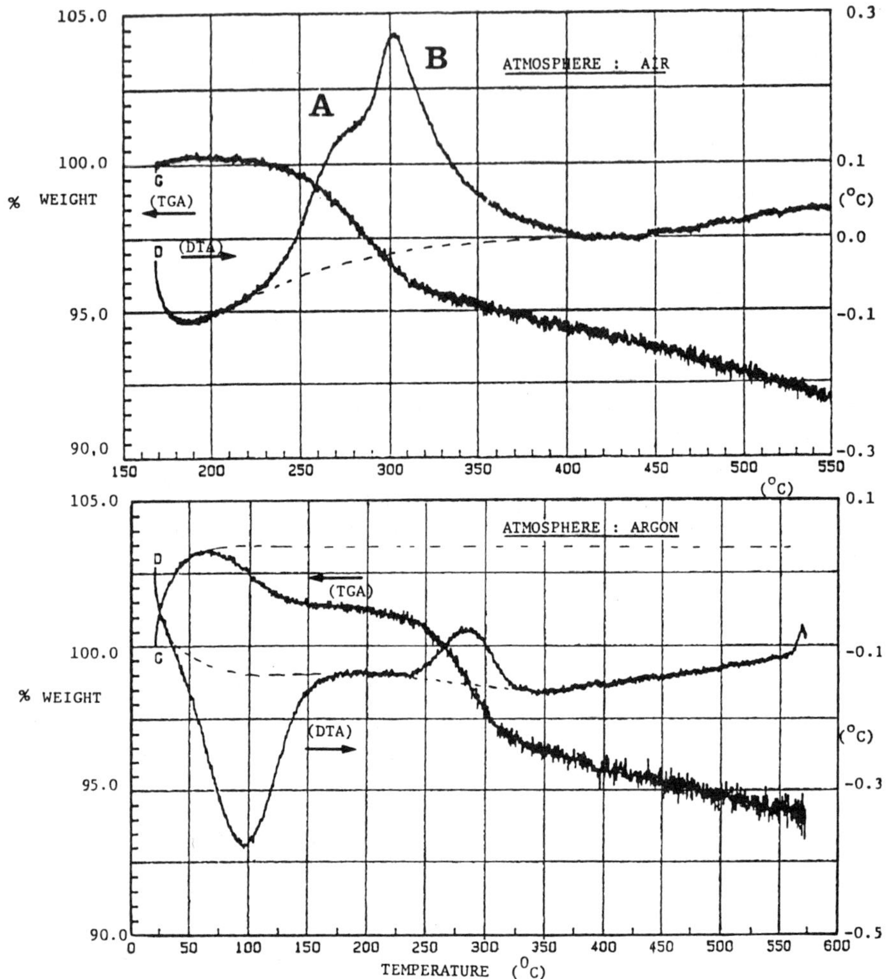

FIG. 3 *TGA and DTA curves with the H-ZSM-5/TFA (4 wt %) in air and in argon.*

IR spectra of the parent zeolite show two vibrations at 3 740 and 3 610 cm^{-1} which are assigned to non-(or weakly) acidic silanol groups and bridged acidic hydroxyl groups, respectively. Upon TFA loading, three additional bands are observed, two at 3 780 - 3 800 cm^{-1} and one at 3 640 cm^{-1}. These are

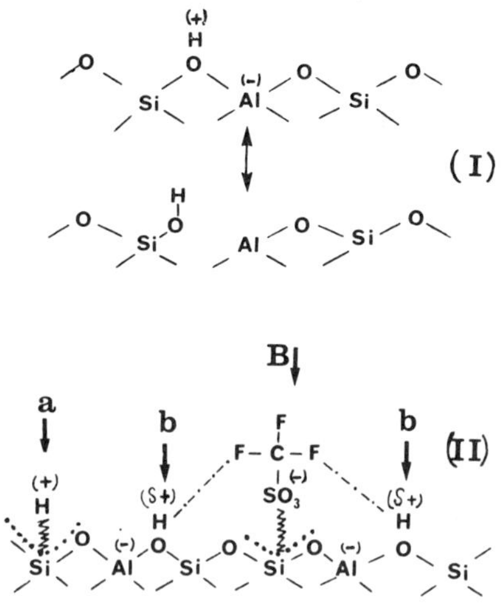

FIG. 4 *Hypothesis on the nature of TFA species on the zeolite surface. (I) = Acid sites on the parent zeolite surface; (II) = dissociative adsorption of TFA and its (eventual) stabilization by formation of hydrogen bonds.*

tentatively assigned to the two types of TFA related OH groups (b and a, respectively, as shown in Figure 4) as previously hypothesized.

Figure 5 (Solid State NMR results) shows the two proton species on the TFA-ZSM-5 catalyst surface: the high inter-exchangeability of these two species may come from the vicinity of the two acidic groups (Zeolite-OH and TFA-OH) which are somehow linked to the zeolite Al sites.

Table 1 reports the results obtained with runs performed over CaO-based catalysts with an aqueous solution of acetic acid (10 wt %). Catalysts obtained by extruding CaO with bentonite gives the highest yield of acetone, more than 97 C-atom % at 425°C.

Table 2 reports the catalytic data obtained in the presence of an aqueous solution of acetone (5 wt %) over the parent ZSM-5, the ZSM-5/La (1 wt %) and the

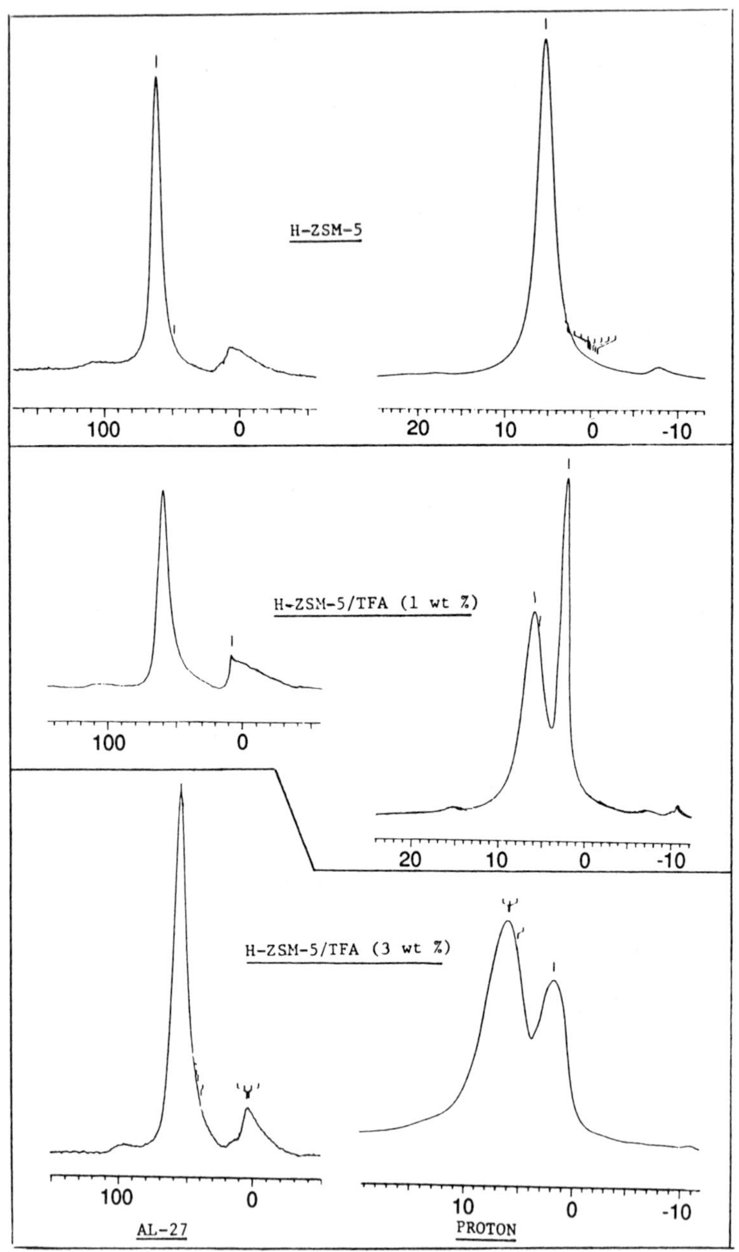

FIG. 5 *Solid State MAS- NMR spectra of the parent H-ZSM-5 and the TFA loaded zeolites.*

TABLE 1 *Conversion of aqueous acetic acid into acetone/hydrocarbons*

Catalyst	Reaction Temperature (°C)	W.H.S.V. (Acetic Acid, h^{-1})	Total Conversion (C-Atom %)	Product Selectivity (C-Atom %)			
				Acetone	Hydrocarbons	O-Compounds	CO_2,CO
CaO (Bentonite)	450	0.06	100	40.81	3.52	3.18	52.49
		0.14	97.55	68.64	3.00	4.90	23.45
		0.20	95.81	71.70	0.89	4.05	23.36
	425	0.05	100	63.78	3.68	0	32.54
		0.12	97.00	69.56	1.07	0	29.37
		0.24	92.36	63.00	0.95	0	36.05
	400	0.05	100	62.40	10.58	0	27.01
		0.13	92.53	58.64	5.07	0	36.29
		0.22	81.11	61.78	1.74	0	36.48
Y-Ca (40 % ion-exchange)	450	0.04	93.05	5.64	29.35	1.65	63.36
	425	0.06	72.63	7.65	11.70	1.91	78.74
	400	0.06	71.45	7.63	16.34	3.62	72.42
H-ZSM-5 (Si/Al:18)	275	0.18	2.93	0	48.65	0	51.35
	333	0.19	17.76	11.77	18.83	0	69.41

TABLE 2 *Conversion of aqueous acetone over ZSM-5, ZSM-5/La and Asb-ZSM-5 catalysts.*
(a) plus some O-compounds

Si/Al ratio	Reaction temperature (°C)	Conversion (C-atom %)				Hydrocarbon yields (C-atom %)			
		Total	To Hydrocarbons	To Ac.Acid (a)	CO_2,CO	Isobutene	Other C_2–C_4 Olefins	C_1–C_4 Paraffins	C_6^+
45	*269*	*72.42*	*29.92*	*6.06*	*36.43*	*6.18*	*15.69*	*2.24*	*5.81*
(ZSM-5	*329*	*88.50*	*36.87*	*2.16*	*49.46*	*9.95*	*13.45*	*4.56*	*11.92*
	402	*76.14*	*42.30*	*1.92*	*31.91*	*0.90*	*31.81*	*2.07*	*7.53*
18	*200*	*45.83*	*16.47*	*4.83*	*24.54*	*8.87*	*2.21*	*0.22*	*5.15*
(ZSM-5)	*272*	*76.70*	*40.41*	*8.22*	*28.06*	*11.62*	*14.68*	*2.93*	*11.17*
	330	*96.72*	*44.95*	*2.36*	*49.41*	*4.03*	*13.66*	*8.87*	*18.39*
18	*250*	*66.60*	*13.10*	*7.43*	*46.06*	*3.24*	*2.54*	*0.94*	*6.38*
(ZSM-5/La)	*300*	*76.67*	*33.18*	*6.79*	*36.69*	*5.71*	*9.92*	*3.95*	*13.60*
	352	*83.66*	*57.19*	*3.33*	*23.14*	*9.55*	*22.60*	*9.18*	*42.34*
15	*273*	*24.06*	*13.60*	*1.71*	*8.76*	*9.77*	*0.41*	*0*	*3.42*
(Asb-ZSM-5)	*329*	*45.92*	*29.61*	*5.06*	*11.25*	*13.80*	*8.98*	*0.24*	*6.58*

TABLE 3 *Conversion of Aqueous Acetone Into Hydrocarbons Over TFA Loaded ZSM-5 and H-Y Zeolites*

CATALYST	REACTION PARAMETERS		CONVERSION (C-atom %)				HYDROCARBON YIELDS (C-atom %)			
	(^{o}C)	W.H.S.V. (acetone, h^{-1})	Total	To Hydrocarbons	To Ac.Acid (a)	CO_2,CO	Isobutene	Other C_2–C_4 Olefins	C_1–C_4 Paraffins	C_6^+
ZSM-5 / TFA (1 wt %)	*200*	*0.09*	*50.54*	*11.11*	*6.09*	*33.34*	*5.29*	*1.53*	*0.16*	*4.12*
ZSM-5 / TFA (2 wt %)	*180*	*0.10*	*16.28*	*7.41*	*1.59*	*7.29*	*4.64*	*0.36*	*0.01*	*2.41*
	200	*0.09*	*45.85*	*24.76*	*6.56*	*14.52*	*11.48*	*3.29*	*0.51*	*9.49*
	211	*0.10*	*61.04*	*39.92*	*10.46*	*10.66*	*15.77*	*7.62*	*1.61*	*14.35*
	213	*0.06*	*66.46*	*45.83*	*9.98*	*10.65*	*18.76*	*8.75*	*1.85*	*16.48*
ZSM-5 / TFA (3 wt %)	*212*	*0.10*	*53.42*	*29.64*	*8.57*	*15.20*	*12.33*	*5.14*	*1.01*	*11.16*
	212	*0.05*	*60.08*	*30.88*	*7.26*	*21.95*	*9.86*	*3.97*	*0.87*	*16.18*
H-Y / TFA (3 wt %)	*210*	*0.05*	*69.34*	*31.18*	*10.97*	*27.19*	*2.95*	*2.99*	*18.94*	*6.30*
	213	*0.06*	*76.17*	*23.23*	*8.85*	*44.09*	*1.93*	*1.79*	*13.96*	*5.56*

(a)plus small amounts of O-compounds

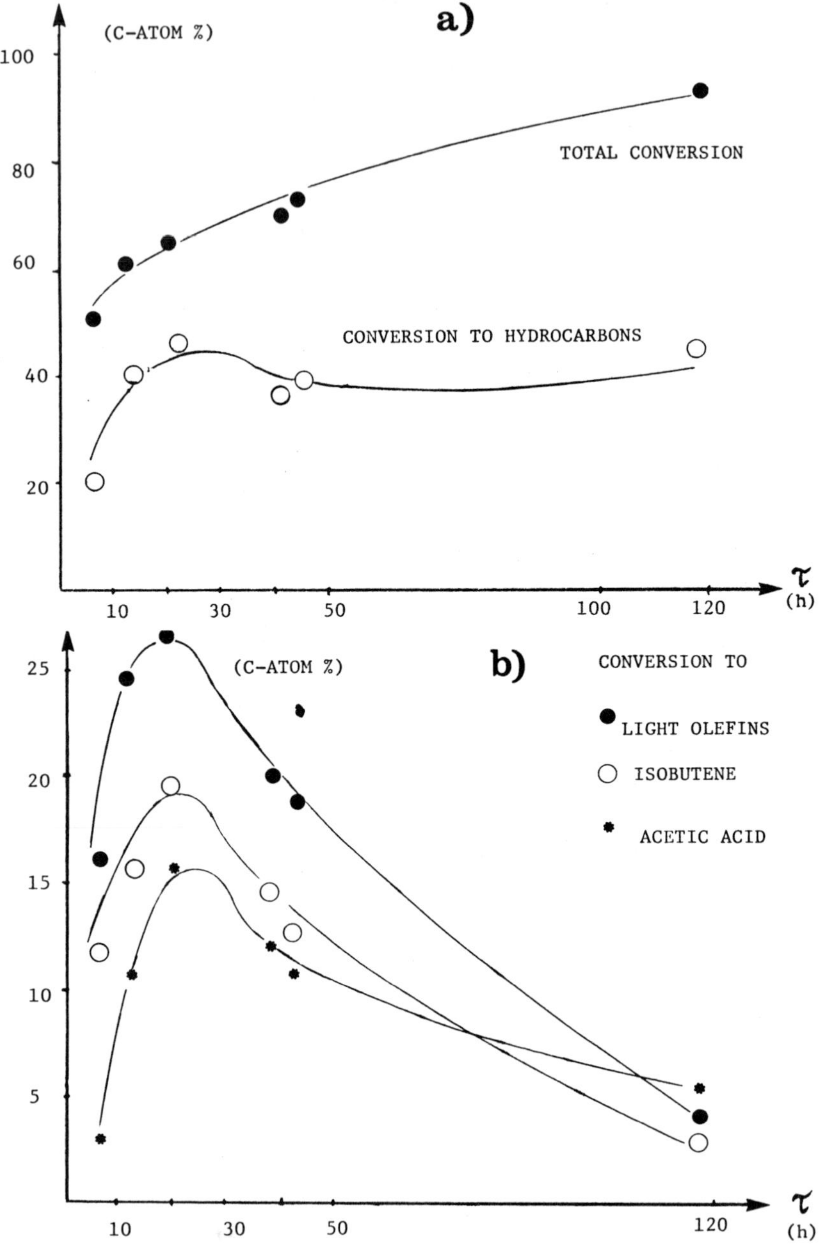

FIG. 6 *Conversions versus contact time (acetone)* τ.

Asbestos-derived ZSM-5 zeolites (W.H.S.V. = 0.09 h^{-1}, acetone). Table 3 reports the results obtained with the ZSM-5/TFA catalysts having different TFA loadings (the Si/Al ratio of the ZSM-5 zeolite used to prepare these catalysts is 18, reaction performed at 210°C). Catalytic data of two Y/TFA samples are also reported in Table 3. While the ZSM-5/TFA catalysts give relatively high yields in isobutene (and heavy hydrocarbons), the larger pore sized Y/TFA produces more paraffins (mostly isobutane). Figures 6 show that the production of isobutene (and C_2 - C_4 light olefins) goes through a maximum at low reaction contact times. Heavy O-compounds which normally result from the aldol-type condensation of acetone (mesityl oxide and diacetone alcohol, mainly) are surprisingly produced in trace amounts. Acetic acid is formed in significant amounts and its yield varies in the same manner as isobutene (see Figures 6). Therefore, the current hypothesis according to which isobutene and other hydrocarbons are derived from products of the classical aldol-type condensation (mesityloxide and diacetone alcohol, for instance) [1, 10] is questionable vis-a-vis the hypothesis of Zaki and Sheppard [11] (see Figure 7). According to the latter, some hydroxyl groups of the catalyst interact with the adsorbed olefinic intermediates to yield isobutene and (adsorbed) acetate species (Figure 7). This indicates that isobutene and acetic acid are probably the primary reaction products. Subsequent acid-catalysed reactions on the product isobutene lead to other light olefins (and paraffins), heavy hydrocarbons and butanols.

Figures 6 also show that the production of hydrocarbons goes through a maximum at lower contact time and exhibits a practically constant value at higher contact time. However, at these contact time values, more carbon dioxide and heavy hydrocarbons are

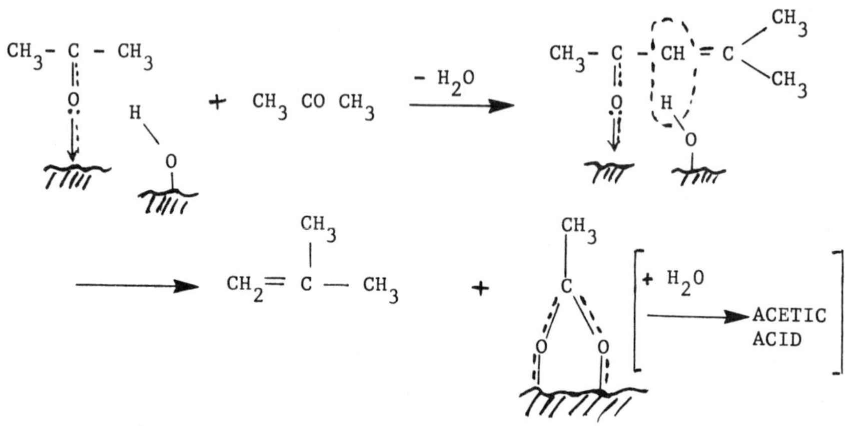

FIG. 7 *Hypothesis for the formation of hydrocarbons, also according to Zaki and Sheppard [11] (permission of J. Catal.).*

formed at the expense of isobutene (and other light olefins) and acetic acid.

Kubelkova et al [10] had observed a rapid decay of the acetone conversion over (pure) ZSM-5 zeolite. Such a loss of activity was attributed to the disappearance of surface acidic hydroxyl groups and a build-up of heavy reaction intermediates. The presence of water in the feed considerably slowed down the decay process [10]. In our case, it was also seen that the catalyst recovered its activity by "in-situ" regeneration with steam which contained some HCl [12].

Finally, Table 4 shows the figure of hydrocarbon yields currently obtained with the bench scale B.A.T.H. reaction system.

CONCLUSION

It is possible to obtain isobutene and other hydrocarbons from a dilute aqueous solution of acetic acid by reacting the latter with a CaO-based catalyst and subsequently the effluents of such a reaction with a TFA bearing ZSM-5 zeolite. Acetic acid and some butanols are the main by-products.

TABLE 4 *Conversion of aqueous acetic acid into hydrocarbons with the B.A.T.H. process.*

FEED:	*AQUEOUS ACETIC ACID (10 wt %)*
REACTOR I:	*Catalyst = CaO- Bentonite; T=425 - 400°C; W.H.S.V. = 1.2 h^{-1} (solution)*
REACTOR II:	*ZSM-5 / TFA; T = 200 - 210°C*
YIELDS OF PRODUCT HYDROCARBONS (C-Atom %):	32%
Gaseous hydrocarbons:	*olefins = 19% (isobutene = 13%), paraffins = 1%*
Liquid hydrocarbons:	*12% (aromatics = 3%)*

ACKNOWLEDGMENTS

The authors thank NSERC-Canada and Actions Structurantes Program of Québec for financial support.

We also thank Mr. Y. Yu, Mr. B. Sjiariel and Mr. P. Aysola for technical assitance.

REFERENCES

1. C.D. Chang and A.J. Silvestri, *J. Catal*, 47 (1977), 249-259.
2. R. Le Van Mao, *Can. Catal. Disc. Group Meeting*, Ottawa (Canada). May 19, 1989.
3. R. Le Van Mao, G. McLaughlin and B. Sjiariel, *Proc. Seventh Can. Bioenergy R & D Seminar - Ottawa (Canada)*, April 1989.
4. R.L. Argauer and G.R. Landolt, U.S. Patent 3 702 886 (Nov. 14, 1972).
5. R. Le Van Mao, P. Levesque, B. Sjiariel and P.H. Bird, *Can. J. Chem.* 63 (1985), 3464.
6. R. Le Van Mao, T.M. Nguyen and G.P. McLaughlin, *Appl. Catal.* 48 (1989), 265-277.

7. a) R. Le Van Mao and T.M. Nguyen, U.S. Patent 4 847 223 (Jul. 11, 1989); b) R. Le Van Mao, U.S. Patent 4 873 392 (Oct. 10, 1989).

8. R. Le Van Mao, T.M. Nguyen, D. Ly and J. Yao, ***Am. Chem. Soc. Spring National Meeting, Boston*** (U.S.A.), April 1990.

9. a) R. Le Van Mao, ***React. Kinet. Catal. Lett.*** 12 (1979), 69; b) R. Le Van Mao, O. Pilati, A. Marzi, A. Villa and V. Ragaini, React. ***Kinet. Catal. Lett.*** 15 (1980), 293.

10. L. Kubelkova, J. Cejka, J. Novakova, V. Bosacek, I. Jirka, P. Jiru, in "Zeolites: Facts Figures, Future", ***Proceedings 8th Intern. Zeolite Conf., Amsterdam***, the Netherlands, Editors: P.A. Jacobs and R.A. van Satem, Elseview Publ. (1989), 1203.

11. M.I. Zaki and N. Sheppard, ***J. Catal.*** 80 (1983), 114.

12. R. Le Van Mao and M.C. Le Van Mao, ***Concordia University***, unpublished data (1988).

24

SAPO-34 CATALYST SELECTIVITY FOR THE MTO PROCESS

G. Pop, G. Musca, D. Ivanescu, E. Pop, G. Maria*,
E. Chirila, and O. Muntean

Department of Catalysis,
Chemical & Biochemical Energetics Institute,
Spl. Independentei 202, Bucharest 77208, ROMANIA

ABSTRACT

In order to investigate the Methanol-to-Olefins (MTO) process on a SAPO-34 zeolite, an experimental program was run in a fixed-bed laboratory-scale reactor. The influence of the operating parameters on the product distribution is used to make some kinetic preliminary considerations, and to examine the catalyst deactivation process.

INTRODUCTION

The MTO process, which converts Methanol into Olefins on a zeolite catalyst, seems to be very attractive because both its two steps (i.e. the methanol synthesis and the MTO) proceed with selectivities exceeding 98%. Economic investigations [1] underlined that this process becomes competitive with the hydrocarbons pyrolysis only if its ethene selectivity is at least 60%, i.e. 25%wt. with respect to the fed methanol. Although the classical Si-Al zeolites don't reach such performances, the new Al-P-Si zeolite class (named SAPO, [2]) brings new prospects to this field: Liang et al. [3] obtained at 450°C (over a SAPO-34 zeolite) a nearly complete methanol conversion

---

* *Author who delivered the paper and to whom all correspondence should be addressed.*

and an 89% total olefins selectivity (the ethene selectivity accounting for 59.9%); Union Carbide reported a 90% total olefins selectivity (at a nearly complete methanol conversion) and an ethene and propene selectivity accounting for 60% [4,5]. Referred to the feed methanol this result corresponds to 39%wt. olefins in products and 26%wt. either ethene or propene respectively.

The aim of this paper is to investigate the performances of the SAPO-34 zeolite under different operating conditions, the catalyst deactivation, and to advance some kinetic considerations about the studied process.

EXPERIMENTAL SECTION

THE CATALYST

The SAPO-34 zeolite was obtained following a modified Lok et al. [6] method.It was crystallized for 100 hours at 190°C in a teflon coated vessel.Thereafter, it was filtered, dried at 120°C and calcinated for 6 hours at 550°C in an air stream.The X-ray diffraction spectra (Figure 1) shows no impurity with respect to the standard SAPO-34 zeolite reported in literature (Lok et al. [6]).The zeolite crystallites (0.3-1.4 μm in diameter) were included in a silica matrix.The extruded particles (2 x 4 mm cylinders) were activated through calcination at 500°C.

APPARATUS AND MATERIALS

The experiments were carried out in a glass tube, laboratory-scale reactor of 22 mm i.d. filled with 10 cm^3 of catalyst diluted with inert material at a 1:1 ratio.An electric heating/cooling system around the reactor maintains the cvasiisothermicity of the catalytic fixed bed. The maximum temperature axial gradient measured in our experiments was less than 7°C.Commercial available methanol (99.8%) was continuously fed and vaporized in a zone with inert material filled preheating.The analytical method for the reaction products has already been reported [7].

EXPERIMENTAL PROGRAM AND DATA AQUISITION

The experimental program, of central-composite type, takes into account only two independent variables: the temperature (varied in the 365-435°C range) and the linear hourly space velocity (LHSV, covering the domain of 0.6-

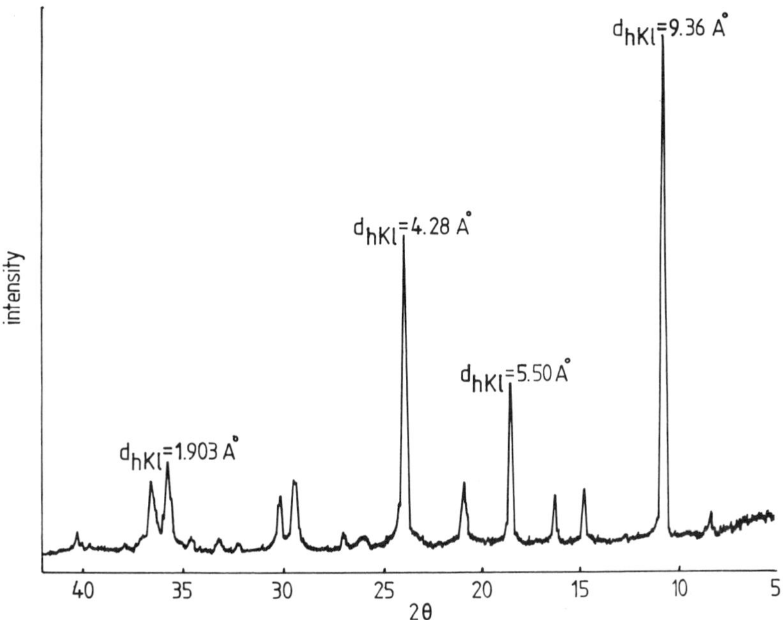

FIG. 1 *XRD- Spectrum of the SAPO-34 zeolite ($Co-K_{\alpha}$ radiation).*

2.4 h).The product analysis and the amount of fraction collected (approximately 2 hours time-on-stream) allow an overall and elementary mass balance for each run.The 8 selected runs, correspond at three different temperature levels (365°, 400°, 435°C), and present a satisfactory reactor input-output mass balance.

PRELIMINARY INVESTIGATIONS

CATALYST PERFORMANCES

The exhaustive product distribution for the 8 runs shows good performances of the catalyst: 36-41 g.olefins/100 g. feed methanol, and 19-27 g. ethene/100 g. feed methanol, for a methanol conversion practically complete (greater than 98%).A typical product distribution is present in Table 1.As shown, the SAPO catalyst is very selective to lower olefins in comparison with the classical zeolite used for the MTO process by Mihail et al. [8](noted MZ; the data are obtained in a stainless steel laboratory-scale, integral reactor).For comparison with the Methanol -to-Hydrocarbons zeolite performances, the reader is re-

TABLE 1. *Product distribution (%wt.) for the SAPO catalyst (435°C, 1.95 h^{-1} run), and for the mordenite zeolite (MZ) catalyst of Mihail et al. [8](370°C, 1.12 h^{-1} run).*

Compound	Catalyst SAPO	Catalyst MZ	Compound	Catalyst SAPO	Catalyst MZ
H_2	0.4290	0.18	C_5H_{12}	0.2979	0.22
CH_4	1.3477	5.00	C_5H_{10} + C_6^+	1.0621	0.28
C_2H_6	0.1671	0.45	CO	-	4.28
C_2H_4	26.6553	7.96	CO_2	-	0.13
C_3H_8	0.4411	3.43	CH_3OH	1.4705	2.97
C_3H_6	10.6310	10.23	CH_3OCH_3 (DME)	0.9074	0.21
C_4H_{10}	0.5445	2.82	H_2O	54.6366	55.12
C_4H_8	1.3898	6.65			

ferred to Pop et al. [9], Mihail et al. [10]. The plotted product concentrations for SAPO-34 catalyst (not presented here) show that the ethene yield presents a maximum value of 27% for high temperatures and LHSV. On the other hand the total olefins yield seems to be less sensitive with respect to the operating parameters, having an approximate value of 41%.

CATALYST DEACTIVATION

Figure 2 presents the variation of methanol conversion, ethene and DME selectivities vs. the time-on-stream. Although the conversion remains very high, the selectivities alteration is important. Therefore, after each 2 hour period, the catalyst was regenerated with air, at 480°C, for 4 hours. The zeolite aging effect was negligible in our experiments.

The exothermic effect of the global MTO reaction is approximately 5 Kcal/mol CH_3OH reacted, instead of 8-10 Kcal/mol CH_3OH reacted for the MTO with the MZ catalyst (Mihail et al. [8]).

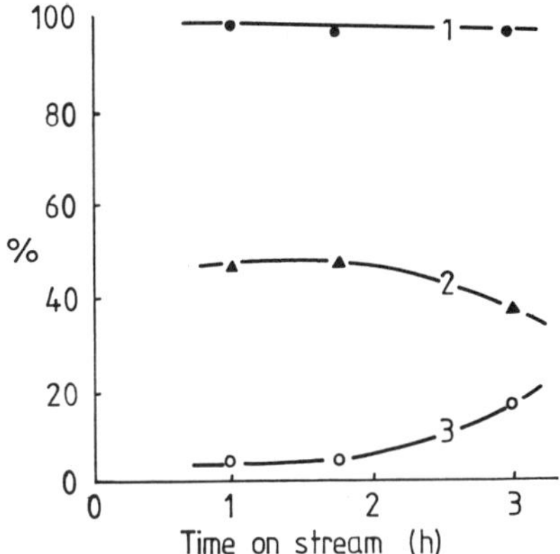

FIG. 2 *The deactivation test: ——1—— methanol conversion; ——2—— ethene selectivity; ——3—— dimethylether (DME) selectivity.*

INITIAL NETWORK FORMULATION

In order to rapidly simulate the process, a reduced and adequate model could be prefered. For the MTO process many extended or reduced kinetic models are proposed in the last years (see for details Mihail et al. [8],[11]; Liu et al. [12]; Iordache et al. [13]). The proposed schemas involve either carbenium or trialkyloxonium ion intermediate (Derouane et al. [14]; Van den Berg et al. [15]; Perot et al. [16]), or the carbene intermediate (Chang and Silvestri [17]; Chang [18]). Among the reduced kinetic models (see the review of Iordache et al. [13]), here we used the schema of Maria [19], derived from the Chang [20] - Anthony's [21] reduced model:

$$A \xrightarrow{k_1} B + H_2O \qquad (1a)$$

$$A + B \xrightarrow{k_2} C + H_2O \qquad (1b)$$

$$C + B \xrightarrow{k_3} P, \qquad (1c)$$

where A notes the oxygenates (without water), B a carbene intermediate, C the olefins, P the paraffins and other products. This model is very robust, i) resisting of some other fictive reaction path added to model (1) (see the special estimation - reduction strategy to identify and to reject fictive reactions of Maria and Musca [22]; Maria [19]), and ii) being also derived from the extended model of Mihail et al. [8](33 reactions) via Maria and Muntean [23] model (14 reactions), by using a lumping analysis (Iordache et al. [13]).

Furthermore, the following model assumptions are made: i) only first order reactions; ii) methanol and DME are at the equilibrium state, constituting the lump A; iii) all species have the empirical formula $(CH_2)_n$, that is a constant average molecular weight value; iv) the B intermediate is constant ; v) the water concentration measurements give few information about the process (see the arguments of Iordache et al. [13]).The

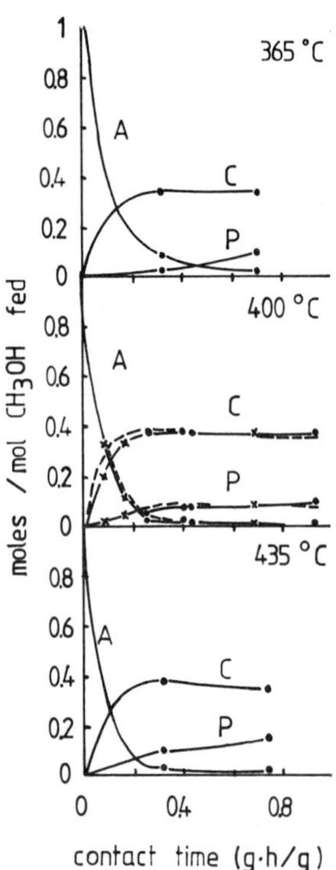

FIG. 3 *Experimental (●) and generated (x) data, spline A, C, P species concentration (moles/mol methanol fed) trajectories (————) and model predictions (------).*

model equations are:

$$dA/dt = - k_1A - k_2AB \quad (2a)$$

$$dC/dt = k_2AB - k_3BC \quad (2b)$$

$$dP/dt = k_3BC \quad (2c)$$

$$B = k_1A/(k_2A + k_3C). \quad (2d)$$

By substituting the B species concentration, the model contains 3 species (A, C, P) and only 2 kinetic parameters: k_1 and $k_{23} = k_2/k_3$. By eliminating lump B, its nature appears to have lower importance in the final reduced kinetic model. The A, C, P species concentration spline trajectories and data are plotted in Figure 3, at three temperature levels: 365°, 400°, 435°C. The catalyst appears to have approximately the same performances (Figure 3) in the investigated temperature domain, so the kinetic parameters were identified only for the mean temperature of 400°C. Additional, spline generated, data are also used in order to increase the estimate quality. The Sum of Squares of Residuals (SSR), computed for all species and runs, was minimized by using the adaptive random procedure of Mihail and Maria [24]; the obtained solution is $k_1 = 6.65$ g/(g.h), $k_{23} = 6.06$, SSR = 0.014. The used contact time t, in the reactor model (cvasiisothermal fixed-bed, integral-type, reactor) is defined as grams of catalyst.h/grams of fed reactant. The obtained model predictions, as shown in Figure 3, are in a very satisfactory agreement with the data. The identified parameters are of the same order of magnitude with those of Maria [19] for the MTO process (370°C), under the MZ catalyst: $k_1 = 5.26$ s^{-1}, $k_{23} = 1.29$, SSR = 0.107 (the used contact time here is defined as m^3 catalyst.s/m^3 gas). As it was to be expected, the (1a, b) reaction rates for the SAPO-MTO process are greater than the corresponding values for the MZ-MTO process.

CONCLUSIONS

The SAPO-34 catalyst performances, i.e. a 90% total olefins selectivity and a 61% ethene selectivity, are above the critical values reported by Sherwin [1] in order to have an economically competitive technology. Recirculating DME and unreacted methanol we can obtain a small improving of the above selectivities by about 1%. The reduced thermal effect of the reaction makes the process more manageable. However, due to the high catalyst deactivation rate, the continuous recirculation of the catalyst in a fluidized bed reactor - regenerator system is more advisable.The proposed reduced kinetic model could satisfactorily correlate the data at 400°C.Further data and kinetic parameters computations could validate the model vs. different operating conditions.

REFERENCES

1. M. B. Sherwin, *Hydroc. Processing, March*: 79 (1981).
2. S. T. Wilson, B. M. Lok, C. A. Messina, T. R. Cannan, and E. M. Flanigen, *J. Amer. Chem. Soc. 104*: 1146 (1982).
3. J. Liang, H. Li, S. Zhao, and W. Guo, Preprint Poster Paper, *7th Int. Zeolite Conf.*, Tokyo, p. 321 (1986).
4. J. Ross, *Appl. Cat. 41*: 391 (1988).
5. *Hydrocarbons Proc. Febr.*: 65 (1988).
6. B. M. Lok, C. A. Messina, R. L. Patton, R. T. Gajek, T. R. Cannan, and E. M. Flanigen, *U. S. Patent 4, 440,871* (1984).
7. G. Pop, G. Musca, E. Pop, D. Hergheleglu, and P. Tomi, *Rev. Chim.(Bucharest), 34*:293 (1983).
8. R. Mihail, S. Straja, G. Maria, G. Musca, and G. Pop, *Ind. Eng. Chem. Process Des. Dev. 22*: 532 (1983a).
9. G. Pop, G. Musca, G. Maria, S. Straja, and R. Mihail, *Ind. Eng. Chem. Product Res. Dev. 25*: 208 (1986).
10. R. Mihail, S. Straja, G. Maria, G. Musca, and G. Pop, *Chem. Eng. Sci. 38*: 1581 (1983b).
11. R. Mihail, S. Straja, G. Maria, G. Musca, and G. Pop, *Ind. Eng. Chem. Res. 26*: 637 (1987).
12. L. Liu, R. G. Tobias, K. McLaughlin, and R. G. Anthony, In: *Catal. Convers. Synth. Gas Alcohols Chem.* R. G. Herman (Ed.), Plenum, New York (1984).
13. O. Iordache, G. Maria, and G. Pop, *Ind. Eng. Chem. Res. 27*: 2218 (1988).
14. G. E. Derouane, J. B. Nagy, P. Dejaifve, J. H. Van Hooff, B. P. Spekman, C. J. Vedrine, and C. Naccache, *J. Catal. 53*: 40(1978).
15. J. P. Van den Berg, J. P. Wolthuizen, J. H. C. Van Hooff, In: *Proc. 5th Int. Conf. on Zeolites*, L. V. C. Rees (Ed.), Heyden, London, p. 649 (1980).
16. G. Perot, F. X. Carmerais, and M. Guisnet, *J. Molec. Catal. 17*: 255 (1982).
17. C. D. Chang, and A. J. Silvestri, *J. Catal. 47*: 249 (1977).
18. C. D. Chang, *Catal. Rev. Sci. Eng. 25*: 1 (1983).
19. G. Maria, *Canad. J. Chem. Eng. 67*: 825 (1989).
20. C. D. Chang, *Chem. Eng. Sci. 35*: 619 (1980).

21. R. G. Anthony, *Chem. Eng. Sci. 36*: 789 (1981).
22. G. Maria, and G. Musca, *Comp. Chem. Eng. 13*: 425 (1989).
23. G. Maria, and O. Muntean, *Chem. Eng. Sci. 42*: 1451 (1987).
24. R. Mihail, and G. Maria, *Comp. Chem. Eng. 10*: 539 (1986).

25

SELECTIVE ALKYLATION OF ISO-BUTENE WITH METHANOL TO PRODUCE ISO-C5 OLEFINS

G. Pop, R. Boeru, O. Muntean and G. Maria*

Department of Catalysis
Chemical & Biochemical Energetics Institute
Spl. Independentei 202, Bucharest 77208, ROMANIA

ABSTRACT

In order to examine the alkylation process of an olefin C4 fraction with methanol to produce iso-C5 olefins, by using a zeolite catalyst, some laboratory-scale experimental data were obtained. The investigated range of reaction temperature was 300-400°C, by using different iso-butene/methanol feed ratios in a reaction time domain of interest. Starting from some competitive reaction routes available from previous experiments, an initial reaction network is formulated. The analysis of the proposed model adequacy allows some considerations concerning the possible significant reaction steps and catalyst performances.

* *Author who delivered the paper and to whom all correspondence should be addressed.*

INTRODUCTION

The iso-C5 olefinic fraction is an important raw material for the petrochemical processing industry and a valuable component of the gasoline. The non-conventional production of it by using an olefinic C4 fraction and methanol is attractive for two reasons: i) the C4 fraction is thus used in a more efficient way, and ii) the process involves the methanol, i.e. a non-petrochemical raw material.

The aim of this paper is to investigate the performances of a zeolitic catalyst for the alkylation of isobutene with methanol, and to use the collected data in order to build and check a reduced kinetic model.

EXPERIMENTAL SECTION

THE CATALYST

The catalyst is a ZSM-12 zeolite, obtained by adapting the method described by Xiang and Li [1]. The Na-zeolite form was successively treated with NH_4NO_3 solution at 95°C, and after drying and calcination, the H-zeolite form was included in a SiO_2 matrix, and used as extruded cylinders 1 mm in diameter and 3 mm in length. Finally the catalyst was activated at 500°C for 8 h. The X-ray diffraction analysis indicated a purity and a crystalline structure in a percent greather than 95%, which corresponds with those of a similar reported catalyst [2].

APPARATUS AND MATERIALS

The experiments were carried out in a stainless steel tube reactor (22 mm i.d.). An electric heating/cooling system around the reactor maintains a constant feed mixture temperature, and forestalls an excessive reactor - ambient medium heat exchange. The liquid, commercially available (99.5%) methanol, continuously fed by means of a dosimetric pump, was vaporized and mixed together with a controlled stream of butene in a zone with inert material filled preheating. The reactor contains 20 cm^3 of catalyst diluted with inert material in a 1:1 ratio. In order to smooth the temperature radial profile, a thin layer of air is present between the reactor and the electric resistance. Special attention was paid to the temperature sensors location around the hot spot location; a six point experimental profile was recorded along the axial coordinate.

PRODUCT ANALYSIS

After cooling at $2^{o}C$, the reaction products were separated into two fractions: the volatile fraction (C1-C6 hydrocarbons and dimethylether (DME)), and the liquid one. The liquid fraction separated into two phases: hydrocarbons (more than C6) and aqueous phase (water, methanol and DME). All fractions were collected and subjected to chromatographic analysis.

The volatile fraction (a 0.5 ml sample) was analysed by using a stream of nitrogen vector gas, in two chromatographic column (FID detector) at the ambient temperature: an external column (EC) of 4 mm in diameter and 6 m in length, containing Chemosorb P60-80 impregnated with 20% bis-2-methoxy-ethyl-adipate, and an internal column (IC) of 4 mm i.d. and 3 m in length containing Chemosorb P60-80 impregnated with 20% tris-cianoethoxy-propane.

The liquid hydrocarbons, stabilized by dilution 1:3 with benzene, were analysed in a column of 4 mm i.d. and 3 m in length containing Chemosorb W60-80, by using a constant heating rate of $8^{o}C$/minute. The C1-C6 hydrocarbons, contained in the liquid phase, were separately analysed in the same manner as the volatile fraction (EC) and then added to the complete compound list of the liquid phase.

The aqueous phase was analysed by using the IC at the constant temperature of $100^{o}C$.

EXPERIMENTAL PROGRAM AND DATA AQUISITION

In order to study the process an experimental program of central-composite type is designed. The independent variables took into account in this program are the feeding temperature, the space time (WHSV), and the molar methanol/iso-butene feeding ratio, varied in the ranges of $300-350^{o}C$, $2-4\ h^{-1}$, and 0.5-2.5 moles/mol respectively. The axial temperature profile of the integral type, fixed-bed, laboratory-scale reactor, is measured for each run (Fig. 1). The product analysis and the amount of fraction collected allow an overall and elementary (C, H, O elements) mass balance for each run. From this point of view we selected 8 runs from the experimental program which presented a relative input-output deviation smaller than 5%. These runs (indexed from 1 to 8) are indicated in Figure 1, i.e. the product species concentrations (in terms of moles/mol CH_3OH fed) and axial temperature profile, and further used for kinetic considerations.

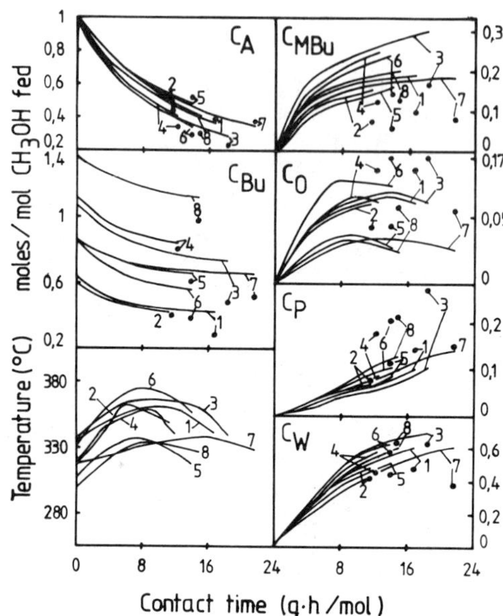

FIG. 1 *Experimental data (●), predicted species A, Bu, MBu, O, P, W concentration (moles/mol CH_3OH fed) trajectories (model 1), and measured axial catalytic bed temperature (°C) vs. the contact time for runs no. 1-8.*

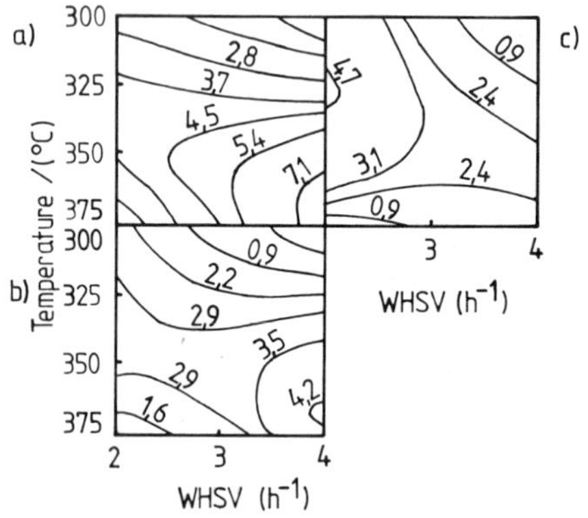

FIG. 2 *Moles CH_3OH reacted/Mol Bu reacted ratio vs. the operating conditions. a) 2.33 moles CH_3OH/mol Bu fed; b) 1.16 moles CH_3OH/mol Bu fed; c) 0.77 moles CH_3OH/mol Bu fed.*

THE KINETIC MODEL

PRELIMINARY INVESTIGATIONS

The alkylation of iso-butene with methanol could be globaly written as:

$$r\ CH_3OH + C_4H_8 \longrightarrow \text{Hydrocarbons + Water,} \quad (1)$$

where r is the CH_3OH/C_4H_8 molar reacted ratio. The experiments indicate that r is strongly dependent on the considered independent variables (see Fig. 2), and generally has a mean value around 2. On the other hand the methanol conversion varies in the 0.5-0.8 range. That could be an indication that, especially for high temperatures, the acid properties of ZSM zeolite could induce, besides the alkylation reactions, a methanol-to-olefins (MTO) path with generation of olefins, paraffins and aromatics.

The thermodynamic computations show that the global reaction (1) is practically irreversible, for all molar feed ratios higher or lower than unity, in the investigated operating domain. The exothermic effect of reaction (1) varies in the range of 13-23 Kcal/mol CH_3OH reacted.

The laboratory-scale used reactor allows the assumption of integral plug-flow type; the deviations from this assumption were minimized by using a suitable particle aspect ratio and Reynolds number and by placing a bed of inert material at the entrance to the catalytic bed.

The each run time is sufficiently large (1-1.5 h) to collect products for a consistent analysis and mass balance calculations. After each run the catalyst was regenerated. The zeolite aging effect was negligible in our experiments.

INITIAL NETWORK FORMULATION

A natural trend to build a kinetic model is to propose a network containing all possible steps based on the product analysis. This tactic ensures that all the feasible reaction steps were considered [3]. However this overparameterization requires not only a steady experimental strategy but leads to a very large computational effort to identify the parameters. A worthy alternative is to build and verify the model step by step; a guiding strategy is presented for instance by McLean et al. [4].

The global initial network, further proposed, takes into account the product distribution and the observation that most olefins are intermediate species. As demonstrated by Behrsing et al.[5], the reaction between methanol and iso-butene is an acid-catalyzed homologation:

$$C_4H_8 + CH_3OH/H^+ \longrightarrow [C_5H_{11}]^+ \longrightarrow C_5H_{10} + H_2O, \quad (2)$$

possible for other olefins too; the oligomerizations of olefins are also possible. On the other hand, the typical MTO reactions are supposed, thus explaining the paraffins and aromatics formation and the great values for r (see *Preliminary Investigations* Section).

By retaining only the relationship (2) as the main alkylation reaction, and by adding a reduced MTO kinetic schema, the denoted model[1] is thus obtained:

$$A + H^+ \underset{k'_1}{\overset{k_1}{\rightleftharpoons}} CH_3^+ + H_2O \quad (3a)$$

$$Bu + CH_3^+ \xrightarrow{k_2} MBu + H^+ \quad (3b)$$

$$A \xrightarrow{k_3} B + H_2O \quad (3c)$$

$$A + B \xrightarrow{k_4} O + H_2O \quad (3d)$$

$$O + B \xrightarrow{k_5} P, \quad (3e)$$

where A notes the oxygenates (without water), B a carbene intermediate, Bu= butene, MBu= methyl-butene, O= C2 and C3 olefins (the C6 and C7 olefins are present in negligible amount in products), P= paraffins and aromatics, W= water. Although for the MTO process many extended or reduced kinetic schema are proposed in the last years (see for details Mihail et al. [6],[7]; Iordache et al. [8]), the adopted equations (3c-e) are derived by Maria [9] from the reduced models of Chang [10] and Anthony [11]. As proved, this MTO reduced model is very robust, also deriving from the extended model of Mihail et al. [6] (33 reactions) via Maria and Muntean [12] model (14 reactions) by using a special estimation - reduction strategy (Maria and Musca [13]; Maria [9]; Iordache et al. [8]).

Furthermore, the following model assumptions are made: i) only first order reactions, and ii) constant B, H^+ and CH_3^+ intermediates concentrations, that is:

$$dC_A/dt = - k_{1m}C_A + k'_1 C_{CH_3^+}C_W - k_3C_A - k_4C_AC_B$$

$$dC_{Bu}/dt = - k_2C_{Bu}C_{CH_3^+}$$

$$dC_{MBu}/dt = k_2 C_{Bu} C_{CH_3^+}$$
$$dC_P/dt = k_5 C_O C_B \tag{4}$$
$$dC_O/dt = k_4 C_A C_B - k_5 C_O C_B$$
$$dC_W/dt = k_{1m} C_A - k'_1 C_W C_{CH_3^+} + k_3 C_A + k_4 C_A C_B$$
$$C_{CH_3^+} = k_{1m} C_A/(k'_1 C_W + k_2 C_{Bu})$$
$$C_B = k_3 C_A/(k_4 C_A + k_5 C_O),$$

where $k_{1m} = k_1 C_{H^+}$ and C notes the species concentrations. Thus, the nonlinear kinetic model 1 contains 6 species (A, Bu, MBu, O, P, W), and only 4 kinetic parameters of Arrhenius type: k_{1m}, $k_{21} = k_2/k'_1$, k_3, $k_{45} = k_4/k_5$, i.e.

$$k_i = A_i^* \exp(-\frac{E_i}{R}(\frac{1}{T} - \frac{1}{T^*})),$$
$$A_i^* = A_i \exp(-E_i/RT^*) \tag{5}$$

(T notes the temperature).The nonlinear transformation (5) of Arrhenius law, proposed by Pritchard and Bacon [14], is useful for the parameter estimation, increasing the numerical procedure convergence rate, and improving the estimate quality (see also Espie and Macchietto [15]).

The kinetic parameters are estimated by minimizing the Sum of Squares of Residuals (SSR), computed for all considered species and runs (the noise is considered the same for all C-observations):

$$SSR = \sum_i^6 \sum_u^8 (C_{iu} - \hat{C}_{iu})^2. \tag{6}$$

The predicted concentrations $\hat{C}$ (moles/mol CH_3OH fed) are computed by successive integration of differential mass balance set of equations corresponding to an integral plug-flow reactor (Froment and Bischoff [16]), by using the experimental temperature profile for each run.The total contact time t, in the reactor model is defined as g. catalyst/(fed mol/h.).

The parameters A_i^* and E_i from (5) are estimated by using a multimodal adaptive random search (i.e. the MMA of Mihail and Maria [17]), and are presented in Table 1. The relative sensitivity coefficients of SSR vs. the activation energies E_i, are small (below 1/100).The sensitivities of SSR vs. the $\ln(A^*)$ (specified in Table 1) are significant, especially for k_{21} and k_3, i.e. the rates of CH_3^+ attack to olefins and of carbene generation which seem to be the less sensitive to temperature in the 300-380°C range.As another observation, and a possible argument for mixed-type process, the value of E_3 = 2.65 Kcal/mol is approximately the same with the correspon-ding carbene generation E-value in the Methanol-to-Hydro-carbons (MTH) extended kinetic schema of Mihail et al. [18], i.e. E_3 = 2.2 Kcal/mol.

Although the measured axial temperature profiles in-volve experimental noise, the obtained model 1 adequacy (Figure 1) is too rough.This is why some supplementary model improvements could be made.

MODEL BUILDING

The model 1 concentration predictions (Figure 1) are un-satisfactory, especially for MBu, O and P species: too high values for C_{MBu} and too low for C_O and C_P vs. the experimental data. The only route to generate more P species is the O species consumption; this way implies however a poorly adequacy because of small C_O values and an implicit increasing of C_W.A simple, but natural modi-

TABLE 1. *The kinetic parameters $\hat{\theta}$.(T^* = 615 K; the relative sensitivity coefficients are computed as $|\partial(SSR/SSR(\hat{\theta}))/\partial(\theta/\hat{\theta})|$).*

Model no.	PARAMETERS ($\hat{\theta}$)	$\ln(A^*)$	E(Kcal/mol)	SSR sensitivities vs. $\ln(A^*)$, (%/%)
1	k_{1m} (mol/g.h)	-1.22	16.978	5/100
	k_{21}	-3.54	4.675	38/100
	k_3 (mol/g.h)	-3.69	2.659	96/100
	k_{45}	-2.54	27.546	1/100
	SSR($\hat{\theta}$)			0.484
2	k_{1m} (mol/g.h)	-1.22	16.978	9/100
	k_{21}	-3.54	4.675	47/100
	k_3 (mol/g.h)	-3.99	2.659	112/100
	k_{45}	-1.23	27.546	2/100
	k_6	-2.96	27.546	7/100
	SSR($\hat{\theta}$)			0.271

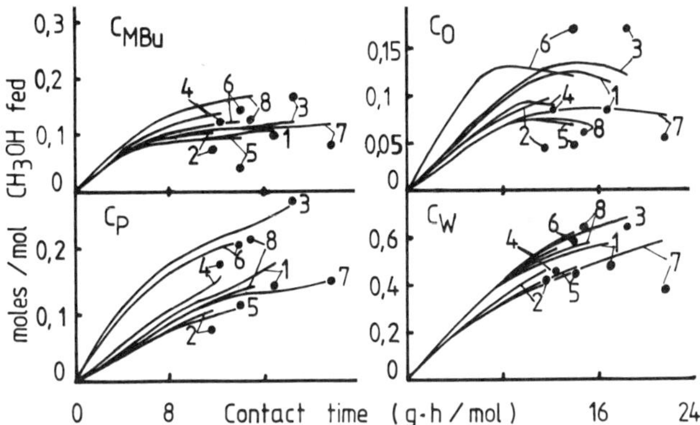

FIG. 3 *Experimental data (●) and predicted species MBu, O, P, W concentration (moles/mol CH_3OH fed) trajectories (model 2) vs. the contact time for runs no. 1-8.*

fication consists in considering MBu as an intermediate species as well as the other olefinic species. Thus, the kinetic schema (3a-e) must be completed with the reaction:

$$MBu + B \xrightarrow{k_6} P, \tag{7}$$

and the model equations (4) should be adequately modified. The model 2 as built contains 5 kinetic parameters. Because of the small sensitivities of SSR vs. the E_i, the activation energies in model 2 are kept at the model 1 values and the equality $E_6 = E_{45}$ is adopted. By applying the same estimation procedure the other parameters are identified and presented in Table 1. As shown in Figure 3, the obtained model 2 adequacy is satisfactory. Supplementary model adequacy tests and parameter precision improvements could be made if supplementary designed runs are performed.

CONCLUSIONS

The mixed kinetic schema for the butene alkylation with methanol appears to be attractive, due to the satisfactory adequacy of the obtained model. However, the simultaneous generation of carbenium ions and of carbenoid species by the acid catalyst could raise some additional questions. A possible approach is to maintain the model 2 equations by considering that the intermediate species

B is a carbenium ion too; the actual data are however insufficient to make one option. It is to notice that the nature of B is of lower importance in the proposed kinetic model because the intermediate considered species are eliminated by means of the quasistationary state hypothesis. Although we do not claim that our model necessarily describes the true reaction mechanism, however it leads to a satisfactory agreement with the experiments.

REFERENCES

1. S. H. Xiang and H. Li, Preprint Paper-Poster, *7th Int. Zeolites Conf.*, Tokyo, p. 242 (1986).
2. E. J. Rosinsky and M. K. Ruber, *U. S. Patent 3,832, 449* (1974).
3. I. F. Boag, D. W. Bacon, and J. Downie, *Canad. J. Chem. Eng. 56*: 389 (1978).
4. D. D. McLean, D. W. Bacon, and J. Downie, *Canad. J. Chem. Eng. 58*: 608 (1980).
5. T. Behrsing, T. Mole, P. Smart, and R. J. Western, *J. Catal. 102*:151 (1986).
6. R. Mihail, S. Straja, G. Maria, G. Musca and G. Pop, *Ind. Eng. Chem. Process Des. Dev. 22*: 532 (1983a).
7. R. Mihail, S. Straja, G. Maria, G. Musca, and G. Pop, *Ind. Eng. Chem. Res. 26*: 637 (1987).
8. O. Iordache, G. Maria, and G. Pop, *Ind. Eng. Chem. Res. 27*: 2224 (1988).
9. G. Maria, *Canad. J. Chem. Eng. 67*: 825 (1989).
10. C. D. Chang, *Chem. Eng. Sci. 35*: 619 (1980).
11. R. G. Anthony, *Chem. Eng. Sci. 36*: 789 (1981).
12. G. Maria and O. Muntean, *Chem. Eng. Sci. 42*:1451 (1987).
13. G. Maria and G, Musca, *Comp. Chem. Eng. 13*:425 (1989).
14. D. J. Pritchard and D. W. Bacon, *Chem. Eng. Sci. 30*: 567 (1975).
15. D. M. Espie and S. Macchietto, *Ind. Eng. Chem. Res. 27*: 2175 (1988).
16. G. F. Froment and K. B. Bischoff, *Chemical Reactor Analysis and Design*, Wiley, New York (1979).
17. R. Mihail and G. Maria, *Comp. Chem. Eng. 10*: 539 (1986).
18. R. Mihail, S. Straja, G. Maria, G. Musca, and G. Pop, *Chem. Eng. Sci. 38*: 1581 (1983b).

26

PRODUCTION OF OLEFINS BY PYROLYSIS OF WASTES

W. Kaminsky, H. Rößler, and U. Bellmann
Institute for Technical and Macromolecular Chemistry
University of Hamburg, *Bundesstr. 45, D 2000 Hamburg 13*
Germany

ABSTRACT

By pyrolysis of hydrocarbon containing wastes in a fluidized bed of sand, it is possible to receive petrochemicals. Feeding materials are suitable plastic wastes, scrap tires, waste oil, or sewage sludge. The investigations are carried out in a continuously working laboratory scale and a small pilot plant with a capacity of 30-70 kg/h. The fluidized bed is heated indirectly by fire tubes up to 600-800 °C. Main products are olefins as ethylene, propene, and butene, if nitrogen as inert gas is used as fluidizing medium. If the pyrolysis gas is recycled and used as fluidizing gas, 30 to 50 percent of aromatics as benzene, toluene, and naphthalene are formed. The influence of the reaction conditions to the product composition is studied.

INTRODUCTION

There are two points for the importance of plastic and rubber recycling. 1. Most plastics are produced from oil and there is an worldwide increase in the price of crude oil. 2. Our booming waste economy has created an increasing demand for recycling processes as they are commonly practiced for scrap metals, glass, and paper in many areas. The recycling of wastes containing hydrocarbons, e.g. mixtures of plastic wastes, used rubber, waste oil, or sewage sludge is far more difficult.

High molecular substances cannot be purified by physical processes like distillation, extraction, or crystallization. They only can be recycled via thermal degradation (pyrolysis of the macromolecules into smaller fragments) (1,2). The pyrolyzate is then suited for the common petrochemical separation processes. As plastic waste consists mainly of polyolefins, a lot of olefins could be recovered by pyrolysis.

The pyrolysis is complicated by the fact that plastics, rubbers, or biopolymers show poor thermal conductivity while the degradation of macromolecules requires considerable amounts of energy.

The pyrolysis of plastic wastes, used tires, and biopolymers has been studied in melting vessels, blast furnaces, autoclaves, tube reactors, rotary kilns, cooking chambers and fluidized bed reactors (3-9). Rotary kiln processes are particularly numerous. They are marked by relatively long residence times of the wastes in the reactor of 20 minutes and more, whereas dwell times in fluidized bed reactors hardly exceed a few seconds with a 1.5 minute maximum. Due to the large temperature gradient inside the rotary kiln the product spectrum is very diverse.

The fluidized bed has a number of special advantages for the pyrolysis because it is characterized by an excellent heat and mass transfer as well as constant temperature throughout the reactor which results in

- fairly uniform product spectra
- absence of moving parts in the hot zone of the reactor
- completely closed system, i.e. the reactor can easily be sealed.

The fluidized bed is generated by a flow of an inert gas that is directed from below through a layer of fine grained material, e.g. sand or carbon black, at a flow rate that is sufficient to create a swirl in the bed. At this stage the fluidized bed behaves like a liquid.

RESULTS AND DISCUSSION

A number of projects that have been conducted at the University of Hamburg in recent years were aimed at determining the suitability of plastic wastes, used tires, and waste oil residues as a source of olefins and other hydrocarbons after pyrolysis in an indirect heated fluidized bed reactor. Laboratory plants with capacities for 60 - 3000 g/h and two pilot plants for 10 - 40 kg/h of plastic and 120 kg/h of used tires were installed and run at the Institute of Technical and Macromolecular Chemistry with financial support from the BMFT, the Verband Kunststofferzeugende Industrie (VKE), the C.R. Eckelmann company (CRE), and Asea Brown Boveri (ABB) (10-14).

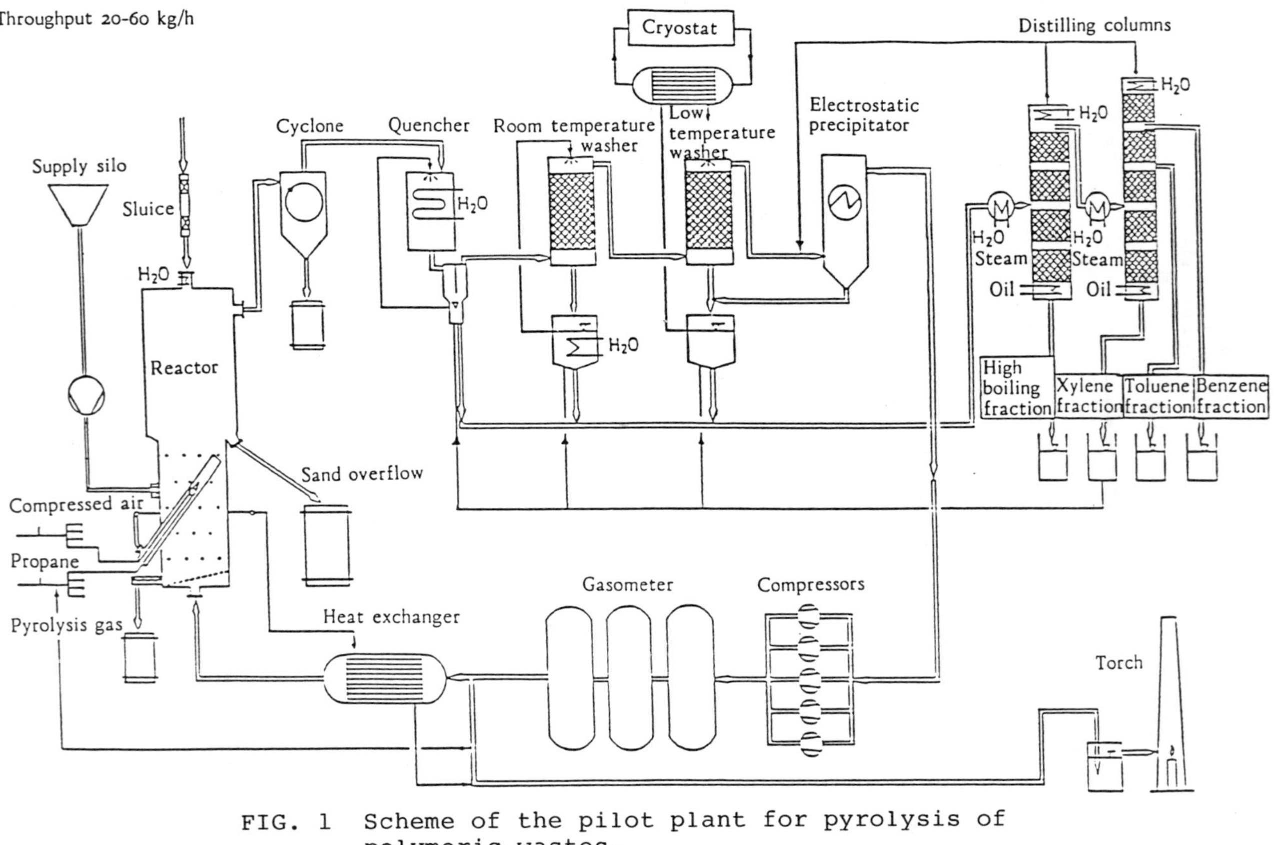

FIG. 1 Scheme of the pilot plant for pyrolysis of polymeric wastes.

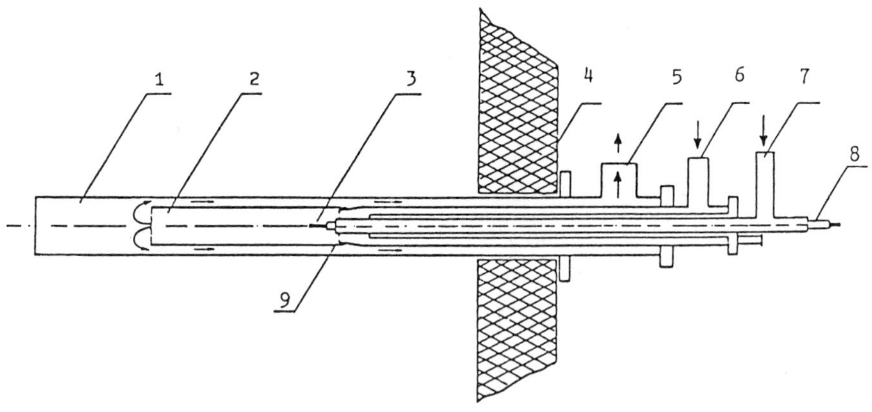

FIG. 2 Scheme of the fire tubes used in the pyrolysis plant: 1 = outer tube closed in front, 2 = inner tube, 3 = ignition electrode, 4 = isolation of the reactor wall, 5 = exhaust gas, 6 = air, 7 = gas, 8 = ceramic isolation tube, 9 = air tip.

Figure 1 shows the scheme of the pilot plant for the pyrolysis of plastics. The core of the plant is a fluidized bed reactor with inner diameter of 450 mm. An auxiliary fluidized bed of quartz sand with a temperature of 600 to 900 °C is used for the pyrolysis of plastics that are fed into the reactor through a double flap gate or a screw. Pyrolysis gas preheated to 400 °C is used to create the swirl in the fluidized bed. The heat input takes place indirectly through heat radiation fire pipes which are heated by pyrolysis gas (Fig. 2) (15).

The exhaust gases are then directed through a heat exchanger. The product gases emerging from the fluidized bed are separated from residual carbon and fine dust in a cyclone and then cooled to room temperature by product oil in condenser. The gas flow is then directed through two packed condensation columns.

The condensed oil fractions are distilled in two distillation columns using the fraction boiling from 135-145 °C (mostly xylenes) as a quenching medium. Also produced are tar with a high boiling range as well as two fractions, one rich in toluene and the other in benzene.

The gas, largely stripped of liquid products, passes to an electrostatic precipitator where it is freed from small droplets. Subsequently, it is compressed in five membrane compressors, connected in parallel, and stored in three gas tanks, each 0.5m^3 in volume. Part of the gas serves as fuel for the radiation heating tubes, while the remainder, preheated by hot fuel gases in the heat exchanger, is used for flu-

TABLE 1 Composition of the gases of the pyrolysis of polyethylene at different temperatures in wt.% and vol.%

Temperature (°C)	650	740	820	820
Compound	wt.%	wt.%	wt.%	vol.%
Hydrogen	0,61	1,26	2,57	21,71
Methane	18,52	39,10	50,27	53,09
Ethene	21,09	32,85	32,45	19,58
Ethane	12,12	11,13	5,42	3,05
Propene	24,27	9,20	2,10	0,85
Propane	2,52	0,39	0,11	0,04
Butene	10,21	0,89	0,08	0,03
Butadiene	1,30	0,42	1,40	0,44
Pentene	2,32	0,01	0,03	0,01
Cyclopentadien	1,80	0,49	0,34	0,09
Other compounds	5,24	4,26	5,23	1,11
Density (g/cm^3)	1,161	0,922	0,704	0,704

idizing the sand bed. The excess gas is burned in a torch.

Any kind of plastic waste can be pyrolyzed even when it is soiled. The products of the pryrolysis of polyolefins were obtained in the following fractions (weight-%):

Gas	20 - 90
Oil	5 - 45

TABLE 2 Products of the pyrolysis of polyethylene (PE) with nitrogene or pyrolysis gas as fluidizing medium at 740 °C

Feed:	PE	PE	PP	PP
Fluidizing medium:	N_2	Pyrolysis gas	N_2	Pyrolysis gas
Part of gas from products (wt-%)	87,7	61,1	83,0	57,5
Hydrogen	0,3	0,5	1,3	0,5
Methane	6,7	16,2	7,8	17,0
Ethene	33,8	25,4	13,3	13,9
Ethane	3,5	5,3	3,9	4,4
Propene	21,8	9,3	35,5	13,7
1-Butene	8,4	1,1	15,2	5,3
2-Butene	3,2	0,5	1,2	0,5
1,3-Butadiene	10,0	2,8	5,3	2,0
Other aliphatic compounds	11,6	8,7	10,6	9,4
Benzene	0,01	12,2	1,5	10,4
Toluene	0,05	3,6	1,7	5,7
Xylene/Styrene	0,05	2,2	0,1	2,7
Other aromatic compounds	0,2	11,3	2,4	13,6
Carbon black	0,4	0.9	0,2	0.9

Carbon black and filling materials 0,2- 10

The main compounds in the gas are methane, ethane, ethene, propene, and butene (Tab. 1).

It is impossible to win back only olefins from polyolefins by pyrolysis in the fluidized bed. So the yields of ethylene and propylene together produced from polyolefins, does not exceed a maximum of 60 %.

The concentration of olefins is influenced by the process parameters in a wide range. If nitrogen is used, a fluidizing medium that has higher amounts of ethylene is formed than in the case when the pyrolysis gas is recycled and used as this medium (Tab. 2). In this case about 10 wt.% of butadiene are formed. The carbon black content is low with 0,2 to 0,9 wt.-%. Even from polypropylene a lot of propene (35 wt.%) could be received.

The process was optimized to achieve the maximum yield of ethylene and other components from polyethene by varying the temperature of the fluid bed, the feed and the flow of the fluidized gas. 16 pyrolysis runs were calculated.

Two plots are shown in Fig. 3 and Fig. 4. The ethylene content shows a minimum at a medium feeding rate and flow rate of the fluidized gas. The propene content increases with decreasing temperature and decreasing feeding rate. Table 3 summarizes the dependence for some points.

In every case there is the formation of methane. In an industrial process it could be useful to separate the gas in the same way which is used in a naphtha cracker plant. The separated methane and ethane is sufficient to be burned in the fire tubes to provide the heat for the pyrolysis process.

Another product fraction which is formed is a pyrolysis oil. This liquid corresponds to a mixture of

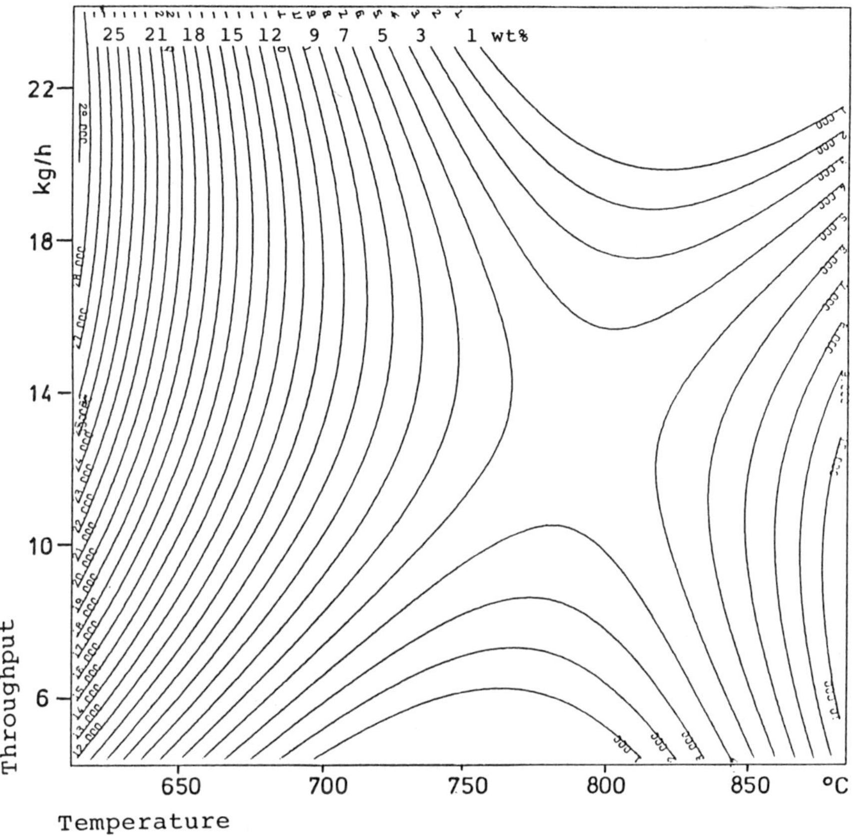

FIG. 3 Optimization of the ethene content. Pyrolysis of polyethylene by 23,8 Nm^3/h flow of fluidizing gas; throughput versus temperature.

light benzene and bituminous coal tar with about 95 % aromatics. The oil may be processed into chemical products according to the usual petrochemical methods. Optimal reaction management aims at high yields of aromatics. Comparable raw material value analysis showed that the chemical processing of this type of high aromatic oil is better converted to valuable raw materials than used for heating or propulsion purposes.

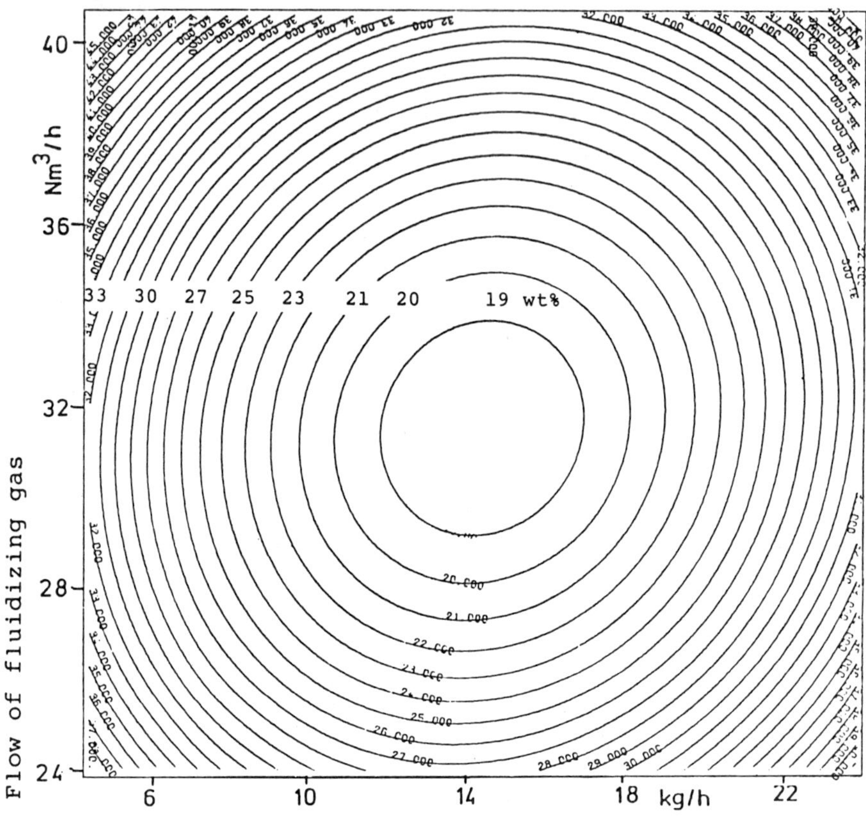

Throughput

FIG. 4 Optimization of the ethene content. Pyrolysis of polyethylene at 750°C. Flow of fluidizing gas versus throughput.

Even mixed plastics gave a high amount of olefins and aromatics (Tab. 4). Carbon dioxide are formed from polyurethanes and cellulose containing materials. The collected plastic fraction from household separation contains to 70 %. Plastic wastes obtained from separate collection contain about 15 % PVC.

By addition of lime to the fluidized bed the hydrogen chloride from the decomposition of PVC can be bound chemically. The calcium chloride that is formed in the process, however, has a tendency to clog the fluidized bed at higher concentrations. Therefore it is attempted to eliminate HCl gas in a preliminary step at 300-400 °C before more extensive cracking reactions take place. The dry HCl gas would then be available for further use.

The residual content of organic chlorine is of great importance for the value of the pyrolysis oils. For processing in existing petrochemical plants, the chlorine content should not exceed 10 ppm. Oils that are obtained from the pyrolysis of plastic mixtures contain between 50 and 200 ppm of organically bound chlorine. Fortunately, no chlorinated dibenzodioxines were found among these chloroorganic compounds. This was confirmed by numerous analyses in other institutes in connection with a BMFT project.

A further dehalogenation of the pyrolysis oil was achieved by dosing sodium vapor into the flow of pyrolysis gas by analogy with the Degussa process in which chlorine containing oils are dehalogenated within a few minutes in the liquid phase using sodium. In the gas phase at 500 °C the same process takes only seconds (16).

TABLE 3 Pyrolysis of Polyethylene. Dependence of methane, ethene, propene, and benzene concentrations of process parameters, feed and temperature

Temperature (°C)	720	780	720	780
Feed (kg/h)	9,2	9,2	19,2	19,2
Methane (wt.%)	23,2	22,0	19,9	25,7
Ethene (wt.%)	26,2	28,0	26,4	17,0
Propene (wt.%)	5,3	1,0	11,3	1,2
Benzene (wt.%)	16,9	23,0	14,3	16,5

TAB. 4 Components (wt.%) by the pyrolysis of different plastic wastes in a fluidized bed process

Feed	Poly-ethylene	Plastics from household separation	Poly-urethanes
Temperature (°C)	740	787	760
Hydrogen	0,75	0,70	0,23
Methane	23,6	17,5	3,5
Ethane	6,7	2,9	0,76
Ethene	19,8	9,8	5,4
Propane	0,08	0,08	0,01
Propene	5,5	1,3	1,7
Butene	0,54	0,13	
Butadiene	1,6	0,44	1,2
CO_2		2,3	14,4
CO		1,2	8,7
HCN			0,31
HCl ($CaCl_2$)		7,2	
Hexene	0,02	0,002	0,2
Benzene	19,1	13,8	5,0
Toluene	3,9	5,6	1,8
Styrene	0,5	3,1	0,01
Indene	0,25	1,0	1,1
Naphthalene	2,8	3,8	1,5
Methylnaphthalene	0,63	0,84	0,69
Diphenyl	0,30	0,41	0,15
Fluorene	0,20	0,22	0,27
Phenanthrene/ Anthracene	0,52	0,66	0,62
Phenylnaphthalene	0,08	0,36	0,01
Fluoranthene	0,08	0,26	0,01
Acetonitrile			0,58
Acrylnitrile			0,95
Benzonitrile			1,8
Ketone, Acetale			9,2
Acetophenone			0,3
Other compounds	11,3	9,0	34,1
Carbon black	1,75	6,4	0,5
Water		2,6	5,0
Inorganic compounds		5,3	

A semi industrial plant that is working according to the Hamburg process of fluidized bed pyrolysis was built in Ebenhausen near Ingolstadt and operated by the Asea Brown Boveri company (ABB) with a throughput of 5000 t/a.

The process has been shown to be successful for the pyrolysis of polyolefins. To process plastics with high PVC content it is necessary to scale down by one step (250 kg/h) and test the plant in continuous operation.

PYROLYSIS OF RUBBER

After experiments on the pyrolysis of plastic wastes had shown how individual pieces that were only slightly smaller than the reactor's inner diameter could be pyrolyzed in the fluidized bed a larger pilot plant with a 90 cm square fluid bed was used to decompose whole used tires. The tires were introduced to the reactor through a lock with pneumatic gates. The steel parts remaining in the reactor, after the pyrolysis was complete, were forked out of the fluidized bed by means of a grate at programmable intervals and deposited in a silo. The solid powdery products were carried out of the reactor by the gas stream and separated in a cyclone.

By pyrolysis of more than 4500 kg of whole used tires in several series of experiments, it could be shown that whole tires with a maximum weight of 20 kg were completely pyrolyzed within 1,5 to 4 minutes at temperatures above 650 °C. The amount of pyrolysis gas produced in the process was sufficient to heat the reactor without additional supply of energy.

TABLE 5 Products of fluidized bed pyrolysis of rubber and scrap tires (wt.%)

Feed Material	EPDM Rubber	Tire Pieces	Whole Tires
Temperature (°C)	750	750	700
Hydrogen	1,92	1,30	0,42
Methane	22,75	15,13	6,06
Ethane	1,08	2,95	2,34
Ethene	8,67	3,99	1,65
Propane	0,02	0,29	0,43
Propene	0,36	2,50	1,53
Butene	0,02	1,31	1,41
Butadiene	0,02	0,92	0,25
Isoprene	0,02	0,34	0,35
Cyclopentadiene	0,14	0,39	0,25
Other aliphatic compounds	0,71	0,36	1,07
Benzene	7,38	4,75	2,42
Toluene	2,33	3,62	2,65
Styrene	0,95	0,17	0,35
Indane/Indene	0,89	0,31	0,48
Naphthalene	1,51	0,85	0,42
Methylnaphthalene	0,54	0,83	0,67
Diphenyl	0,15	0,49	0,39
Phenanthrene	0,21	0,29	0,19
Other aromatic compounds	4,52	10,20	16,71
Carbon monoxide	1,25	3,80	1,48
Carbon dioxide	0,75	1,95	1,74
Water	1,27	0,10	5,11
Hydrogen sulphide	0,1	0,23	0,02
Thiophene	0,05	0,15	0,25
Carbon soot, fillers	41,85	40,59	40,0
Steel cord		1,62	11,30

The products of the pyrolysis were obtained in the following five fractions (weight %):

Gas	15-20
Oil	20-30
Water	5-10
Carbon black and filling materials	30-40
Steel cord	5-20

Main components next to the solids carbon black and steel cord are methane, ethylene, ethane, benzene, and toluene (Table 5).

EPDM-rubber gave the highest ethylene value while scrap tires more propene.

It is also remarkable about the product spectrum that the hydrogen sulfide content was found to be below 0,3 wt-% in all experiments although some 2,0 % of sulfur are incorporated in the rubber. After the pyrolysis the main portion of sulfur is found in the carbon black fraction in the form of zinc or calcium sulfide. During the pyrolysis a reaction of sulfur containing fission products with alkaline filling materials takes place. The carbon black that is obtained is suitable for recycling as filling materials.

Also dried sewage sludge can be polymerized with the Hamburg process. At 600 °C the pyrolysis yields (weight-% referring to dry organic substance)

Tar and light oil	44
Water	16
Gas	18
Coke	22

The gas contains 6,7 wt.-% of ethene and 2,1 wt.-% of propene at a pyrolysis temperature of 750 °C.

CONCLUSION

The necessity of economizing on the amount of raw materials taken from natural deposits to preserve sufficient amounts of resources for future generations and the need to minimize the production of wastes calls for an extensive recycling of raw materials.

Through pyrolysis of plastic wastes, used tires, waste oil residues and other organic waste materials, it is possible to obtain olefins and other hydrocarbons which can be used as petrochemicals. Separate collection and sorting of wastes as they are carried out in many industrialized nations at the moment produce increasing amounts of materials that are suitable for recycling through pyrolysis. Since olefins and oil are generated as valuable substances, the profitability of the process depends largely on the price of mineral oil. The combination of a refinery together with a pyrolysis plant would optimize the use for the products. It is to be expected that some of the processes can be established in the market in the near future.

LITERATURE

1. W.L. Hawkins, *Polymer degradation and stabilization*, Springer Press, Berlin (1984)
2. L.F. Albright, B.L. Crynes, W.H. Corcoran, *Pyrolysis theory and industrial practice*, Academic Press, New York (1983)
3. H. Sueyoshi, Y. Kitaoka, *Hydrocarbon Process* 161 (1972)
4. B. Schulmann, P.A. White, *Pyrolysis of Scrap Tires Using the Tosco II-Process, ACS Sympos. Ser.* 76 (1978)

5. G.L. Ferrero, K. Maniatis, A. Buekens, A.V. Bridgewater, *Pyrolysis and Gasification*, Elsevier A. Science, London (1989)
6. G.P. Bracker, in Ullmanns Enzyklopädie der Technischen Chemie, *Pyrolysis*, 4th ed., Vol. 6, Verlag Chemie, Weinheim, p. 553
7. K.J. Thomé-Kozmiensky, *Pyrolyse von Abfällen*, EF-press, Berlin 1985
8. H. Sinn, W. Kaminsky, J. Janning, *Angew. Chem.* 88: 737 (1976); *Angew. Chem. Int. Ed. Engl.* 15: 660 (1976)
9. J. Piskorz, D. Radlein, D.S. Scott, *J. Anal. Appl Pyrol.* 9: 121 (1986)
10. W. Kaminsky, H. Sinn, J. Janning, *Chem. Ing. Tech.* 51: 419 (1979)
11. W. Kaminsky, Ressourc. *Recovery Conserv.* 5: 205 (1980)
12. W. Kaminsky, H. Sinn, *Hydrocarbon Process* 59: 187 (1980)
13. W. Kaminsky, *J. Anal. Appl. Pyroly.* 8: 439 (1985)
14. W. Kaminsky, A.B. Kummer, *J. Anal. Appl. Pyroly.* 16: 27 (1989)
15. Th. Schmidt, *Gas Wärme Int.* 25: 487 (1976)
16. W. Kaminsky, H. Sinn, *Nachr. Chem. Tech. Lab.* 38: 333 (1990)

27

THE FLASH PYROLYSIS AND METHANOLYSIS OF BIOMASS (WOOD) FOR PRODUCTION OF ETHYLENE, BENZENE, AND METHANOL

M. Steinberg, P. T. Fallon, and M. S. Sundaram

Process Sciences Division
Department of Applied Science
Brookhaven National Laboratory
Upton, New York 11973

ABSTRACT

The process chemistry of the flash pyrolysis of biomass (wood) with the reactive gases, H_2 and CH_4 and with the non-reactive gases He and N_2 is being determined in a 1 in. downflow tubular reactor at pressures from 20 to 1000 psi and temperatures from 600 to 1000°C. With hydrogen, flash hydropyrolysis leads to high yields of methane and CO which can be used for SNG and methanol fuel production. With methane, flash methanolysis leads to high yields of ethylene, benzene and CO which can be used for the production of valuable chemical feedstocks and methanol transportation fuel. At reactor conditions of 50 psi and 1000°C and approximately 1 sec residence time, the yields based on pine wood, carbon conversion are up to 25% for ethylene, 25% for benzene, and 45% for CO, indicating that over 90% of the carbon in pine is converted to valuable products. Pine wood produces higher yields of hydrocarbon products than Douglas fir wood; the yield of ethylene is 2.3 times higher with methane than with helium or nitrogen, and for pine, the ratio is 7.5 times higher. The mechanism appears to be a free radical reaction between CH_4 and the pyrolyzed wood. There appears to be no net production or consumption of methane. A preliminary process design and analysis indicates a potentially economical competitive system for the production of ethylene, benzene and methanol based on the methanolysis of wood.

INTRODUCTION

Flash pyrolysis is a technique for pyrolyzing carbonaceous materials at high heat-up rates, in the order of 10^3 to 10^6 $^oC/sec$, maintaining the reactants at temperature for short periods of time, in the order of seconds, followed by rapid cool-down of the products in the order of 10^2 to 10^3 $^oC/sec$. Conventional pyrolysis involves heat-up rates in the order of 1° to $10^oC/sec$ and reacting at elevated temperature for long residence times. The total yield of volatile matter in flash pyrolysis of carbon containing material such as coal and wood is significantly higher than in conventional pyrolysis (Belt and Bissett, 1978). This comes about because secondary reactions involving primary volatile material are prevented from further cracking or repolymerization on hot surfaces which lead to tar and char formation. The primary volatile species tend to be preserved thus forming valuable lighter hydrocarbon fuels and chemical feedstocks. The objective of this work is to develop a process chemistry data base for the flash pyrolysis of biomass, particularly wood.

The investigation involves the flash pyrolysis with non-reactive or inert gases, which includes helium (He), argon (Ar), and nitrogen (N_2). The purpose of studying these gases is to determine the differences in the yield and distribution of products as a function of the rate of heat and mass transfer in the absence of chemical reactions. The differences in the heat capacity and diffusivity of the inert gases He, N_2 and Ar causes differences in mass and heat transfer rates of the particulate carbon containing material, which could lead to varying product yields.

Flash pyrolysis with gases that could react with the biomass particles will yield a different slate and distribution of product yields. The potential reactive gases under study include hydrogen (H_2), carbon monoxide (CO), carbon dioxide (CO_2), water (H_2O) and methane (CH_4). The choice of these gases is made on the basis that they are low in cost, readily available and abundant in supply. Furthermore, these gases are usually found in the products from the pyrolysis of wood so that any economically viable biomass conversion process would require recycling of these gases. It thus becomes important to understand the affect on the yield and distribution of products when reacting with these process gases.

When the reactive gas H_2 is used, the reaction is termed hydropyrolysis. When the reaction takes place with water, the reaction term is referred to as hydrolysis. We have also found that methane gas can be reactive towards biomass so that we have termed this reaction methanolysis.

EXPERIMENTAL EQUIPMENT AND PROCEDURE

The experimental equipment is a highly versatile and instrumented 1 in. I.D., 8 ft. long entrained downflow tubular reactor system, which has previously been described in detail (Fallon and Steinberg 1977). A schematic flowsheet of the system is shown in Figure 1. Wood particles in the size range between 300 and 1000 microns are fed by gravity into the top of the reactor from a wood feeder enclosed in a high pressure vessel. The process gas enters through a preheater, contacting the wood particles at the top of the tubular reactor. The gas to solid heat transfer, pyrolysis and reaction takes place as the wood and entraining gas meet and flow concurrently down the reactor tube. Below the 8 ft. reactor section is a 3 ft. forced air cooled section in which the product gases and solids are quenched to approximately 250° to 300°C. The residual char is separated from the effluent gas stream in a vessel at the bottom of the reactor, which is also maintained at 250°-300°C to prevent liquid product condensation. The products, in the effluent

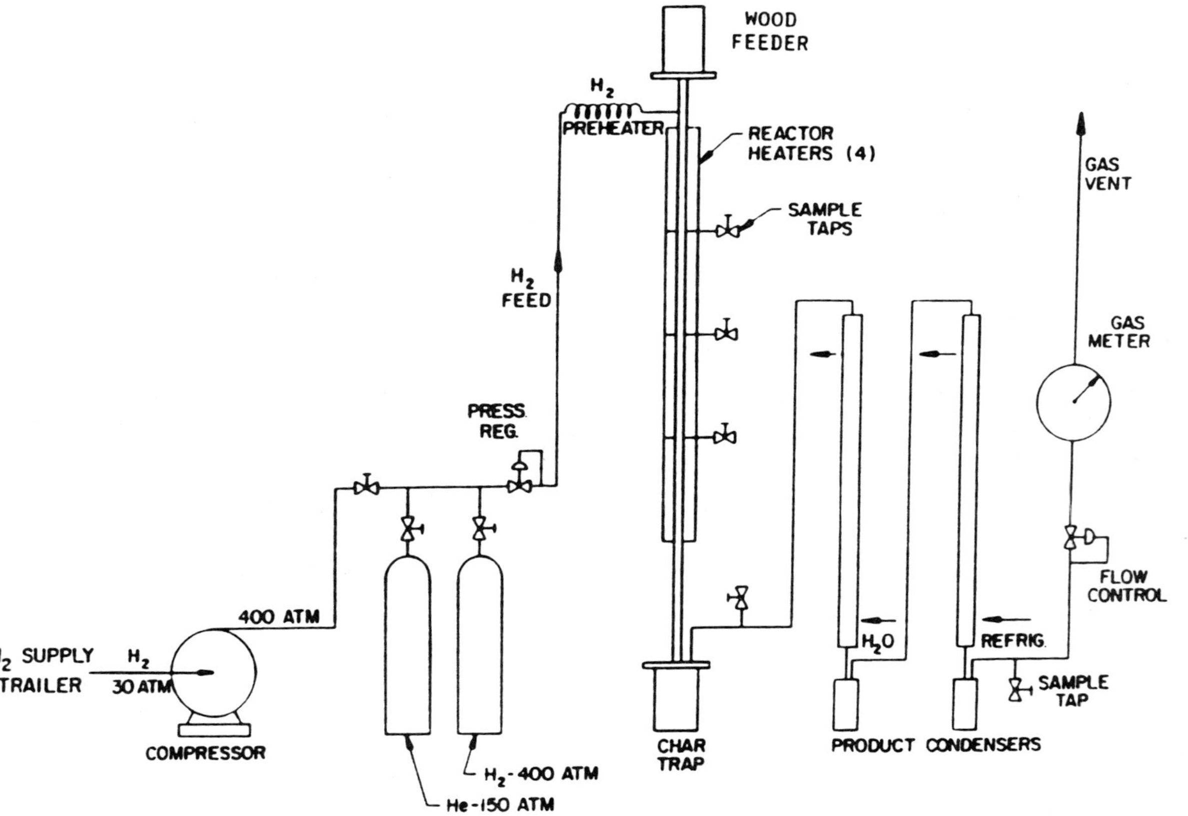

Figure 1. Schematic flowsheet of entrained tubular experiment.

gas, are then removed in two condensers, one water cooled (~15°C) and the other freon cooled (~-40°C). The remaining gases are then reduced to atmospheric pressure, measured with a positive displacement meter and vented to the atmosphere. Wood particle feed rate is usually in the order of 1 lb/hr (500 gm/hr) and gases up to 5 lbs/hr (2500 gas/hr).

On-line product analysis is accomplished via a programmable gas chromatograph (GC) which determines CO, CO_2, CH_4, C_2H_6, C_2H_4 and BTX (benzene, toluene, xylene) concentrations every 8 minutes. Products heavier than BTX (>C_9) are not measured on-line due to condensation in the sample lines and the temperature limit (230°C) of the gas chromatograph. These products are collected in the condensers and measured at the end of each experiment. Product yields, as a function of wood particle residence time, are determined by sampling from taps located at 2 ft. intervals along the length of the reactor. Wood particle residence times are calculated by the addition of the wood particle free fall velocity and the gas flow rate. Calculated values are in reasonable agreement with calibrated values (Fallon and Steinberg, 1977).

The wood feed usually consists of 300-1000 micron particles of oven dried Douglas fir having an elemental analysis of 50 wt% C, 44% O and 6% H or oven dried pine containing 50.7% C, 43% O and 6.2% H. The ash content is <0.1%. For purposes of preventing caking and producing a free flow of wood particles through the reactor, approximately 15% to 30% of a fine silica flour (Cab-O-Sil) is mixed in with the wood. The silica has been shown to have no affect on the product yields in blank and control experiments.

Yields are based on the percent of the carbon in the feed wood converted to product. The heavier hydrocarbons (naphthalene, anthracene, phenols, etc.) are collected, measured and the composition determined on a second GC at the end of each experiment. From these yields and a carbon analysis of the char produced, a complete material balance can be made, which under controlled conditions, can be determined within 10% of the carbon feed.

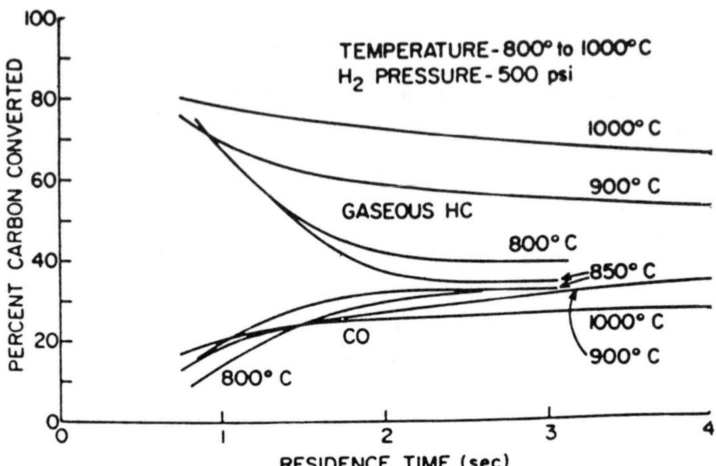

Figure 2. Rapid hydropyrolysis of fir wood, percent carbon converted to gaseous HC & CO vs residence time.

A summary of the major findings (Steinberg and Fallon, 1980, 1981; Sundaram, Steinberg, and Fallon, 1982; Steinberg, Fallon, and Sundaram, 1983) are as follows.

FLASH HYDROPYROLYSIS OF BIOMASS WITH HYDROGEN

In the flash hydropyrolysis of wood particles, yields of hydrocarbon gases (methane + ethane + ethylene) and CO are affected significantly by the residence time of the wood particles in the reactor. As shown in Figure 2 at all temperatures investigated (800°C to 1000°C) at a pressure of 500 psi, increasing the residence time, decreased the gaseous hydrocarbon yield and increased the CO yield, although not to an equal extent.

The available data indicate that the primary devolatilization and attendant hydrogenation are complete within 1 sec, and that increasing the residence time causes secondary reactions. The flash hydropyrolysis of wood appears to produce mainly methane and water at residence times of 1 sec or less. On the basis of calculated H and O balances, starting with the initial composition of wood, the following stoichiometric equation approximates the short residence time flash pyrolysis.

For <1-sec residence time:

$$CH_{1.4}O_{0.7} + 2H_2 \longrightarrow CH_4 + 0.7H_2O \qquad (1)$$

As the reactants and products (wood plus hydrogen, methane, and water) proceed further down the tubular reactor and the residence time becomes longer, the methane appears to react with the water in a reforming type of reaction, which increases the amount of CO and reduces the methane and water in the product gases:

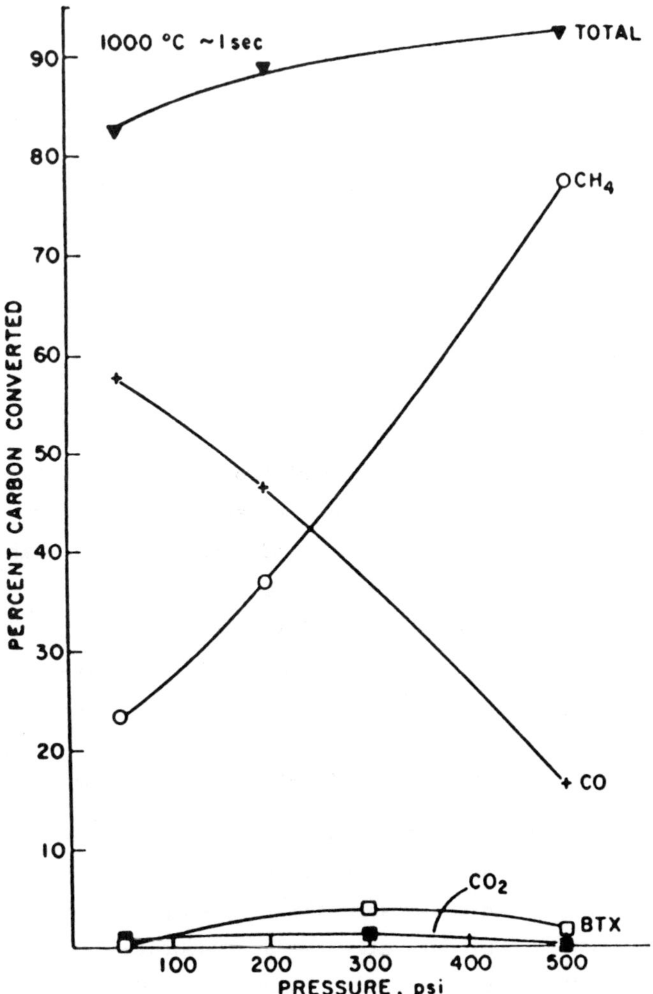

Figure 3. Flash pyrolysis of fir wood in hydrogen, effect of pressure on product yields.

$$CH_4 + H_2O \longrightarrow CO + 3H_2 \quad (2)$$

The stoichiometry of the overall reaction for longer residence times can be approximated as follows.
For >4-second residence time:

$$CH_{1.4}O_{0.7} + 0.5H_2 \longrightarrow 0.5CH_4 + 0.5CO + 0.2H_2O \quad (3)$$

It also appears from Figure 2 that at higher temperatures (1000°C), it takes a longer period of time for the secondary reforming reaction to take place. At lower temperatures, in the order of 800°C and 500 psi, significant quantities of benzene form with yields up to 13% C conversion.

Decreasing pressure at short residence time (~1 sec) causes an increase in CO yield and a decrease in methane. Figure 3 indicates that a yield reversal occurs at 250 psi at 1000°C. This indicates that higher pressures of hydrogen favors both the direct hydrogenation of the wood carbon and the methanation reaction.

$$CO + 3H_2 \longrightarrow CH_4 + H_2O \quad (4)$$

PROCESS DESIGN AND ANALYSIS-FLASH HYDROPYROLYSIS OF BIOMASS

The above data have resulted in the design of two process concepts. One process converts wood to high BTU pipeline gas (SNG), as shown in Figure 4, and the other to methanol fuel and chemical feedstocks, benzene and ethylene, as shown in Figure 5. It is noted that at 1000°C, 500 psi and 1 sec residence time, flash hydropyrolysis converts over 90% of the carbon content of Douglas fir to volatile HCs, of which 80% consists of methane.

FLASH PYROLYSIS WITH THE NON-REACTIVE GASES HELIUM AND NITROGEN

Figures 6 and 7 give the product yields for the flash pyrolysis of wood in the inert gas helium as a function of pressure and temperature, respectively. In this case, the largest yield consists of CO. As the pressure is increased at constant temperature of 1000°C, the CO and total yield go through a maximum at 200 psi. The total yield also increases with increasing temperature mainly

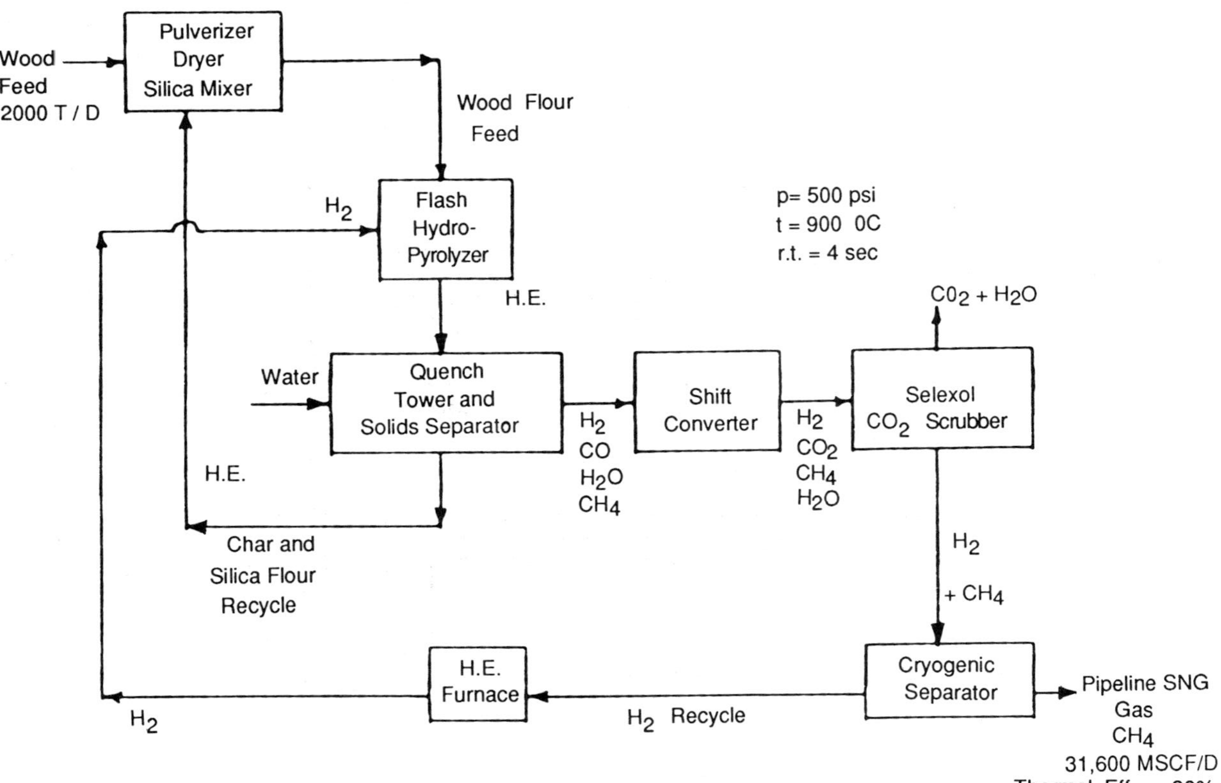

Figure 4. Flash hydropyrolysis of wood for pipeline gas.

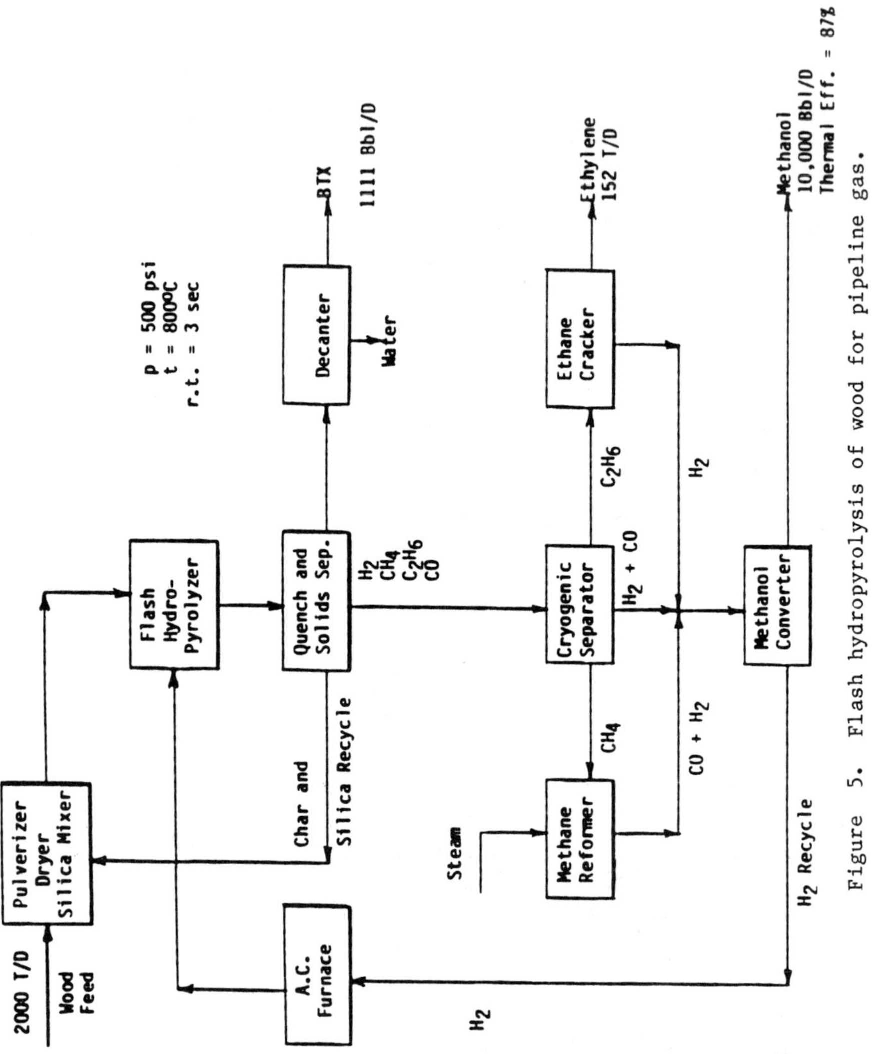

Figure 5. Flash hydropyrolysis of wood for pipeline gas.

because the yield of methane, benzene and ethylene increase with temperature. The maximum HC conversion occurs at 900°C and 50 psi pressure. At this point, about 10% is converted to CH_4, 10% to C_2H_4, and 5% to C_6H_6, which results in a total of 25% conversion of the C to hydrocarbons.

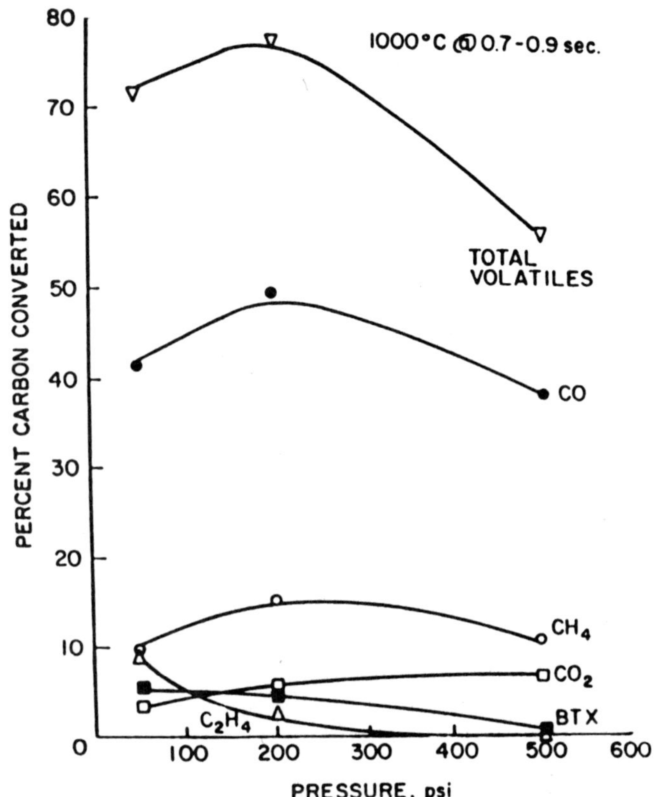

Figure 6. Flash pyrolysis of fir wood in helium, effect of pressure on product yields.

Figures 8 and 9 present hydrocarbon yield data on the flash pyrolysis of wood with the inert gas nitrogen at 900°C as a function of pressure and at 50 psi as a function of temperature, respectively. The total HCs decrease with increasing pressure mainly due to a decrease in ethylene yield from 12% at 50 psi to zero at 500 psi. The total HC yield tends to increase from 900°C to 1000°Cmainly due to improved BTX formation. About the same total hydrocarbon yields (25%) are obtained with nitrogen flash pyrolysis as with helium.

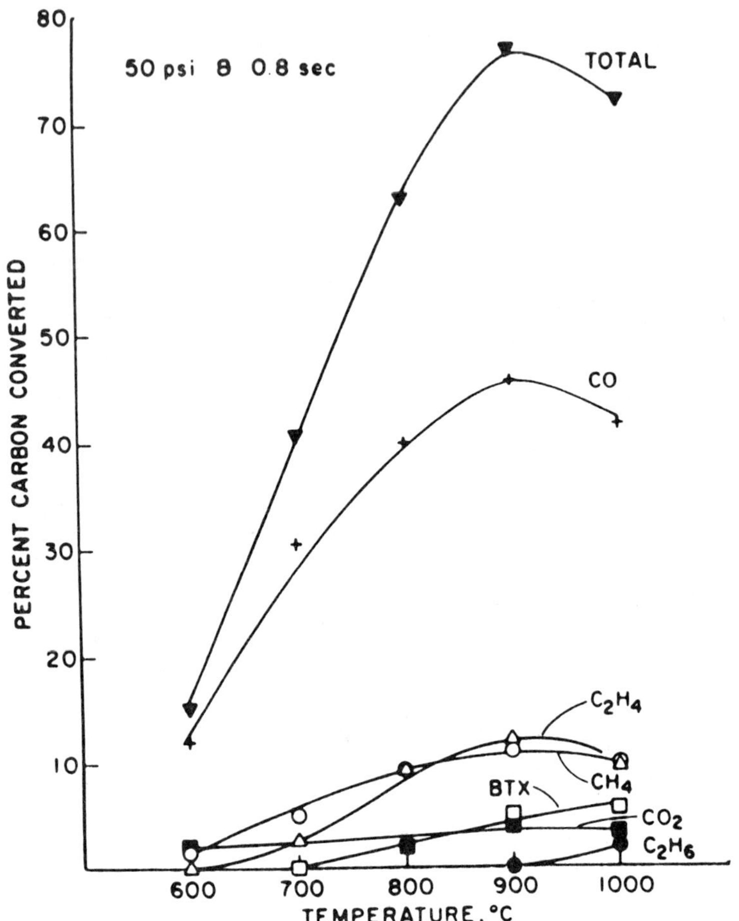

Figure 7. Flash pyrolysis of fir wood in helium, effect of temperature on product yields.

FLASH METHANOLYSIS OF BIOMASS WITH METHANE

A series of experiments were performed on the flash pyrolysis of fir wood with methane. The following are the major findings based on Douglas fir wood and a sugar pine wood.

1. The yield in terms of conversion of carbon in the biomass to products are given in Figure 10 as a function of pressure from 20 to 200 psig at a constant temperature of 1000°C. The major product was

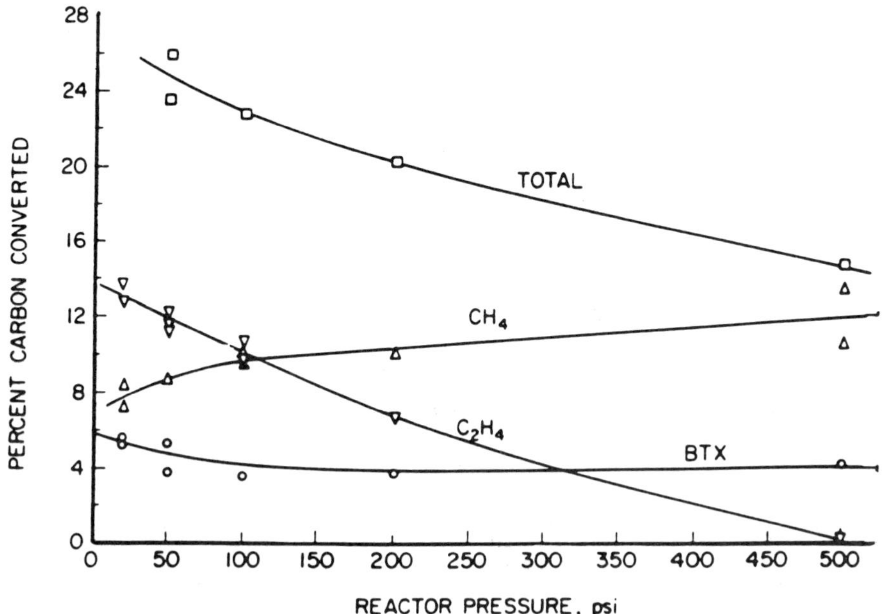

Figure 8. Flash pyrolysis of fir wood with nitrogen, hydrocarbon yield vs reactor pressure, reactor temperature = 900°C.

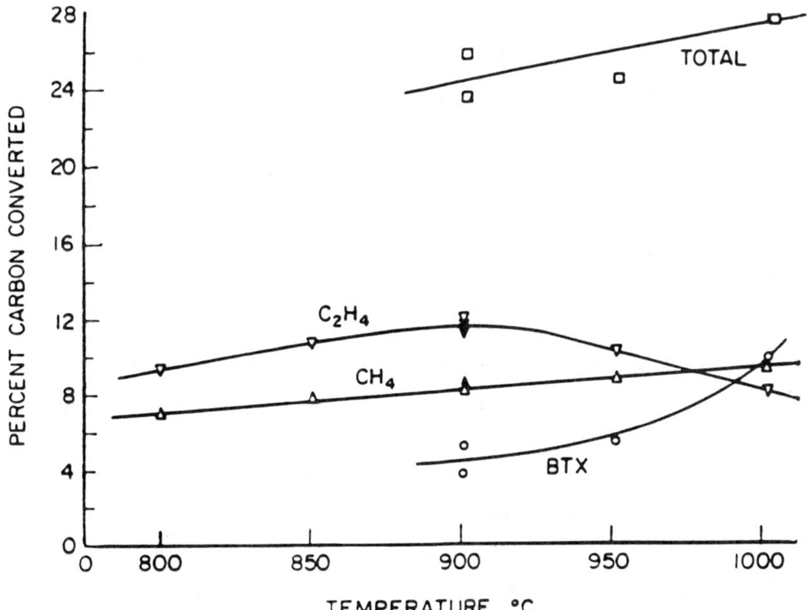

Figure 9. Flash pyrolysis of fir wood with nitrogen, hydrocarbon yields vs pressure, reactor pressure - 50 psi.

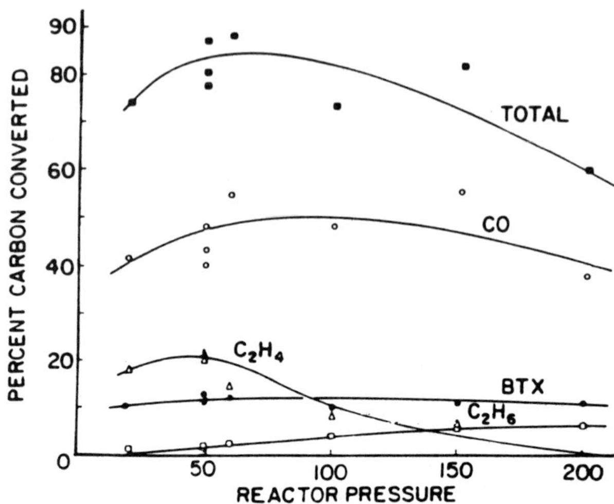

Figure 10. Flash methanolysis of fir wood, product yields vs reactor pressure, reactor temperature - 1000°C, methane/wood ratio - ≈5, wood residence time - ≈0.9 - 2.8 sec.

CO reaching a peak of about 55% at 100 psi. The BTX remains almost constant at 10 to 12%. The ethane yield increases linearly from 1.5% at 20 psi to 6% at 200 psi. The ethylene yield is most interesting in that there is very little ethylene formation at a pressure of 200 psi and above. The ethylene yield increases with decreasing pressure reaching a maximum of 22% at 50 psi. Pressures below 50 psi do not appear to improve the yield of ethylene.

2. The yields as a function of temperature are shown in Figures 11 and 12. At the optimum pressure of 50 psi, the yields increase with increasing temperature. The total volatiles excluding any possible methane formation reaches 88%, with the CO increasing to 48% at 1000°C. There is very little ethylene formation at 600°C and the ethylene increases almost linearly with temperature to 22% at 1000°C. The BTX also increases from a small value at 600°C to 12% at 1000°C. The ethane and CO_2 yields remain fairly constant at about 2 to 3%. There appears to be be a peak in heavier liquid

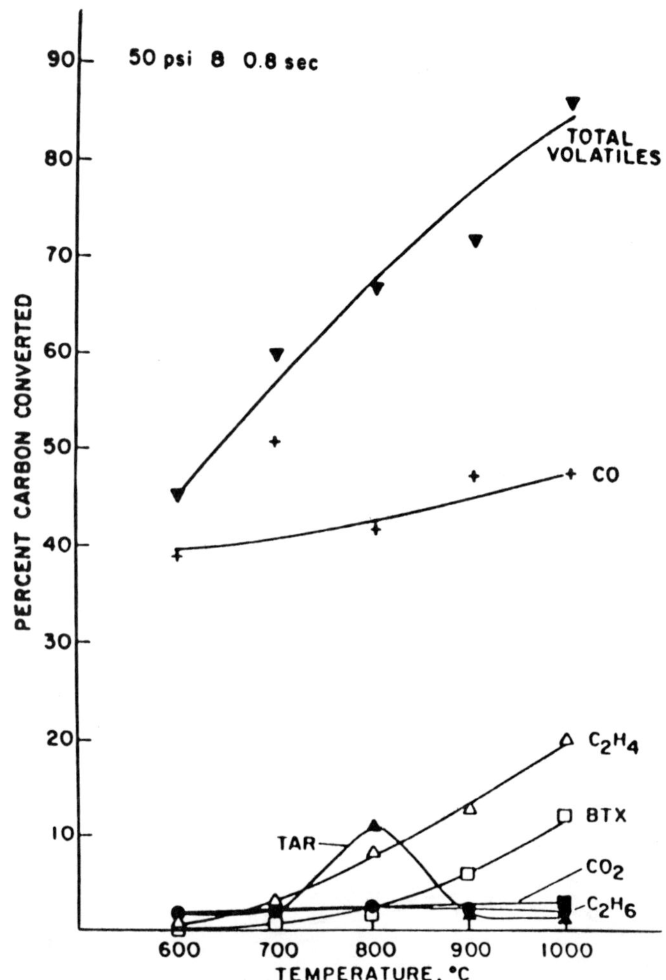

Figure 11. Flash pyrolysis of fir wood in methane, effect of temperature on product yields.

hydrocarbon tar formation at 800°C to the extent of 11% C conversion. The liquids are identified as mainly oils and asphaltenes. The oils are predominantly naphthalenes with a small amount of phenol.

3. A comparison of the yields of ethylene under flash pyrolysis with CH_4, He, N_2 and H_2 are shown in Figure 13. An increased

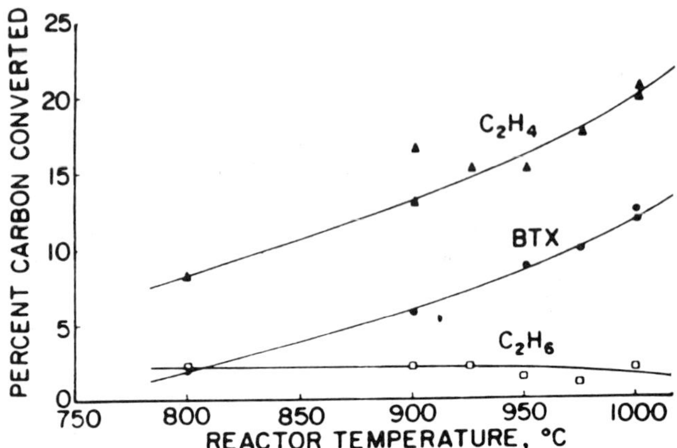

Figure 12. Flash methanolysis of fir wood, hydrocarbon yields vs reactor temperature, reactor pressure - 50 psi, methane/wood ratio - ≈5, wood residence time - ≈0.9 sec.

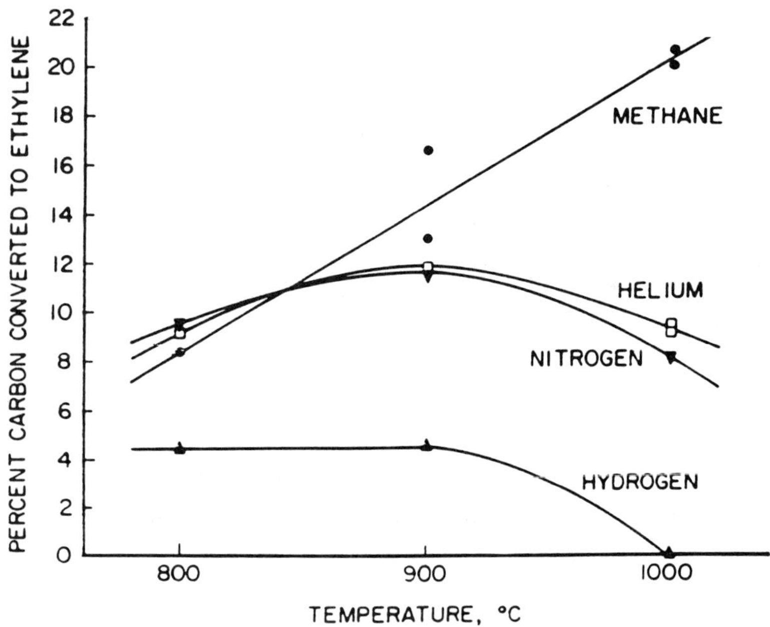

Figure 13. Flash pyrolysis of fir wood, ethylene yield vs reactor temperature reactor pressure - 50 psi.

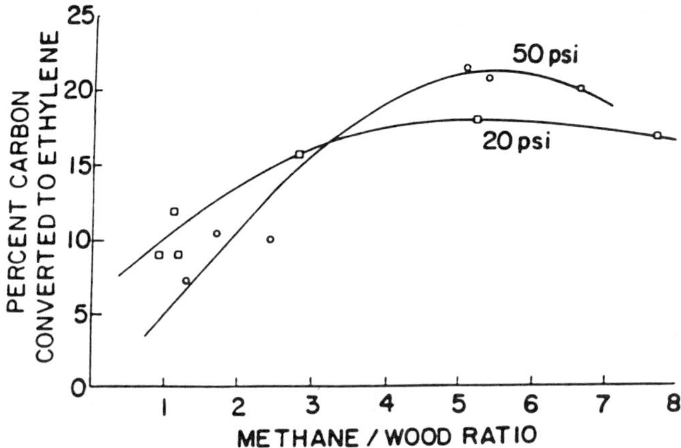

Figure 14. Flash methanolysis of fir wood, ethylene yields vs methane/wood ratio, temperature - 1000°C, reactor pressure - 20 and 50 psi, wood residence time - ≈0.9 sec.

ethylene yield with CH_4 over inert He and N_2 begins at about 850°C and reaches a factor of about 2.3 at 1000°C. At lower temperatures,
4. The ethylene yield decreases with increasing flow rate of wood particles at approximately constant flow rate of methane. As shown in Figure 14, the ethylene yield increases with increasing methane/wood ratio and becomes constant at a ratio above 5. This indicates a concentration saturation effect of reactive species originating from the wood interacting with the methane pyrolysis gas. A continuing linear increase of ethylene yield with methane/wood ratio would have indicated an independent additive yield of methane pyrolysis and wood pyrolysis (Steinberg, Fallon, and Sundaram, 1984).
5. A series of runs were made with sugar pine wood. The experimental results are presented in Table I and are compared to the results obtained with Douglas fir wood in Figures 15 and 16. A Kraft processed pine lignin was also run and is shown in Table I. It is noted that the total HC yields were from 23 to 43% higher for

TABLE 1 Flash Methanolysis of Wood and Lignin at 50 psi

Wood Type	Fir(1			Pine(2			Pine Lignin(3
Run No.	590	593	592	746	746	746	737
Reactor Temp. (°C)	900	950	1000	900	950	1000	900
Wood Feed Rate (lb/hr)	0.89	0.80	0.92	0.55	0.55	0.55	0.55
Methane Feed Rate (lb/hr)	4.21	5.23	4.89	3.84	3.84	3.84	3.99
Particle Res. Time (sec)	0.9	0.8	0.8	0.9	0.9	0.9	0.9
% Carbon Conv. to Prod.							
C_2H_4	13.0	15.2	20.7	15.7	20.6	27.4	5.1
C_2H_6	2.2	1.5	2.0	0	0	0	0
Total Gas H.C.	15.2	16.7	22.7	15.7	20.6	27.4	5.1
BTX	5.8	8.9	12.7	11.9	13.4	24.6	7.3
Oils ($<C_9$)	1.4		1.0				
Total Liquid H.C.	7.2	8.9	13.7	11.9	13.4	24.6	7.3
Total H.C.	22.4	25.6	36.4	27.6	34.0	52.0	12.4
CO	47.7	49.1	48.1	35.6	36.1	38.7	10.2
CO_2	2.5	3.3	3.3	2.4	2.4	2.7	1.9
Total CO_x	50.2	52.4	51.4	38.0	38.5	41.4	12.1
Total Conv.	72.6	78.0	87.8	65.6	72.5	93.4	24.5

(1 Fir fed with 15% Cab-O-Sil
(2 Pine fed with 30% Cab-O-Sil
(3 Lignin fed with 20% Cab-O-Sil

pine wood than for fir wood. At the same time, the CO yields were 25% lower for pine than for fir. Since the elemental analysis for each of the woods are about the same, these data indicate less methane formation in the case of pine.

The results of the lignin experiment indicate much lower yields of hydrocarbon products (more than 50% lower), which leads to the conclusion that the cellulose component of the wood is the major source of the HC products. The yield of ethylene from pine at 900°C

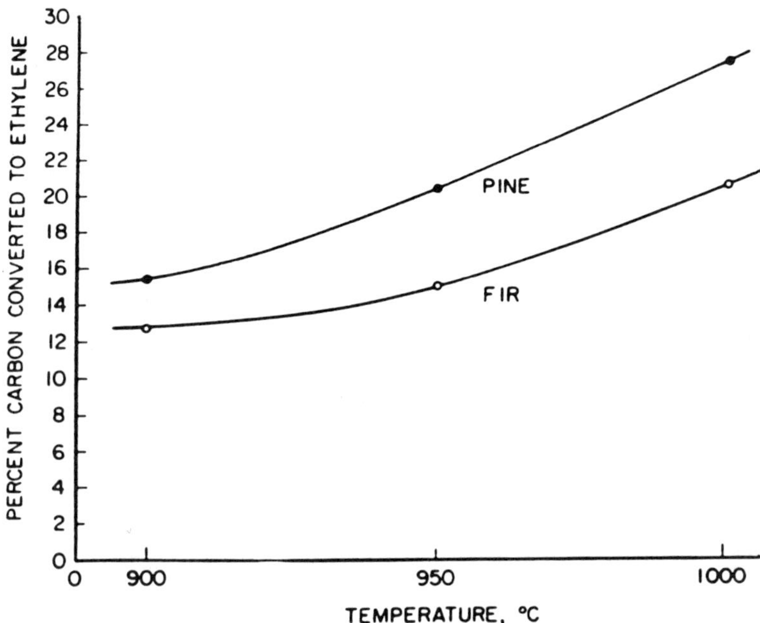

Figure 15. Flash methanolysis of fir and pine woods with methane, ethylene yield vs temperature, reactor pressure - 50 psi.

and 50 psi (15.7%) is over three times higher than for pine lignin (5.1%), while the benzene yield in pine (11.9%) is only 50% higher than for lignin (7.3%). This would indicate that the major source of ethylene is the cellulose fraction of the wood, while the benzene is derived more from the aromatic structure of the lignin.

It is also noted that for pine at 50 psi and 1000°C, as much as 52% of the carbon in the wood is converted to BTX (mainly benzene) and ethylene. Yields of 29.6% conversion to ethylene, 24.6% to benzene, and 45% to CO are the highest obtained for the flash methanolysis of pine wood. Figures 15 and 16 compare the higher yields of ethylene and BTX for pine and fir, respectively as a function of temperature. The structure of pine is evidently more conducive to producing unsaturated and aromatic hydrocarbons in

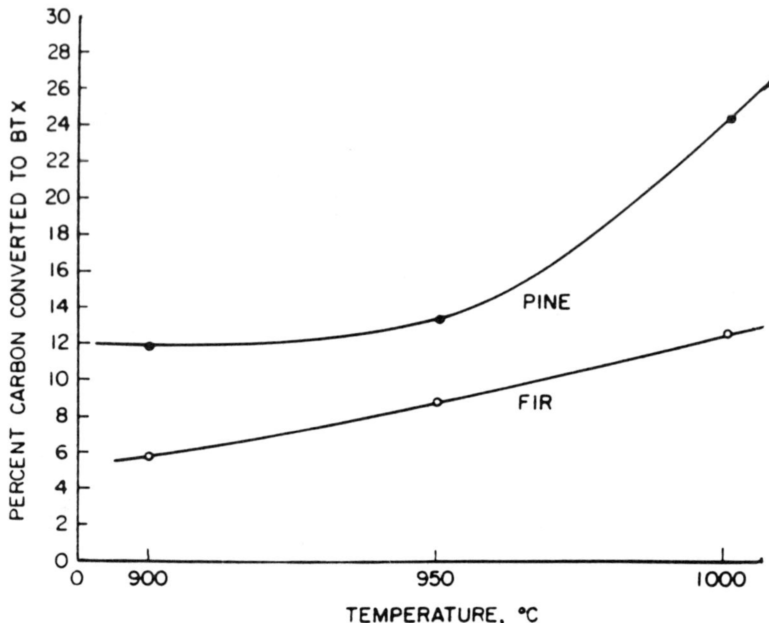

Figure 16. Flash methanolysis of fir and pine woods with methane, BTX yield vs temperature, reactor pressure - 50 psi.

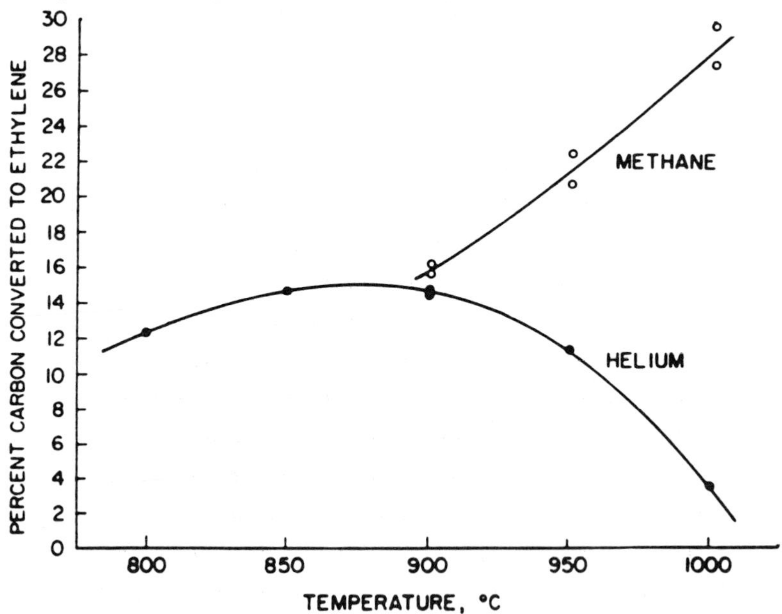

Figure 17. Flash pyrolysis of pine wood, ethylene yield vs reactor temperature, reactor pressure - 50 psi.

conjunction with methane. This is further indicated in Figure 17, which shows that there is a 7.5 fold increase in ethylene when pyrolyzing at 1000°C with methane as compared to straight pyrolysis with helium.

Experiment with flash methanolysis performed at temperatures above 1000°C and up to 1100°C, appeared to indicate a leveling off and no further improvement in ethylene and benzene yields. Pressures lower than 50 psi also tend to decrease yields. The maximum volumetric concentration of ethylene obtained in our experiments was 3.2%.

6. A series of blank runs were made to assure that the ethylene does not arise from the methane cracking alone or with other substrates in the reactor as indicated in the following:

a) Methane flowing through the tubular reactor alone before and after every run at the same temperature and pressure as during a run does not produce any ethylene. Thus, there is no cracking of methane to ethylene on the alloy Inconel 617 surfaces of the reactor.

b) Running silica flour (Cab-O-Sil) and sand under the same conditions as in the methane/wood runs did not produce any ethylene.

c) Running silica flour (Cab-O-Sil) and sand wetted with 23% water with the same flow rate of methane as used in the wood run, as well as pressure and temperature did not produce any ethylene.

d) Running recycled wood char residue from previous methanolysis runs and coconut charcoal under the same conditions as with wood, do not produce additional ethylene yields.

These blank and control runs strongly indicate that one of the species responsible for producing ethylene arises from the initial volatilization of the wood. Furthermore, the improvement in yield with methane above 900°C compared to helium pyrolysis under the same conditions strongly indicates a reaction between the species originating from the wood and methane.

7. Methane balances were performed by measuring the flow rate into and out of the reactor. Experiments were also performed with inert

tracer He in the methane measuring the ratio of He/CH_4 in and out of the reactor. Within experimental error, no appreciable net consumption or production of methane was observed. It should be noted that in the flash pyrolysis runs with inert He alone, up to 15% CH_4 is formed in addition to ethylene. Thus, methane is probably also formed from the wood carbon during flash methanolysis and, therefore, tends to keep the total methane in balance. Another indication in the methane balance can be obtained when the C balance adds up to less than 100%, then methane must be produced from the carbon. When the C balance adds up to more than 100%, then CH_4 must be consumed. Additional measurements are needed to obtain precise methane balances.

8. One explanation for the enhanced ethylene and benzene yields is that free radical reactions take place between the biomass and the high concentration of high temperature methane. Flash heating of biomass particles produces free radicals (thermal fragments, e.g. CH_2 and CH_3), which are either stabilized by the high partial pressure of methane or abstract hydrogen from methane producing other HC radicals which subsequently form C_2H_4 and C_6H_6. Ethane may also form as an intermediate precursor to ethylene formation (Steinberg, Fallon, and Sundaram, 1984).

$$\text{wood} \longrightarrow \text{HC free radicals} + CO + H_2O \qquad (5)$$

$$\text{HC free radicals} + CH_4 \longrightarrow C_2H_4 + C_6H_6 \qquad (6)$$

However, we cannot unequivocally eliminate a possible catalytic conversion of methane on the wood surfaces until a ^{13}C tracer experiment is performed. A ^{13}C tracer experiment has been performed on the methanolysis of coal and it has been found that there indeed is a reaction between the C in methane with the C in coal (Sundaram and Steinberg, 1986). The fact that there is a non-linear function of yield with the methane to wood ratio also indicates a free-radical saturation effect in the mechanism of formation of ethylene in the interaction of methane with wood.

9. A number of other experiments were performed pertinent to the development of the process for the flash methanolysis of wood.

a) Addition of increasing concentrations of wood ash prepared from previous pyrolysis experiments indicates that there is little effect on the hydrocarbon yields. This indicates that recycling of partially converted wood in a large scale plant to achieve higher wood conversion efficiency would be satisfactory.

b) The effect of fresh wood and dried wood were determined. Fresh pine containing approximately 47% moisture indicated approximately the same conversion of wood carbon to ethylene as with dry wood containing less than 35% moisture. However, the BTX yields were lower with the fresh wood.

c) The lower temperature flash pyrolysis of fresh pine wood containing 46% moisture with methane, produced a maximum of 11% carbonaceous liquids at a temperature of 650°C. The liquids appear to be oxygenated organic materials including phenols, acetone, and other organic acids.

d) Some experimental work was performed with hard wood on which most of the work was performed. At a temperature of 1000°C and 50 psi pressure, there appears to be a slightly lower yield of ethylene and benzene than pine.

Additional experiments are required to establish these results with other hard woods.

e) An interesting experiment was performed on the flash methanolysis of a mixed feedstock of wood and coal. The purpose of this experiment was to determine whether the higher H/C ratio in wood would affect the conversion of the lower H/C ratio in coal to produce additional gaseous and liquid hydrocarbons. In the two runs made, it was found that the yields obtained were additive. Additional runs are needed to determine whether a synergistic enhancing effect on yield can be obtained when co-processing wood with coal.

PROCESS DESIGN AND ANALYSIS-FLASH METHANOLYSIS OF BIOMASS FOR FUEL AND CHEMICAL FEEDSTOCK PRODUCTION

The flash methanolysis of wood in the presence of methane indicates that significant yields of valuable fuels and chemical feedstocks

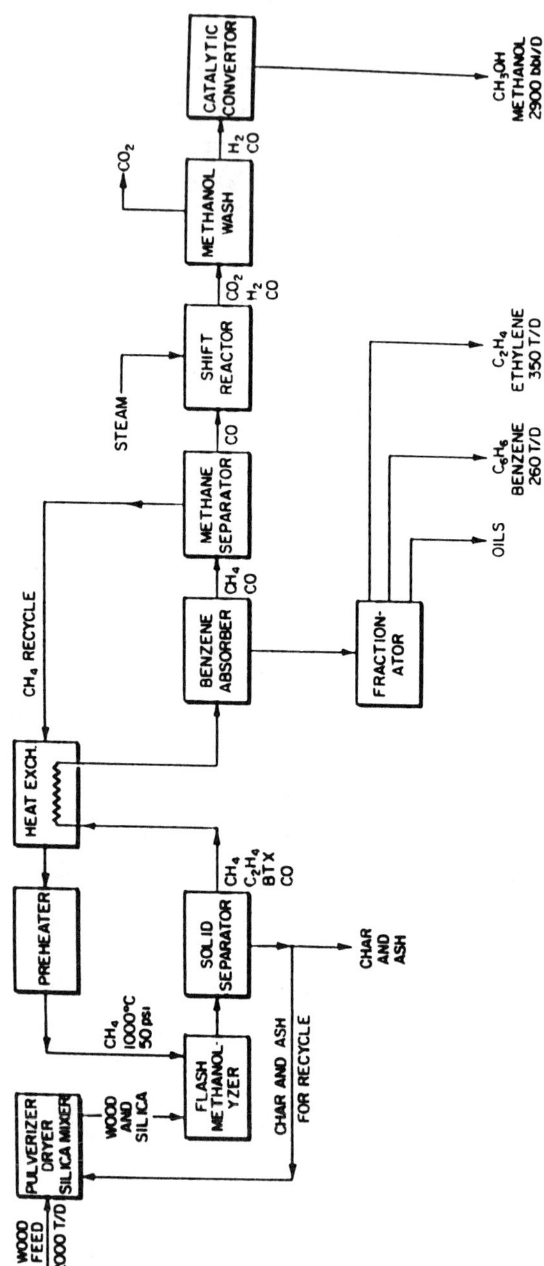

Figure 18. Flash methanolysis of wood for production of chemical feedstock and methanol.

are produced. Benzene is valuable in blending with gasoline. Ethylene and benzene are valuable chemical feedstocks for the large polymer and plastics market. These materials are conventionally produced from petroleum and higher HCs extracted from natural gas. The direct production of these products from biomass therefore becomes a valuable alternate and renewable source of supply of these materials for the U.S. In addition, our experiments indicate that a large yield of CO is produced and this can be converted to methanol, which is a valuable coproduct, as a transportation fuel.

A conceptual flow sheet based upon the experimental results is given in Figure 18. Preliminary economic analyses indicate that a reasonably competitive process (Steinberg and Fallon, 1981; Wan and Price, 1983) can be developed for the production of ethylene, benzene and methanol fuel and feedstocks based on the flash pyrolysis of wood with methane (flash methanolysis).

CONCLUSIONS

1. Flash hydropyrolysis of wood with hydrogen in an entrained down flow reactor leads mainly to methane and water at very short particle reaction residence times (<1 sec). At longer residence times (>4 sec), the products tend mainly to methane and carbon monoxide. At 1000°C and 500 psi, over 90% of the wood carbon can be converted to volatile hydrocarbons, mainly methane.
2. Decreasing pressure in flash hydropyrolysis of wood tends to favor CO yields and decreased CH_4 yields.
3. Flash pyrolysis of wood particles with inert helium gas, mainly yields CO (~40-50%) and significant amounts of methane (15%), ethylene (10%), and BTX (5%, mainly benzene). Similar hydrocarbon yields are obtained with nitrogen under similar flash pyrolysis conditions of 50 psi and 900°C.
4. The flash methanolysis of Douglas fir wood particles with methane at 1000°C and 50 psi produces 2.3 times as much ethylene as the flash pyrolysis of wood with helium. A free radical originating from the volatilization of the wood interacting with the methane appears plausible for the mechanism of the reaction.

5. The flash methanolysis of pine wood produces higher hydrocarbon yields than Douglas fir. The maximum yields for pine were obtained at 50 psi, 1000°C and ~1 sec residence time; the ethylene yield amounted to 29.6%, the benzene is 24.6% and the CO is 45.8%. With pine, ethylene is 7.3 times higher than that for the flash pyrolysis of wood with an inert gas. Thus, over 90% of the carbon in pine is converted to valuable products.

6. Preliminary process designs and analysis indicates a reasonably competitive process for production of the chemical feedstocks, ethylene and benzene, and methanol transportation fuel by the flash methanolysis of wood, thus indicating an alternative source for these commodity products.

ACKNOWLEDGMENT

This research was performed under the auspices of the U.S. Department of Energy, Division of Materials Sciences, Office of Basic Energy Sciences under Contract No. DE-AC02-76CH00016.

REFERENCES

1. Belt, R. J., and L. A. Bissett. Assessment of flash pyrolysis and hydropyrolysis, METC/RI-79/2. Morgantown Energy Technology Center, Morgantown, West Virginia (1978)
2. Fallon, P., and M. Steinberg. Flash hydropyrolysis of coal experimental unit, BNL 50698, Brookhaven National Laboratory Upton, New York (1977)
3. Steinberg, M., P. Fallon, and M. Sundaram. Flash pyrolysis of biomass with reactive and non-reactive gases, summary report, BNL 35607, Brookhaven National Laboratory, Upton, New York (1984)
4. Steinberg, M., and P. Fallon. Flash Pyrolysis of biomass with reactive and non-reactive gases, BNL 51560, Brookhaven National Laboratory, Upton, New York (1980)
5. Steinberg, M., P. Fallon, and M. Sundaram. The Flash methanolysis of wood for the production of fuels and chemicals, BNL 32400, Brookhaven National Laboratory, Upton, New York Presented at the VII Symposium on Energy from Biomass and Wastes, Lake Buena Vista, Florida (1983)

6. Steinberg, M., and P. Fallon. Flash pyrolysis and hydropyrolysis of biomass, progress report no. 2, BNL 30263, Brookhaven National Laboratory, Upton, New York (1981)

7. Sundaram, M., M. Steinberg, and P. Fallon. Flash pyrolysis of biomass with reactive and non-reactive gases, BNL 32280, Brookhaven National Laboratory, Upton, New York. Presented at the International Conference on the Fundamentals of Thermo chemical Biomass Conversion, Estes Park, Colorado (1982)

8. Wan, E. I., and J. D. Prince. Technical and economic assessment of emerging advanced biomass thermochemical conversion techniques. Proc. of the 15th Biomass Thermochemical Conversion Contractor's Meeting, Atlanta, Georgia, CONF 830323, PNL-SA-11306 (1983)

9. Sundaram, M. S., and M. Steinberg. Flash methanolysis of coal: A mechanistic study, BNL 37935, Brookhaven National Laboratory, Upton, New York and in FUEL Journal (in press) (1986)

10. Steinberg, M., P. Fallon, and M. Sundaram. Flash pyrolysis of biomass with reactive and non-reactive gases, BNL 39113, Brookhaven National Laboratory, Upton, New York (1986)

28

PRODUCTION OF BTX AROMATICS FROM LIGHT ALKANES AND OLEFINS ON HYBRID CATALYSTS: AN EXAMPLE OF SYNERGY EXPLAINED BY SPILLOVER

L. Dufresne, J. Yao and R. Le Van Mao *

Department of Chemistry and Biochemistry
CONCORDIA UNIVERSITY, 1455 De Maisonneuve Blvd. W., Montreal (Quebec) Canada H3G-1M8

ABSTRACT

Selective conversion of light olefins and/or alkanes into BTX aromatics has been demonstrated to occur on hybrid type catalysts. The catalysts are made-up of two components. The major component is a ZSM-5 zeolite in its acid form and the minor component may be an oxide, mixed oxide or a supported oxide. Here it is shown that the enhanced aromatic production on hybrid type catalysts, is due to a long distance hydrogen transfer from the zeolite acid sites to the co-catalyst surface (the so-called Long Distance Hydrogen Back-Spillover action or LD-HBS in acronym form). This concept allows considerable flexibility in the catalyst formulation since the different components may be modified separately.

INTRODUCTION

BTX aromatic hydrocarbons have a high commercial value. They can be used as octane boosters in gasoline, as feedstock for plastics, as commodity chemicals, etc. Great efforts have been devoted in the last decade to finding new up-grading

----

(*) To whom correspondence should be addressed.

processes for the production of such monoaromatics from lower value fractions.

Aromatics can be synthesized from olefinic and/or paraffinic feedstocks. Enhanced aromatization of ethane [1] was demonstrated early on when a synthetic zeolite of the pentasil family, a ZSM-5 ($SiO_2/Al_2O_3 = 70$), was modified by ion exchange with small amounts of zinc (2.1wt% Zn) or copper-zinc (0.25wt% Cu, 1.0wt% Zn) mixtures. For the Cu/Zn- ZSM-5 catalyst, the best conversion of ethane was 34.5% and the total aromatic yield 21.5%. The composition of the aromatic fraction was mostly monocyclic due to the transition state and product shape selectivity of the zeolite channel network. Gallo-alumino-silicates or gallium (0.5wt% Ga) exchanged ZSM-5 zeolites have also been studied in the conversion of ethane [2]. Gallium containing catalysts were seen to have greater aromatization capacity than their Cu/Zn analogs. In the CYCLAR process [3] developed jointly by BP/UOP, gallium is introduced by reflux in the ZSM-5 lattice. The resulting solid may incorporate a certain quantity of Ga^{3+} ions as well as Ga_2O_3 on the surface. It is observed that propane and n-butane are converted to aromatics and a significant amount of hydrogen is also produced. Several other examples of similarly modified catalysts exist in the literature [4-8]. In all these cases, the aromatization reaction is enhanced via a bifunctional type of catalysis, since acid sites and metal ion or oxide active sites are in very close proximity, both located inside or at the openings of the zeolite channels.

The hydroconversion of n-heptane over a Pt/Al_2O_3- erionite system [9] indicates that it is possible to increase the catalytic activity even in a case where the two types of active sites are not in adjacent positions, by a hydrogen spillover action from the Pt/Al_2O_3 to the erionite lattice.. Recently, it has been demonstrated by Le Van Mao et al. [10-13] and Fujimoto et al. [14] that this concept of long distance transfer of hydrogen atoms can be applied successfully to explain the increased aromatization activity.

The hybrid catalysts developed in this work show a synergistic effect for BTX aromatic hydrocarbon production when a metal oxide is positioned "close" to the openings of the zeolite channels. Therefore, by mechanically mixing a ZnO or a ZnO/Al_2O_3 co-precipitate powder with H-ZSM-5 zeolite fine particles and then immobilizing their respective positions within a bentonite binder matrix, the BTX production is greatly favored [10].

From a series of experiments designed specifically to probe certain catalytic behaviors of our system, the importance of hydrogen transfer has been clearly established [12]. The results have permitted the developement of a new generation of LD-HBS hybrid catalysts which exhibit very high activity toward aromatic production [13]. This paper presents a collection of the latest results obtained.

EXPERIMENTAL

ZSM-5 zeolite synthesis

Parent ZSM-5 zeolites were synthesized according to the well known method of Argauer and Landolt [15]. The preferred Si/Al ratio was between 36 and 43 to obtain both high thermal stability and a great acid site density which is a prerequisite for high aromatization activity. The acid form of the zeolites was obtained by successive ion exchange with a 5wt% ammonium chloride solution at 80ºC, under mild stirring. The solid was then filtered, washed, dried overnight at 120ºC and activated in air at 550ºC for at least 12 hours.

The zeolite powder in its acid form was characterized by X-ray powder diffraction for identification of the ZSM-5 structure and for measurement of its degree of crystallinity [16]. Atomic absorption provided the exact chemical composition and ensured that the amount of sodium ions was less then 0.2wt%. The catalytic activity of the resulting solid was tested against a chosen standard reaction to confirm a good uniformity between individual batch syntheses.

Preparation of the catalysts

Synthesis of the co-catalysts is described elsewhere [10-13]. For the preparation of the hybrid catalysts, the H-ZSM-5 zeolite powder (mesh size number higher than 60) (75wt% or 80wt%) and the co-catalyst powder (mesh size higher than 60) (5wt% or 13wt%), were mechanically and intimately mixed at room temperature. The bentonite binder (20wt% or 7wt%) was subsequently added and also intimately mixed with the solid catalyst. Water was added dropwise until a malleable paste was obtained. The latter was extruded into 1 mm O.D. "spaghettis". The final extrudates were dried at

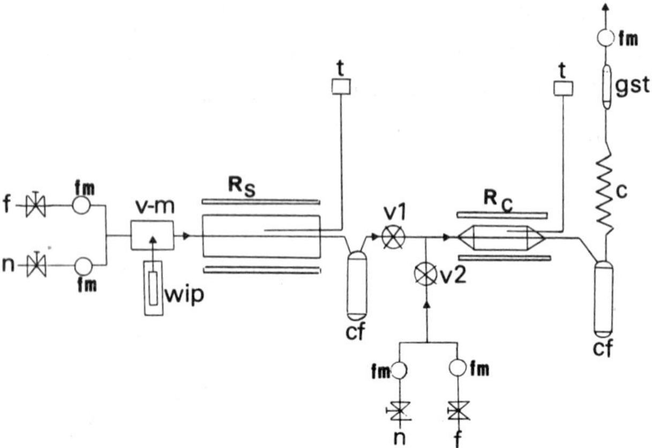

FIGURE 1: EXPERIMENTAL DESIGN
LEGEND: Outstream gases of steam-cracking obtained when **V1** is open **V2** is closed.**c**, condensers; **cf**, collecting flasks; **f**, feed (propane or n-butane); **fm**, flow-meters; **gst**; gas sampling tube; **n**, nitrogen (carrier gas); **Rc** or **RII**, catalytic reactor; **Rs** or **R1**, steam-cracking reactor; **t**, thermocouples; **v-m**, vaporizer-mixer; **wip**, water infusion pump.(Reprinted with permission of Applied Catalysis).

120ºC for several hours and activated in air at 550ºC for about 10 hours. The final length of the extrudates used in catalytic testing was about 5 mm.

Catalytic testing

The experimental set-up for catalytic testing is as shown in Figure 1. $\mathbf{R_S}$ in series with $\mathbf{R_C}$ was utilized in the case where the reaction feedstock was generated in a laboratory scale propane steam-cracker (through **V1**). The resulting feedstock was a mixture of mostly light olefins and light paraffins smaller than C_6 (because of the condensing flask between the two reactors). The product distribution of the gaseous mixtures is given in Table 1. A blank reaction with pure bentonite, the matrix binder, is included for comparison and it can be seen that its aromatization activity is negligible. The **V2** configuration was utilized in the case where pure olefinic or paraffinic feeds were used. The WHSV for all reactions presented in this paper was arbitrarily chosen as $0.6h^{-1}$.

TABLE 1: *Total product distribution of a propane steam-cracker and resulting gaseous outstream. Results of a reaction on the bentonite binder (blank)*

TEMPERATURE R1 / PRODUCT DISTRIBUTION (C atom %)	820ºC		840ºC		Bentonite
	Total	Gas	Total	Gas	820ºC
Methane	18.2	19.0	21.7	23.3	20.4
Ethylene	33.3	34.8	38.3	41.1	37.7
Ethane	2.0	2.2	2.4	2.6	2.3
Propylene	17.1	17.8	14.8	15.9	2.1
Propane	21.9	22.8	13.0	13.9	30.3
Butanes	0.5	0.5	0.1	0.1	0.7
Butenes	2.8	2.9	2.9	3.1	3.2
$C5^+$ Aliphatics	1.6	-	2.0	-	1.7
Aromatics	2.6	-	4.8	-	1.7
Total	100.0	100.0	100.0	100.0	100.1
BTX/Arom. (%)	61.8		64.8		100.0

COMPOSITION OF THE BTX AROMATICS

	820ºC	840ºC	Bentonite
Benzene	89.9	84.0	100.0
Toluene	2.8	5.3	-
Ethylbenzene	0.6	0.9	-
Xylenes	1.1	1.5	-
Styrene	5.6	8.3	-
Total	100.1	100.0	100.0

C_5^+ Aliphatics: Hydrocarbons having at least 5 C atoms

Residence time: 0.5sec.

Typical reaction temperatures for R_c were 500ºC and 540ºC. 4 grams of catalysts were used and reaction duration was 4-5 hours, showing no activity decay during that period. The amount of carbon atoms and the product distribution of the gaseous effluents was determined against an internal standard (2,2-dimethylbutane) and analyzed on a 2.5m packed column of 15wt% squalane coated Chromosorb P, by FID gas chromatography (HP 5790). The breakdown of the liquid products was also acquired by GC on a 50m PONA capillary column.

In the following data, the total conversion of feed is defined as:

$$C_t \text{ (C atom \%)} = \frac{(NC)_F - (NC)_P}{(NC)_F} * 100$$

where $(NC)_F$ and $(NC)_P$ are the number of C atoms of reactant fed and in the reactor outstream respectively.

The selectivity for product i is defined as:

$$S_i \text{ (C atom \%)} = \frac{(NC)_i}{(NC)_F - (NC)_P} * 100$$

where $(NC)_i$ is the number of C atoms of product i in the reactor outstream.

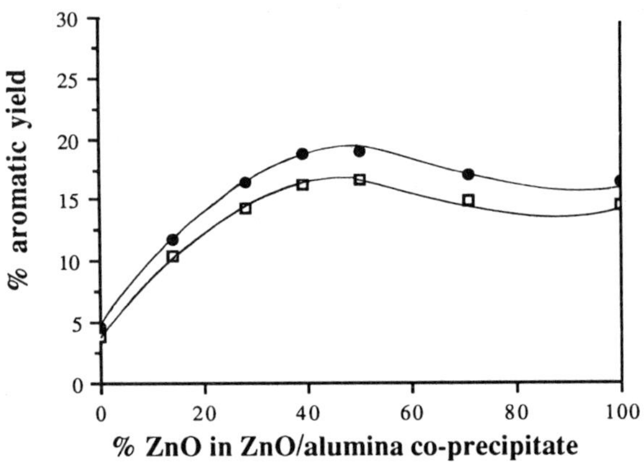

FIGURE 2: EFFECT OF ZnO CONTENT IN THE CO-PRECIPITATE
Temperature R1:820ºC; RII: 500ºC; WHSV: $1.6h^{-1}$.
Catalyst composition: H-ZSM-5(37)(40wt%) + ZnO/alumina(40wt%)
LEGEND: ● TOTAL AROMATICS
□ BTX AROMATICS
(Reprinted with permission of Applied Catalysis).

The yield of product i is defined as follows:

$$Y_i \text{ (C atom \%)} = S_i * C_t * 1/100$$

RESULTS AND DISCUSSION

HBS phenomena in the catalytic aromatization of light olefins and paraffins

At the onset of this work, from results obtained with a gaseous olefinic/paraffinic feedstock, it was demonstrated that hybrid catalysts obtained by mechanically mixing a ZnO/Al_2O_3 co-precipitate powder with the H-ZSM-5 zeolite particles yielded a substantial increase of the aromatization propensity of the zeolite lattice. As can be observed in Figure 2, the maximum total aromatic yield is obtained when the ratio Zn/Al=1.0, but remains relatively constant after as the amount of ZnO increases in the

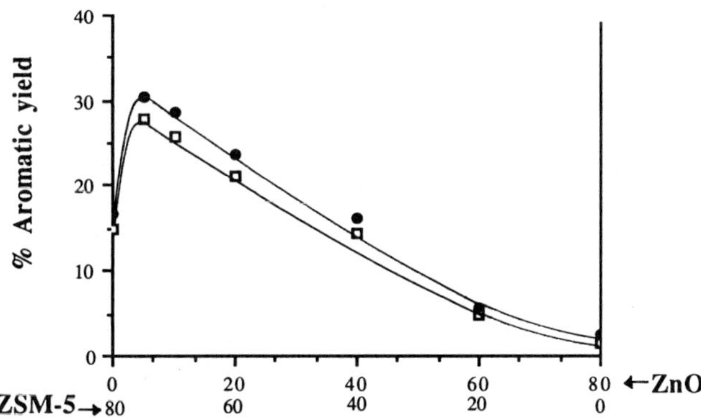

FIGURE 3: OPTIMIZATION OF CATALYST COMPOSITION
Same conditions , catalyst composition and legend as in FIG. 2.
(Reprinted with permission of Applied Catalysis).

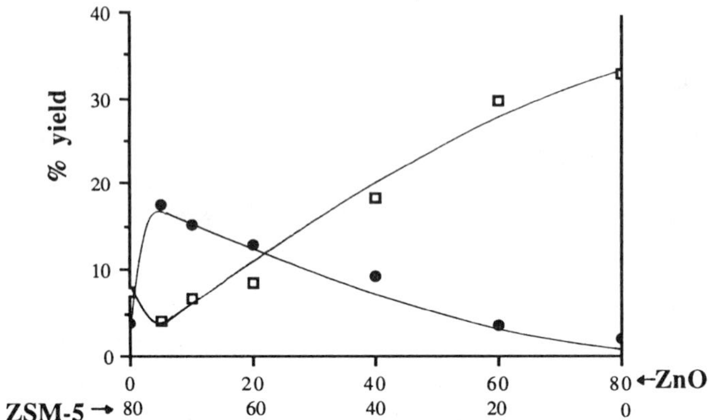

FIGURE 4: OPTIMIZATION; CORRELATION GRAPH
Temperature R1: 820ºC; RII: 500ºC; WHSV: $1.5h^{-1}$.
Catalyst composition: H-ZSM-5(37)(0-80wt%)+ZnO(80-0wt%)
LEGEND: ● ETHANE ■ ETHYLENE

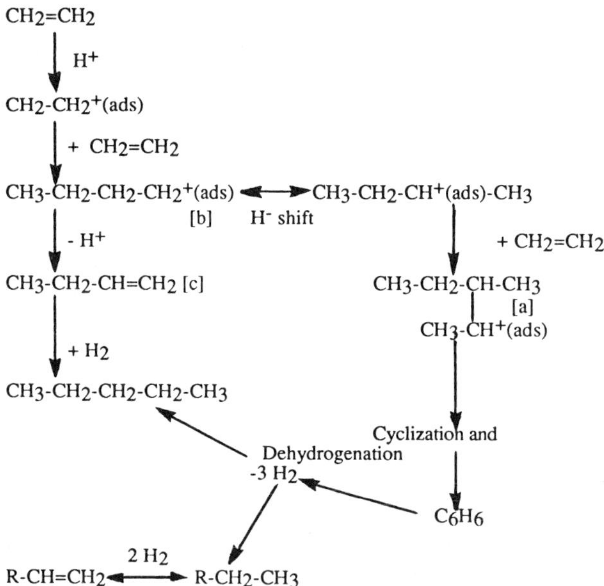

FIGURE 5: MECHANISM FOR OLEFIN AROMATIZATION IN H-ZSM-5
LEGEND: [a] Olefinic oligomers (2º) [b] Adsorbed carbenium ion [c] Olefin

co-precipitate, or even in the case of the pure component oxide. Optimization of the catalyst formulation was performed with pure ZnO containing hybrid (Figure 3). It can be noted that when the catalyst is made up of 5wt% ZnO and 75wt% H-ZSM-5 (Si/Al=38), the greatest aromatization activity is recorded. This effect can in fact be qualified as synergy since the resulting aromatic production of the hybrid system is much more than the sum of the individual components. Figure 4 shows that the trend in increased aromatic production is paralleled by an increased ethane production and a greater conversion of the ethylene fraction in the feed. This suggests that ZnO has a strong reducing ability which is fueled by the adsorption of hydrogen species released by the zeolite component. In short, as more aromatics are produced, more hydrogen must be evacuated from the zeolite channels due to the cyclization/ dehydrogenation step in the aromatization mechanism [17] . ZnO acts as an acceptor for the hydrogen produced and in turn also has a catalytic activity of its own for the reduction of ethylene in ethane.

Experimental evidence of LD-HBS in the aromatization of light alkenes and alkanes

It is assumed that for the H-ZSM-5 zeolites which are not modified by any metallic ions, the mechanism which leads to the formation of aromatic products from ethylene is as shown in Figure 5 [10] . It is suggested by Mole, Anderson et al. [18] that the dehydrogenation of the saturated cyclic molecule occurs by a series of hydride abstraction steps. H^- may combine with the Brønsted acid sites on the internal

TABLE 2: *Catalytic activity of the pure and hybrid catalysts*

CATALYSTS	TOTAL AROMATICS (C atom %)
ZnO(80wt%)	2.50
H-ZSM-5(37) (80wt%)	16.9
H-ZSM-5(37)(75wt%)+ ZnO(5wt%)	30.5

Temperature RI:820ºC; RII:500ºC; WHSV:1.5h^{-1}.

surface of the zeolite channels and thus it is possible that if the resulting hydrogen species are not removed efficiently, the acid catalyzed aromatization reaction can not reach its equilibrium position. The following tests show the involvement of the hydrogen species released during the reaction.

(a) When the feed is olefinic in nature, is the presence of the co-catalyst alone responsible for the enhanced aromatic production?

Table 2 shows that it is not, because in the case where a pure ZnO catalyst is used, no appreciable amount of aromatic product is formed. Furthermore, as shown in Figure 3 the amount of co-catalyst needs be very low in the hybrid system to exhibit the highest activity for aromatization.

(b) Does an alkane feed (n-butane) undergo dehydrogenation over the co-catalyst before it enters the zeolite pores?

Table 3 shows that it does not, even when the ZnO has been reduced and saturated with hydrogen prior to the reaction. These results are in agreement with Fujimoto et al. [14].

(c) Is the physical contact between the zeolite component and the co-catalyst necessary?

TABLE 3: *Catalytic activity of pure ZnO and reduced ZnO*

CATALYST / PRODUCT DISTRIBUTION (C atom %)	Bentonite (100%)	ZnO (80wt%)	ZnO reduced (80wt%)
C_1-C_4 Paraffins	96.7	97.4	96.7
C_2-C_4 olefins	3.3	2.6	3.2
C_5^+ non-aroma.	-	-	-
Aromatics	-	-	0.1
Conversion (%)	21.0	31.4	22.4

Temperature RII:540ºC; WHSV: 0.5h^{-1}; Feed: n-butane

TABLE 4: *Effect of the distance between the zeolite and co-catalyst components*

CATALYST / PRODUCT DISTRIBUTION (C atom %)	A	B	C	D	E	F
C_1-C_4 Paraffins	63.0	47.3	46.0	62.7	63.8	65.5
C_2-C_4 olefins	17.2	6.4	8.9	13.5	16.8	18.0
C_5^+ non-aroma.	3.2	0.5	1.3	2.8	3.2	2.6
Aromatics	16.6	45.8	43.8	21.0	16.2	13.9
BTX/Arom. (%)	92.0	89.0	92.0	93.0	92.0	93.0
Conversion (wt%)	81.4	95.5	88.8	70.9	75.4	78.8

Temperature RII:540ºC; WHSV: $0.5h^{-1}$; Feed: n-butane

A: H-ZSM-5(36)(80wt%)
B: H-ZSM-5(36)(75wt%) + ZnO(5wt%) dry impregnation
C: H-ZSM-5(36)(75wt%) + ZnO(5wt%) hybrid
D: H-ZSM-5(36)(80wt%) + ZnO(80wt%) mixed catalytic bed
E: H-ZSM-5(36)(80wt%) --> ZnO(80wt%) 2 catalytic beds
F: ZnO(80wt%) --> H-ZSM-5(36)(80wt%) 2 catalytic beds

Table 4 shows that when the H-ZSM-5 and ZnO are at a close distance such as in the catalyst prepared by dry impregnation (where ZnO is on the external surface of the zeolite particles or incorporated in the bentonite binder , in the hybrid catalyst configuration or when extrudates of the pure components are mixed together in the same catalyticbed),the aromatic yield is enhanced considerably with respect to the pure H-ZSM-5 catalyst. When two beds separated by a porous disk are used, the presence of ZnO does not influence the aromatic yield, indicating that the distance becomes too large for HBS to occur between the two components. Therefore, LD-HBS is dependent on the distance which exists between the zeolite and the co-catalyst.

(d) Is there any possibility that Zn^{2+} or Ga^{3+} ions undergo solid state ion exchange and be incorporated into the zeolite lattice [19] (which could maybe explain the enhanced aromatic formation)?

TABLE 5: *Catalytic results obtained with hybrid catalysts where the co-catalyst is an inert oxide*

NAME OF HYBRID	CO-CATALYST	TOTAL AROMATICS (C atom %)	BTX/ AROM. (%)
H-ZSM-5(36)	nil	13.5	92.0
H-ZSM-5///Q	Quartz	27.7	90.0
H-ZSM-5///Si	Silica Gel (Grace)	29.8	91.0
H-ZSM-5///Si	Ludox Colloidal Silica (Dupont)	34.2	93.1
H-ZSM-5///Al	γ-Alumina (Merk)	32.6	87.8

Temperature RII:540ºC; WHSV:0.6h^{-1}; Feed:n-butane

Table 5 shows that for n-butane feed, even when small amounts of alumina, silica gel, colloidal silica or quartz in their pure form are added as co-catalyst, without any metallic ions , the increase in aromatic yield is 2-fold or more. Therefore, the solid state ion exchange of metal ions alone cannot explain the synergy seen for aromatization.

(e) Does the introduction of hydrogen or oxygen in the feed disturb the aromatic formation activity of the pure zeolite?

Table 6 shows that in the case of n-butane feed, when O_2 is co-fed more aromatic are formed, as also seen by Fujimoto et al. [20] and Centi et al. [21]. It is postulated that the oxygen molecule can react with some hydrogen species adsorbed on the surface and remove it as water from the zeolite lattice (the formation of H_2O was indeed observed). In effect the oxygen acts as a chemically reactive hydrogen removal entity in a similar fashion to the co-catalyst which itself is active by virtue if its physical properties (surface, potential energy...). When hydrogen is co-fed, the aromatic yield is somewhat depressed.

TABLE 6: *Effect of added oxygen or hydrogen to the catalytic activity*

FEED / CONVERSION (wt%) / PRODUCT DISTRIBUTION (C atom %)	n-butane + N_2	n-butane + air	n-butane $+N_2+H_2$
CONVERSION (wt%)	81.4	85.3	75.2
C_1-C_4 Paraffins	63.0	54.7	64.0
C_2-C_4 olefins	17.2	19.4	18.3
C_5^+ non-aroma.	3.2	3.6	2.9
Aromatics	16.6	22.3	14.8
Water (g)	-	0.5	-
BTX/Arom. (%)	92.0	91.0	95.0

Temperature RII:540ºC; WHSV: $0.5h^{-1}$; Feed: n-butane
Flow-rates: n-butane:14ml/min; N_2:10ml/min; air:12.5 ml/min; H_2:5ml/min
Catalyst: H-ZSM-5(36)(80wt%)

In the case where ethylene is used as feed, Table 7 reports the aromatic yields and hydrogen production in the gas phase for the pure H-ZSM-5(42) catalyst and the hybrid systems with ZnO and ZnO/Al_2O_3 co-precipitate (Zn/Al=1.0) when the hydrogen partial pressure is varied. It can be seen that for the pure system, as the amount of hydrogen co-fed is increased, the aromatic production decreases slowly while the production of gaseous molecular hydrogen remains relatively constant. This probably means that the reaction mechanism slowly shifts to another pathway when the hydrogen partial pressure is increased. In the case of the ZnO hybrid system, the aromatic yield and hydrogen consumption decrease are dramatic. Conversely, the production of ethane increases very rapidly. Therefore, it appears that even if there is only a small amount of ZnO in the hybrid system, as the hydrogen partial pressure increases the reaction tends to shift from aromatization to hydrogenation with a great driving force. For the ZnO/Al_2O_3 co-precipitate, the same trends are observed but the effects of added hydrogen are not so radical.

TABLE 7: *Effect of co-fed hydrogen on the catalytic behavior of pure and hybrid catalysts. Partial pressure experiments*

CATALYST	H_2 PARTIAL PRESSURE (atm)	AROMATIC YIELD(ethane) (C atom %)	HYDROGEN (+) or (-) (mmol/hr)
H-ZSM-5(42) (80wt%)	0	41.9(4.2)	+14.7
	0.25	38.3(4.0)	+15.3
	0.42	36.6(3.7)	+12.3
	0.63	31.5(4.0)	+18.9
	0.76	31.8(4.0)	+19.3
H-ZSM-5(42)///ZnO	0	51.7(26.0)	+16.2
	0.25	41.3(36.5)	+ 6.4
	0.42	32.0(45.0)	- 1.8
	0.63	22.7(58.2)	-10.9
	0.76	13.4(66.6)	-21.7
H-ZSM-5(42)/// Zn/Al=1.0	0	58.5(14.8)	+30.5
	0.25	54.5(16.5)	+30.5
	0.42	50.8(14.2)	+30.9
	0.63	38.3(23.5)	+16.6
	0.76	34.7(35.0)	+ 4.4

Temperature RII: 500°C; WHSV:0.6h^{-1};Feed: Ethylene

Figure 6 is a schematic representation of the proposed LD-HBS reaction network. The zeolite is responsible for the aromatization of light alkenes or alkanes. The co-catalyst, in contact or positioned at close distance from the zeolite particles, is the porthole for the hydrogen species formed during the reaction and released at the zeolite pores. Once adsorbed on the surface of the co-catalyst, hydrogen can desorb molecularly (sink action), or react with another absorbed unsaturated molecule (ex. ethylene) reducing it to the corresponding saturated molecule (ex. ethane) (scavenging action).

New generation of LD-HBS catalysts

It can be observed in Table 5 that hydrogen removal can be done by "inert" oxides, thus increasing appreciably the aromatic yield with respect to the pure zeolite catalyst.

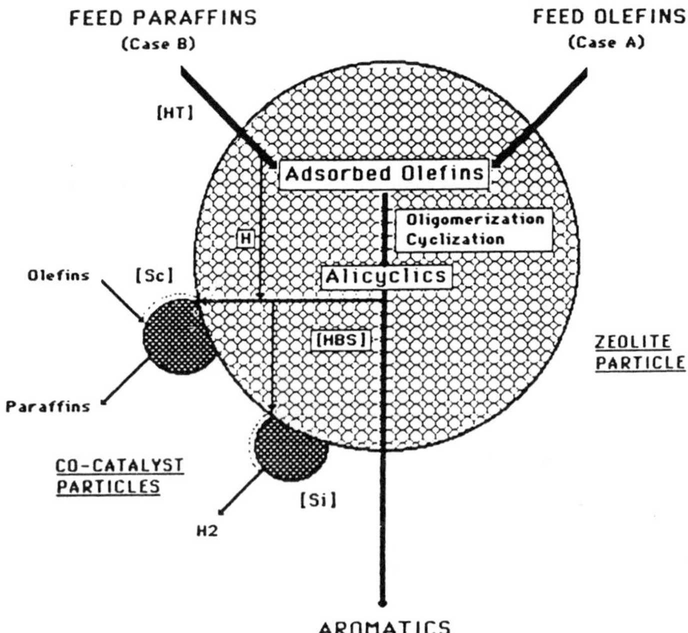

FIGURE 6: SCHEMATIC REPRESENTATION OF THE PROPOSED REACTION NETWORK OF OLEFIN (CASE A) AND PARAFFIN (CASE B) AROMATIZATION.

LEGEND: [HT] Hydride transfer

[HBS] Hydrogen back-spillover

[Sc] Scavenging action

[Si] Sink effect

(Reprinted with permission of Applied Catalysis).

In Table 8 it can be observed that when gallium oxide is impregnated or co-evaporated onto these supports, the increase in aromatization activity of the hybrid catalysts is again significant. Furthermore, if the H-ZSM-5 is previously modified with a small amount of sulfate ions, the aromatic yield reaches its highest level, maybe because SO_4^{2-} ions have a favorable effect on the rate of diffusion (out) of spilt-over hydrogen species from the zeolite lattice.

Finally, tests of hydrogen chemisorption (pulse technique) performed at the rection temperature show that the gallium oxide/ silica Ludox system solid has the

TABLE 8: *Catalytic results obtained with the new generation of hybrid catalysts*

NAME OF HYBRID	CO-CATA.	ZEOLITE	TOTAL AROM. (C atom %)	BTX/ AROM. (%)
H-ZSM-5(36)	nil		13.5	91.7
H-ZSM-5///Ga-Q	3wt% Ga oxide Quartz	ZSM-5	48.5	88.7
H-ZSM-5///Ga-Si	3wt% Ga oxide/ Silica Gel (Grace)	ZSM-5	50.0	87.6
H-ZSM-5///Ga-SiL	3wt% Ga oxide/ Ludox Colloidal Silica (Dupont)	ZSM-5	63.4	87.3
H-ZSM-5///Ga-Al	3wt% Ga oxide/ γ-Alumina (Merk)	ZSM-5	42.2	87.7
H-ZSM-5///Ga-SiL	3wt% Ga oxide/ Ludox Silica	ZSM-5/ SO_4^{2-}	69.8	89.5

Temperature RII:540ºC; WHSV:0.6h^{-1}; Feed:n-butane

highest inital rate of hydrogen up-take (it is also the most performing co-catalyst). This may be due to a favorable configuration (dispersion...) of gallium species onto the evaporated Ludox silica.

CONCLUSION

Evidence for the active participation of the Long Distance Hydrogen Back Spillover phenomena has been confirmed with various hybrid catalyst formulations, in the aromatization of both light olefins and paraffins or a mixture thereof, and for variable catalytic testing conditions. Due to the concept of preparation of the hybrid system, separate chemical modification of the zeolite and co-catalyst components prior to their combination becomes possible and has been demonstrated to lead to higher performances for the hybrid catalytic systems.

ACKNOWLEDGMENTS

The authors thank NSERC of Canada, the "Actions Structurantes" program and FCAR (scholarship for L. Dufresne) of Quebec for financial support.

REFERENCES

1. Pochen Chu; U.S. Patent 4, 120, 910 (Oct. 17, 1978).
2. A.W. Chester; U.S.Patent 4, 350, 835 (Sept. 21, 1982).
3. J.A. Johnson, J.A. Weizman, G.K. Hiller and A.H.P. Hall; Paper presented at NPRA *Annual Meeting*, San Antonio TX (USA), March 1984.
4. T. Inui, Y. Makino, F. Okazumi et al.; *J. Chem Soc., Chem. Commun.* 7: 571 (1986).
5. T. Inui, Y. Makino, F. Okazumi et al.; *Ind. Eng Chem. Res.* 26: 647 (1987).
6. N.S. Gnep, J.Y. Doyemet, A.M. Seco, F.R.Ribeiro and M. Guisnet; *Appl. Catal.* 43: 155 (1988).
7. T. Inui, K. Kamachi, Y. Ishihara et al.; *Proc. 2nd Int'l. Conf. on Spillover*, K.H. Steinberg ed., Leipzig GDR, p.167 (1989).
8. Y. Ono, H. Nakatani, H. Kitagawa and E. Suzuki, Successful Design of Catalysts, T. Inui ed., Elsevier Sc. Publishers, p.279 (1988).
9. A. El Tanany, G.M. Pajonk, K.H. Steinberg and S.J. Teichner; *Appl. Catal.* 39: 89 (1988).
10. R. LeVanMao and L. Dufresne; *Appl. Catal.* 52: 1 (1989).
11. J. Yao, R. LeVanMao and L. Dufresne; *Appl. Catal.* 65(2):175 (1990).
12. (a) R. LeVanMao, L. Dufresne and J. Yao, *Appl. Catal.* 65: 143 (1990); (b) ibid. Can. Catal. Disc. Group Meetings, Ottawa (Canada) May 1989 and Kingston (Canada) November 1989.
13. R. LeVanMao, J. Yao and B. Sjiariel; *Catal. Lett.* 6(1): 23 (1990).
14. K. Fujimoto, I. Nakamura and K. Yokoda; *Proc. 2nd Int'l. Conf. on Spillover*, K.H. Steinberg ed., Leipzig GDR, p.176 (1989).
15. R.L. Argauer and G.R. Landolt; U.S.Patent 3, 702, 886 (Nov. 14, 1972).
16. R. LeVanMao, P. Levesque, B. Sjiariel and P.H. Bird; *Can. J. Chem* 63: 3464 (1985).
17. J.P. Dejaifve, J.C. Vedrine et al.; ACS/CSJ Meeting, Honolulu p.286 (1979).
18. T. Mole, J.R. Anderson et al.; *Appl. Catal.* 17: 141 (1985).
19. *Appl. Catal.* 55(1): N2 (1989).
20. K. Fujimoto, I. Nakamura, K. Yokota; *Zeolites* 9: 120 (1989).
21. G. Centi and G. Golinelli; *J. Catal.* 115: 452 (1989).

29

NEW APPROACH TO THE CATALYTIC SYNTHESIS OF AROMATICS FROM LOWER (C_2-C_3) ALKANES AND ALKENES OVER MODIFIED PENTASILS

O.V. BRAGIN[a]

N.D. Zelinsky Institute of Organic Chemistry Academy of Sciences
Leninsky Prospekt, 47
Moscow, 117913, USSR

Dehydrocyclooligomerization of low molecular weight olefins and alkanes to produce aromatics is obviously an important area to investigate. One of the most important problems at present and through the year of 2000 consists in developing new catalysts to obtain economically sound ways of utilizing non-traditional feedstocks such as natural gas and associated refinery gases. This problem is close to the rational utilization of tail gases from large refineries and chemical productions. Substantial amounts of low-molecular hydrocarbons formed in these processes are, as a rule, not optimally used: part are used as process fuel; part are incinerated in flares.

Ethane and ethylene aromatization reactions that occur in the presence of metallicalumina catalysts were discovered at our Institute about 15 years ago [1]. At that time, these conversions were of only scientific interest but indicated the possibility of converting C_2's to aromatic hydrocarbons. The studies conducted within the last few years have shown, that to convert C_2–C_4hydrocarbons to high-molecular hydrocarbons certain high-silica zeolites of the pentasil-type acid catalysts are more efficient than metallicalumina catalysts. It should be emphasized that elucidation of the main mechanisms of oligomerization-aromatization reactions on pentasils is at present an important scientific problem. Practical aspects of this work have already been discussed.

Conversion of C_3 to C_6 paraffins into aromatics seems to be much more easier than conversions of ethane. Chen described the "M2-Forming" process [2] for the conversion of C_3 to C_6 hydrocarbon fractions into gasolines rich in aromatics using ZSM-5 type zeolite catalysts. Other authors [3-6] reported that Ga and Zn supported H-ZSM-5 zeolites are effective catalysts for the conversion of lower paraffins to aromatics.

---

[a] Deceased.

TABLE 1 *Catalytic conversion of ethylene on Pt/Al_2O_3 and zeolites (500-550°C)*

Catalyst	Conversion	Yield of liquid products (%)	
		Aliphatic compounds C_5-C_8	Aromatic C_6-C_{12}
Pt/Al_2O_3 (0.6% Pt)	40 - 50	2 - 3	10 - 12
H-ZVK	90 - 98	1 - 2	45 - 55
H-ZVM	95 - 99	traces	56 - 75

H-ZVK (SiO_2/Al_2O_3 = 40) and H-ZVM (SiO_2/Al_2O_3 = 70) are pentasils with structure of ZSM-5.

RESULTS

The catalytic activities of the new generation zeolites were compared with those of platinum-alumina catalysts. Table 1 shows comparisons of ethylene aromatization. As indicated for zeolites the aromatic hydrocarbons yield is 5-6 times higher than on a platinum-alumina catalyst, even though pentasils are free of noble metals. The investigation of the activity and selectivity of homemade ZSM-type pentasils (ZVK, ZVM, ultrasils) has shown that when converting C_3–C_4 olefins the process can be directed to production of either aliphatic hydrocarbons or aromatic hydrocarbons (Fig. 1).

The lack of appreciable conversions of alkenes, particularly ethylene, on zeolite Na-forms proves to be an important feature of the catalytic systems under discussion. In this case, a 75% yield of aromatic hydrocarbons was obtained on ZSM H-form under optimal conditions. These results are indicative of the important role played by the Bronsted acidic sites in ethylene oligomerization and aromatization. The composition of the active sites proves to be the main problem and hence needs to be carefully investigated. The activity of pentasil-containing catalysts (as already shown) is to a large extent dependent on their acidic properties. In this connection, the mechanism with pentasils acid catalysts was comprehensively studied during the current investigation by both catalytic and spectrometric methods.

In Figures 2 and 3, corresponding data are presented, not only for zeolites as such, but for a number of zeolite-alumina formulations within a wide range of compositions where the pentasil content varied from 0 to 100%. Typical spectra obtained with one of these formulations (30%

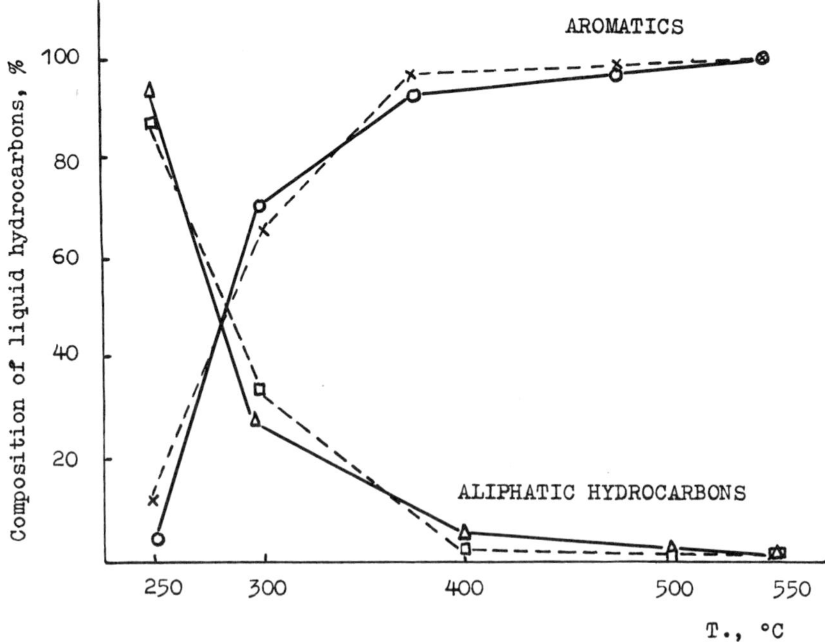

FIG. 1 *Conversion of ethylene over pentasils (solid line - ZVK, dashed line - ZVM).*

alumina) are compared with the spectrum of an individual zeolite (Fig. 2a). It is clearly seen that the incorporation of alumina caused "dilution" of the zeolite system. For formulations of other compositions, no shift of the corresponding bands was observed indicating that there are no intimate interactions between pentasil and Al_2O_3 or no intrusion of alumina into the zeolite framework.

The next important point deals with quantitative determination of Bronsted and Lewis acidic sites (Fig. 2b). It is shown that the HZSM-γ-Al_2O_3 system is characterized by the availability of both L- and B-sites; the relation between these sites depends on the composition of the formulation. Finally, concerning the catalytic properties of this system (Fig. 3), high activity and high selectivity for ethylene aromatization occurred over a wide range of compositions. A sharp increase in activity is observed when a very small amount of zeolite (below 5%) is introduced into alumina.

These results permit two conclusions. The first conclusion: for ethylene aromatization not only B-sites (as previously assumed), but a combination of B- and L-sites are necessary. And, the second conclusion: the relation between B- and L-sites in the catalysts can easily be controlled by introducing a specific amount of alumina into the system.

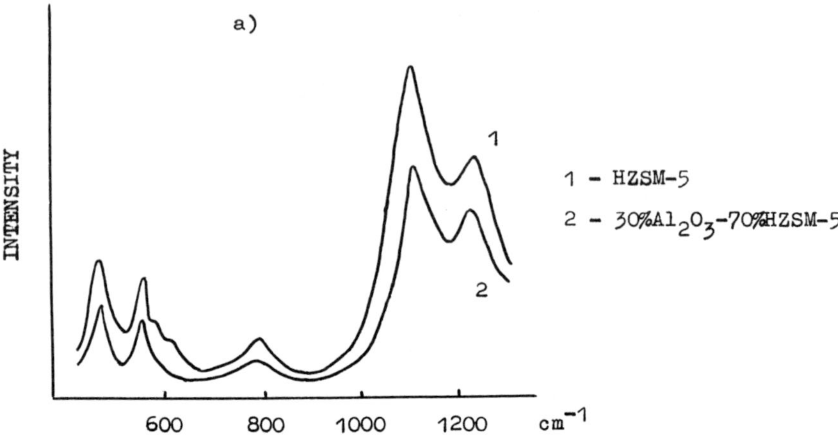

FIG. 2 *a) IR-spectra of HZSM-5 and compositions 30% Al_2O_3- -70% HZSM.*

b) Separate determination of B- and L-sites in different compositions Al_2O_3-HZSM.

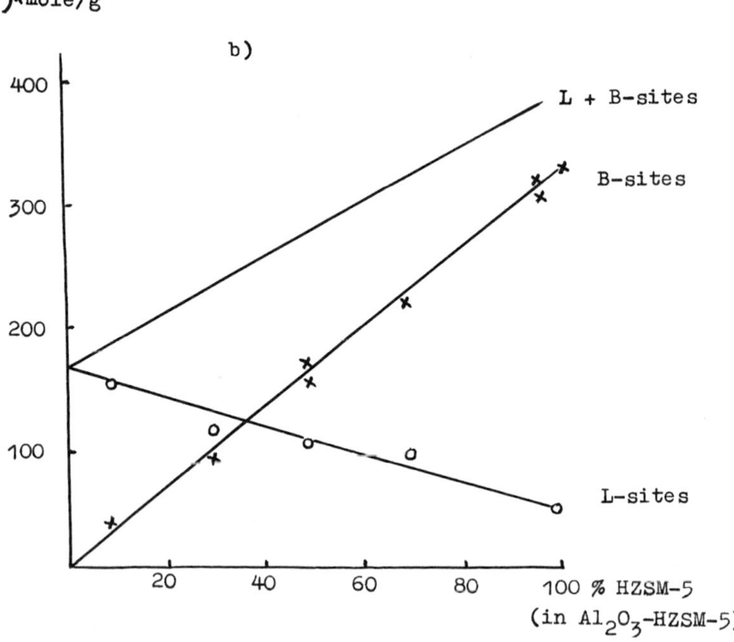

FIG. 3 *Conversion of C_2H_4 over Al_2O_3-HZSM-5 (500°C).*

TABLE 2 *Aromatization C_2-C_4-alkanes*

Hydrocarbons	Yield aromatics (wt. %)		
	H-ZSM	M_I-ZSM	M_{II}-ZSM
C_2C_6	8-10	20-30	20-30
C_3H_8	37-39	26-28	55-60
i-C_4H_{10}	40-43	-	48-52

M_I and M_{II} - modifiers of I and II types

Let us now discuss the activation of alkanes. At this moment, it should be emphasized that this task is much more complex because of the lower reactivity of alkanes as compared to that of olefins, but specific information have also been revealed in the present investigation. The pronounced catalytic activity of the zeolite H-form for aromatization of ethane and other alkanes seems to be unusual and important (Table 2). Modification of the zeolites can result in a considerable increase in their aromatizing ability. Modifiers of the first type (metals from VIII group) sharply increased the yield of aromatic hydrocarbons from ethane - by two and even by three times. C_3–C_4 alkanes had higher reactivities than ethane; the yield of aromatic hydrocarbons from them on modified pentasils reached 50% and even greater. In conclusion, certain modified zeolites result in the aromatization of appreciable amounts of C_2–C_4 alkanes. While investigating the aromatizing of propane (Fig. 4) on metal-containing pentasils, it was found that catalytic activity is determined by the metal component of the catalyst.

Actually it has been shown [7] by XPS analysis that a change in valency of a metal portion can occur (Fig. 5). On the upper spectrum, the Pt valency was recorded when the maximum activity was attained by the catalyst. Here it is possible to isolate two types of platinum sites: first zero-valency Pt and second Pt with an electronic density deficiency, arising in the course of the reaction. The latter type was apparently responsible for the enhanced catalytic activity of metal-containing zeolites in the reaction under discussion.

Recently non-platinum systems such as pentasils modified with zinc or gallium oxides have been successfully used as aromatization catalysts. While preparing and testing a number of Zn- and Ga-containing catalysts, we were convinced of their high activity. As seen from Table 2, zinc

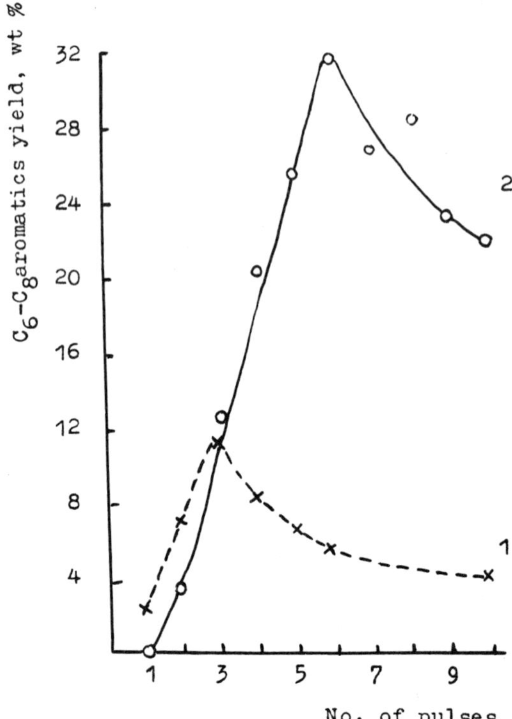

FIG. 4 *Dependence of the yield of C_6-C_8 aromatic hydrocarbons over 1.8% Pt/HZSM subjected to air-hydrogen pre-treatment on the number of alkane pulses at 550°C from 1 - ethane; 2 - propane.*

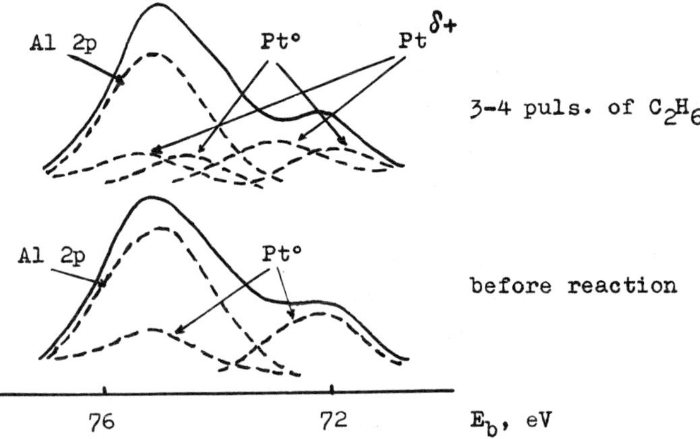

FIG. 5 *The changes of Pt-electron state in Pt/HZVM in the process of ethane aromatization.*

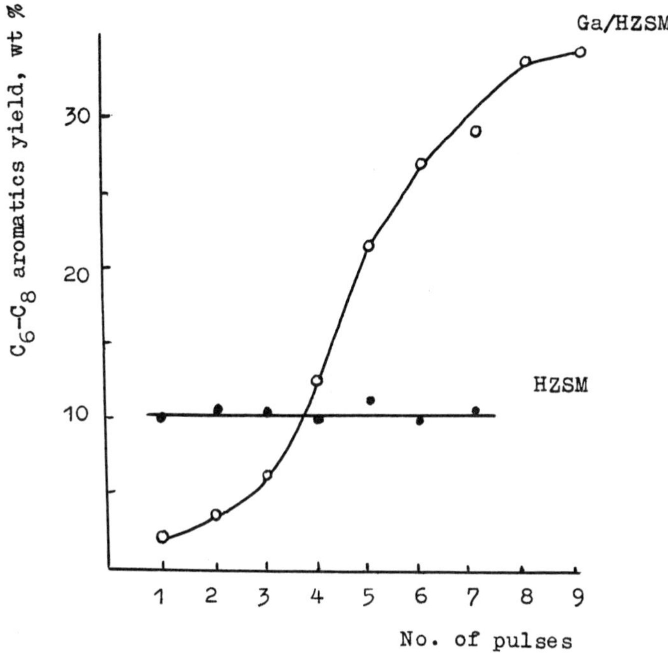

FIG. 6 *Dependence of C_6-C_8 aromatics yield on the number of propane pulses at 500°C over Ga/HZSM.*

and gallium (only 1-2%) and combinations with other elements belong to modifiers of the second type. As an example of the aromatization of light paraffins, Figure 6 indicates results for propane aromatization over Ga-pentasil. The reasons that such catalysts are effective are apparently different as compared to catalysts containing Group Eight metals. Figure 7 presents results obtained for several modified pentasils containing from 0.5 to 5% Zn when used for C_2–C_3 alkanes aromatization. The greatest yield of aromatic hydrocarbons was achieved on catalysts containing 1-2% Zn.

A question is raised: what causes the promoting action of zinc and gallium in alkane aromatization? To compare the catalytic and acidic properties of pentasils, a spectrometric study of the interaction of pyridine (Py) with the zeolite Bronsted and Lewis acidic sites was performed [8,9]. Based on the assumption that while introducing Zn, one Zn^{2+} cation is substituted for two OH-groups, the absolute number of B- and L-sites in a series of Zn-containing catalysts was calculated. As can be seen in Figure 8, the experimental data on the change in number of L- and B-sites are in good agreement with the calculated ones when up to 2% Zn is introduced into the catalyst. We wish to emphasize that the most efficient catalysts contain 1-2% Zn.

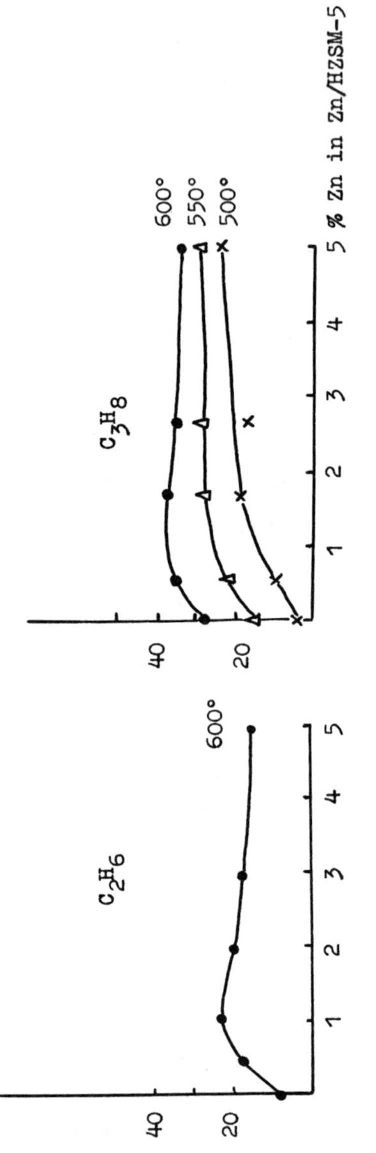

FIG. 7 *Conversion of C_2H_6 and C_3H_8 over Zn/HZSM-5.*

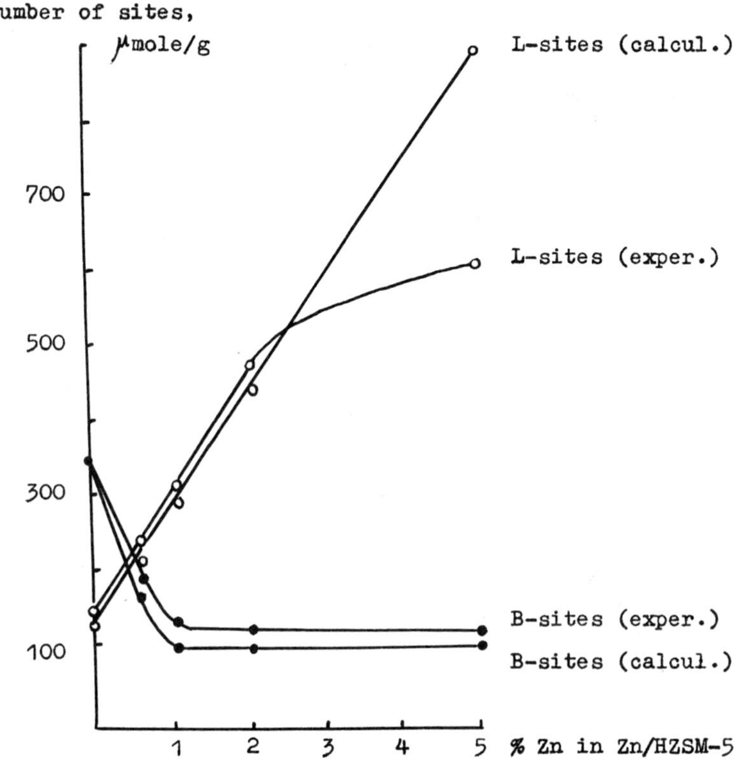

FIG. 8 *The dependence of the number of B- and L-sites on Zn/HZSM-5 systems.*

The main conclusion of the present investigation is that when pentasils are properly modified, strong aprotic acidic sites are formed capable of activating and dehydrogenating lower alkane molecules due to electron-acceptor properties. Of special importance, methods for controlling the total amount and the L:B site ratio in the catalyst system under study have been revealed; these methods consist in introducing a pre-calculated amount of modifier, thereby permitting control of the activity and the selectivity of the catalyst.

The introduction of gallium into pentasil (0.5-2%) accelerates aromatization of ethane and propane as compared to the original pentasil. In the presence of 2% gallium, yields of aromatics from ethane increase by three times; when propane is used, yields arc as high as 60% (Fig. 9). Further increase of the gallium content to 3% in the zeolite however leads to a considerable decrease of the aromatizing properties of the catalyst.

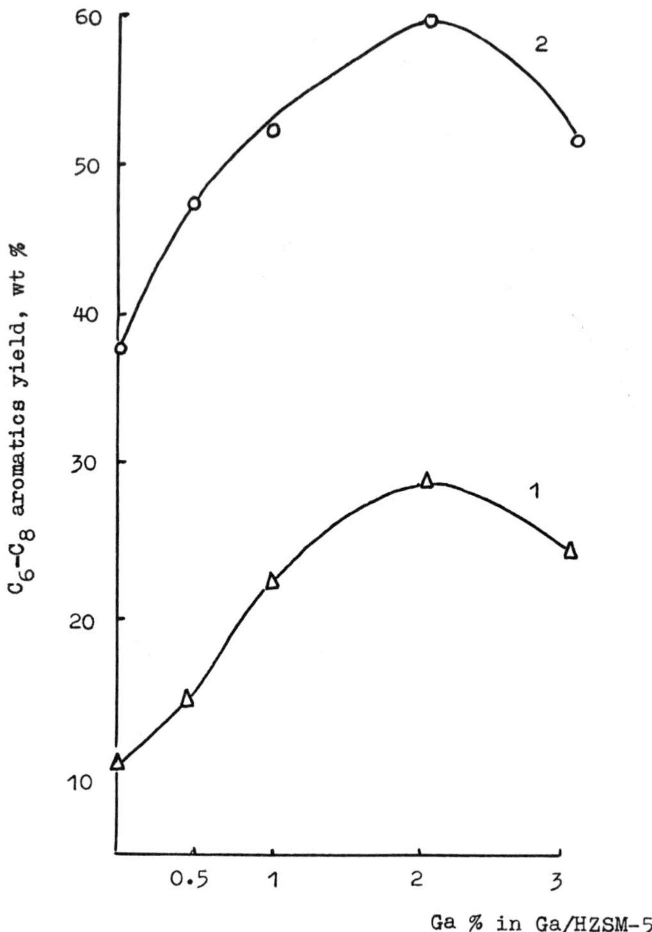

FIG. 9 *Aromatization of C_2H_4 (1) and C_3H_8 (2) over Ga/HZSM-5 systems.*

For the Ga^{3+}/HZSM system, the concentration of B_1-sites, which can be characterized by a.b. at 3610 cm^{-1}, does not depend to any large degree on the Ga content in the range 0.5-5% (Fig. 10). The number of L-sites increases slightly with increased Ga concentrations; this increase of Ga content leads to appearance of a second type of sites, L_{11}, which are absent in the initial zeolite. In the Ga^{3+}-pentasil, there is a constant concentration of L_1-sites, but the number of L_{11}-sites increases almost linearly with the Ga content. However, the increase of the number of L_{11}-sites is insignificant. This shows that the most Ga does not interact with Py as do L-sites. It can

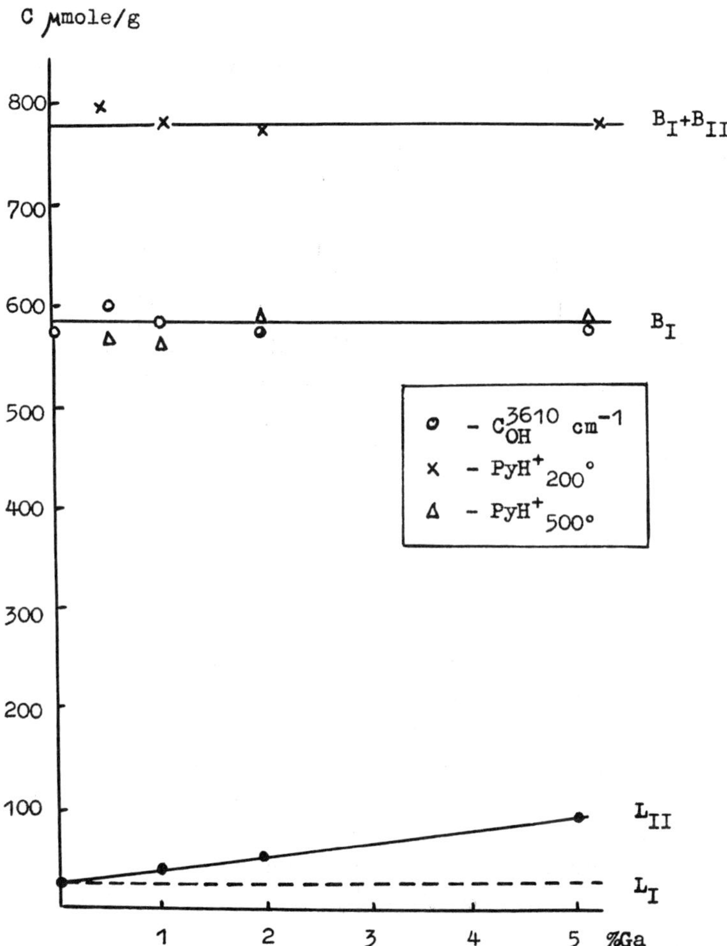

FIG. 10 *The dependence of the number of B- and L-sites on Ga/HZSM-5 systems.*

be also concluded that there are two types of B-sites in the catalyst: i) stronger sites, B_1, which are characterized by a.b. at 3610 cm^{-1} and ii) less strong ones, B_{11}, which do not exhibit any clear a.b. of OH-group but can be detected by thermal desorption of Py.

The Ga does not interact with OH-groups and the behavior and the concentration of the latter do not depend on the presence of Ga. Unlike Zn, Ga ions probably do not occupy cation positions and are not located in zeolite channels. Most probably they are located on the outer surfaces of the zeolite in aggregated forms as gallium oxide.

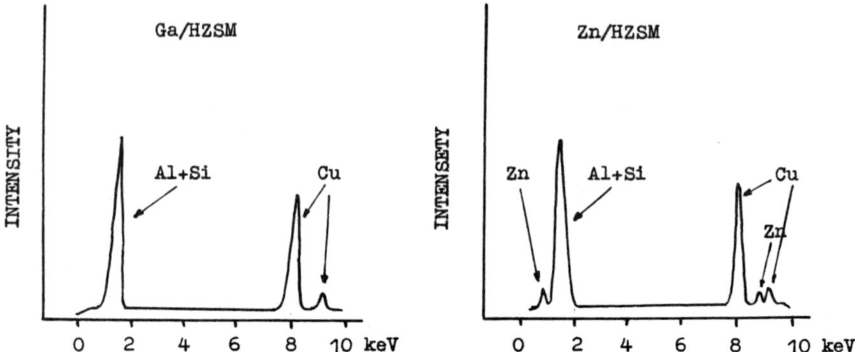

FIG. 11 *Energy dispersive spectrum of characteristic X-ray emission for the plate of Ga- and Zn-containing zeolites (Al and Si lines correspond to zeolite framework, Cu - to copper grid of holders).*

Additional experimental evidence for above-mentioned assumption concerning the distribution of Zn^{2+} and Ga^{3+} ions in the pentasil matrix were determined using an electron microscopy study. In the case of the Ga^{3+}/HZSM system, Ga was not detected in pseudohexagonal plates which are characteristic for zeolite (Fig. 11). Ga can be seen only as an individual morphological formation located on the outer surface of zeolite which are extended microcrystals. Using an electron microprobe, Zn is shown to always be present in pseudo hexagonal plates of Zn^{2+}/HZSM zeolite. The content of Zn in relation to other elements (Si, Al) is nearly constant indicating that Zn tends to distribute itself uniformly in the zeolite channels. Thus the distributions of activators - Zn and Ga - after their introduction into zeolites are different.

DISCUSSION OF RESULTS

Considering all the data on catalytic and acidic properties of studied systems, we note two important features. The first one is the positive effect of introduction of Zn and Ga into zeolites, that leads to considerable increases of the rates of lower alkanes aromatization. The second one is the different distribution of the mentioned additives into the pentasil matrix. The difference in Zn and Ga distribution may be caused by the fact that the latter additive is a multicharged ion and therefore in order to compensate the charge, the Ga ions should be distributed in the zeolite structure as aquacomplexes and bridged structures. The distribution of such structures in zeolite channels is hindered because of steric reasons. In spite of the different distribution of Ga as compared to Zn, the introduction of Ga also shows a positive effect in aromatization and this effect depends significantly on the gallium content. There can be several reasons for this effect:

i) it is possible that the reaction media or pretreatment in different gases that have strong influence on catalytic properties may cause dispersion of the Ga-containing phase and migration of the dehydroxylated Ga ions into the zeolite channels; ii) it cannot be ruled out also that activation of alkane molecules proceeds directly on the surface of the Ga-containing phase followed by conversion of intermediates in zeolite channels.

CONCLUSIONS

While evaluating methods of activating lower (C_2–C_4) hydrocarbons, considerable progress has been achieved in the last few years in this field of heterogeneous catalysis. New catalysts for low-molecular hydrocarbons activation, and in particular catalysts for oligomerization, aromatization and condensation have been developed and reported; significant achievements have been made in studying the catalytic action of zeolite-containing and other catalysts.

ACKNOWLEDGMENT

Dr. Olga V. Chetina helped with the final editing of this chapter following Dr. Bragin's untimely death on January 7, 1991.

REFERENCES

1. O.V. Bragin, A.V. Preobrazhensky and A.L. Liberman, Izvestia Acad. Nauk SSSR. Ser. khim., 1205, 1670, 2751 (1974) (Russian).
2. N.Y. Chen and Y. Tsoung, Ind. Eng. Chem. Proc. Des. Dev. 25: 151 (1986).
3. T. Mole and J.R. Anderson, Appl. Catal. 17: 141 (1985).
4. G. Sirokman, Y.Sendoda and Y. Ono, Zeolites 6: 299 (1986).
5. T. Inui, Y. Makino, F. Okazumi, S. Nagano and A. Miyamoto, Ind. Eng. Chem. Res. 26: 647 (1987).
6. N.S. Gnep, I.Y. Doyemet and M. Guisnet, J. Mol. Catal. 45: 261 (1988).
7. O.V. Bragin, E.S. Shpiro, A.V. Preobrazhensky, S.A. Isaev, T.V. Vasina, B.B. Dyusenbina, G.V. Antoshin and Kh. M. Minachev, Applied Catalysis 27: 219 (1986).
8. V.I. Yakerson, T.V. Vasina, L.I. Lafer, V.P. Sytnyk, G.L. Dykh, A.V. Mokhov, O.V. Bragin and Kh. M. Minachev, Catalysis Letters 3: 339 (1989).
9. T.V. Vasina, V.P. Sytnyk, A.V. Preobrazhensky, L.I. Lafer, V.I. Yakerson, and O.V. Bragin, Izvestia Akad. Nauk SSSR. Ser. Khim. 3: 528 (1989) (Russian).